LUNATRICK?

Graham Frank

APOLLO 11—WAS IT ALL A

LUNATRICK?

WAS THE **EAGLE** *MERELY A SKYLARK?*

100 MILLION AMERICANS NOW BELIEVE THAT ONE GIANT
LEAP WAS ONE GIANT MOON, SO, BECOME A MEMBER OF
MY JURY AND LET US TOGETHER PUT THE GREAT MOON
HOAX CONSPIRACY THEORY ON TRIAL

Some original research by Keith House
HE STILL WANTED TO BELIEVE…

✛ **Strategic Book Publishing**
New York, New York

Strategic Book Publishing
An imprint of AEG Publishing Group
845 Third Avenue, 6th Floor — #6016
New York, NY 10022
www.StrategicBookPublishing.com

ISBN: 978-1-60860-113-4
SKU: 1-60860-113-7

Printed in the United States of America

To Keith

Contents

Introduction

On July 20 1969, something quite beyond extraordinary happened. A dream mankind had nurtured for centuries finally came true. The United States landed two living, breathing human beings on the surface of the Moon. Or so it is claimed.

In the words of the first man to set foot upon dusty alien soil, it was "…one small step for a man; one giant leap for mankind." That man, of course, was Neil A. Armstrong. He was followed down the ladder of an all too ungainly-looking spacecraft called the Lunar Module, or LM for short, by another man. That man was fellow astronaut Edwin "Buzz" Aldrin.

Orbiting above them in a larger vehicle called the Command Module, and awaiting their safe return from the Moon's surface, was a third astronaut named Michael Collins. All too briefly, or so it now seems, these were the three most famous men in the world. Or, more precisely, *off* it!

Their successful expedition, which ended with their safe return from outer space, was named Apollo 11. It was soon followed by another expedition that was equally as successful, Apollo 12. There were a five other successful Moon landings, Apollo missions 14 through 17, with only the aptly named Apollo 13 mission failing to accomplish what it originally set out to do. This was due to an unfortunate accident while in space. Thankfully, though, no loss of life was incurred.

But the incredible adventure that was Apollo 11 is still the most memorable by far, and of the twelve men that history records as having walked upon the surface of the Moon, the name Neil Armstrong remains way out in front. Measured against any yardstick whatsoever,

Armstrong should continue to be recognized the world over as not only the most famous man in the world, but the most famous man in all of history.

Strangely, though, once all the initial razzamatazz had died down and all the obligatory media appearances were done and dusted, Armstrong was seemingly allowed to vanish into almost total obscurity. It is alleged that he shunned TV appearances, radio talk shows, all manner of later interviews, and further invasions into his privacy by the media. Armstrong was nothing like his Soviet predecessor, first man in space Yuri Gagarin, who basked in the glory of it all and enjoyed his celebrity status until the day he tragically died in an airplane crash in 1968.

The recalcitrant attitude, not only of Armstrong but of NASA, which seemed a little backward in coming forward in informing its former employee that whether he personally liked it or not, he was public property now, of course led to questions being asked. Not so much by the media, curiously enough, although "explanations" for his behavior were occasionally published. A certain amount of unease regarding not just Apollo 11, but the whole Apollo program eventually began to find a home with some members of the American public, especially when the alarming number of astronaut deaths in shuttles merely ascending to, or descending from Earth orbit began to grow and grow. To them, when that they thought about it, Armstrong's attitude throughout had been un-American at best and downright suspicious at worst! All anyone had ever wanted to do—maybe each year on the anniversary of that first momentous Moon landing—was to have a chance to shower a great American hero with the adulation and acclaim they thought he thoroughly and richly deserved and to put the more recent tragic failures to one side.

But no Apollo 11 Day was ever declared to commemorate it all anyway. And even if it had been, it seems unlikely that Armstrong himself would even have bothered to show up, let alone lead the official parade through Times Square. Was there a problem? Was Armstrong unwell? Had his awesome experience made him ill?

No. He was quite well, apparently.

But still we were fobbed off with lame excuses from his old bosses at NASA, who claimed, amongst other things, that Armstrong had always been a very private man, and that he was someone who, having done his duty not only for his country, but for all mankind, then simply wanted to be left alone to get on with the rest of his life. And for him, it seems, that meant "far from the madding crowd."

But those who now numbered many and who harbored certain suspicions about it *all* were quick to dismiss this "defense" of Armstrong's behavior since Apollo as arrant nonsense. They continued to labor the point by asking obvious questions like what manner of shy, retiring, "shrinking violet" is it, who—at least as far as we had all been led to believe—knowingly stepped onto the surface of the Moon for the first time in history and on live TV in front of an audience of hundreds of millions?

More than a few eyebrows were again raised when the normally affable and approachable Buzz Aldrin, attending a function at which he was the guest of honor, was seen suddenly bursting into tears and running from the banquet hall sobbing. Why? Well, it seems that after a little alcohol had been consumed by all present, a fellow guest had asked him to describe his feelings while walking upon the Moon on that mind boggling day in 1969.

Sadly, the upshot of all this is that it has provided ammunition for what has now become an army—its ranks ever-swelling—of reluctant skeptics who, to their dismay, find themselves forced to deduce that there can only be one logical reason why a man like Neil Armstrong, but for a brief, abortive foray into politics, continues to bury himself away from all those adoring, hero-worshipping, overly inquisitive admirers. There can only be one explanation that makes any sense at all, why a grown man with seemingly everything to be proud of, would run away from a function with tears in his eyes after having been asked a perfectly natural and ordinary question that he must have been asked hundreds of times before…

Heartbreak.

Shame. With the benefit of hindsight, both men now realized that against their better judgment as otherwise honorable, courageous, honest, law-abiding, decent human beings, they had once upon a time allowed themselves to be "bamboozled" into "doing something for their country," which they now deeply regretted. Persuaded. Cajoled. Perhaps even browbeaten. Maybe even bullied and threatened into accepting that "to play along" would enhance the security of the United States of America and the entire free world by delivering a "killer blow" to the rival Soviet space program.

So, this book, then, will ask the all too pertinent question: *Were* fine men forced, with threats of disgrace and the loss of their pension rights after a genuine course of intensive training as astronauts, to accept that it was *their duty* to play out roles in a…

A hoax to end all hoaxes?

You can fool all of the people some of the time
You can fool some of the people all of the time
But you can't fool all of the people all of the time

—Abraham Lincoln

The Case of the People versus NASA and the United States
Government
Opens with the Preliminary Hearing

COURT NOW IN SESSION

Before the Grand Jury of the People:
The Case of the People versus NASA and the U.S. Government
The Preliminary Hearing

1

An Outline of the Case by the Author

At the very outset of embarking upon this project all by myself—it was originally supposed to be with the help of my researcher and friend Keith House, who died whilst we were merely in the process of gathering together material for a book we hoped would fly in the face of an increasingly vociferous group claiming that the Apollo Moon landings of the late sixties and early seventies were faked—I had no idea where my solo efforts would eventually lead me.

I still owe much of *LUNATRICK?* to Keith's infectious interest in the entire American space program *and* to the amount of research he was able to do for me before becoming too ill with lung cancer to continue. We had previously worked together on another project but, sadly, he was called upon to serve a much greater master than myself before I even managed "to put pen to paper"—metaphorically speaking in this word processing age—this time around.

Despite the fact that the original intention of the book was to "cock a snoot" at "The Moon Landings Were A Hoax" lobby, at the back of my mind was a little voice all the while warning me that I had to be fair and at least hear these people out. But the trouble was that once I did that, I found myself not only understanding but *agreeing* with a lot of the thinking behind their arguments! As a result, this same little voice was soon nagging that because there was just so *much* logic in what these people were attempting to say, should I not perhaps consider picking up *their* banner instead and try saying it a little better?

But old loyalties do die hard and besides, I am not *that* fickle! No, I decided. If I was going to continue at all, I would do my very utmost to try to be fair to *both* sides of the debate. I would be picking up the pro-hoax banner of course, waving it about and marching along stridently with it. And I am especially inclined to do so when I see people more stumbling along than marching, and struggling to keep such a banner aloft. People with no real axe to grind one way or the other, but feel they have been duped. Even more so if they express such feelings often and loudly enough to have been joined by many millions more. I am nothing if not a coward and wholeheartedly subscribe to the philosophy that there *is* a certain safety in numbers! And, speaking of numbers, one is even more inclined to suspect that something might be "up" (or maybe *not up* in this particular case!) when 99 percent of those protesting are clearly not your average, stereotypical "tree-sitters" or animal rights activists. Not that I personally have anything against such people. Their hearts are undoubtedly in the right place even if their bodies often aren't! It is also a rather sad fact that the trees and animals these people are always seeking to protect often look a lot cleaner than *they* do!

But anyway, that much done, I would then, with an equivalent amount of gusto, verve, and vigor, grab the banner of those who believe there to have been no hoax, no fakery, and that all had been perfectly genuine and above board about Apollo, and march *back* along that earlier well-trodden path, now using my new banner like a broom to sweep away all those earlier doubts, preposterous allegations, and antipathy bordering on the fanatical towards Apollo.

And, speaking of fanatics, one must always be very careful indeed, especially in a book of this nature, to avoid over-representing the opinions of those who either possess unshakeable beliefs or who are clearly in the grip of an obsession. As sincere, expert, and devoted to a particular subject as such people undoubtedly often are, they see with an impaired vision. The trouble is, though, that they are generally far too obsessed to realize this. Already at a disadvantage as far as seeing any broader picture might be concerned, they insist upon looking at things only through a very narrow tunnel while wearing spectacles tinted only with the color of their own favorite brand of skepticism. By doing so, they invariably not only do themselves and their cause great damage, but also cause an otherwise serious business to be cast in a less than serious light. Because of this, and in no small measure, the media, which we all rely upon for information and prior warning of anything nasty about to appear over the horizon, often "dismiss" potentially mind-blowing subjects with undisguised disdain. What might have been better off being treated as a serious news story becomes worded in a "tongue-in-cheek" fashion and presented as the "and finally" item in a news program together with a witty comment from a newsreader, intended to leave no one in any doubt that the matter is *not* to be taken seriously.

One such "and finally" item was recently aired on News At Ten (ITV, U.K.). In fact, not only was it the *last* item, but it even came after the sports news and weather report! It was footage of a protestor in the United States being assaulted by someone he had just loudly denounced—in front of the flashing cameras of the world's press—as having participated in a hoax.

The denouncer left holding his jaw was Bart Sibrell, a well-known member of "The Moon Landings Were A Hoax" fraternity. The man he had accused was one of the most famous men in the world, Colonel Edwin "Buzz" Aldrin. As far as the record shows, Aldrin was the second human being in all history, to set foot upon the surface of the Moon. The first, of course, was his crewmate and commander on that same unforgettable occasion, astronaut Neil A. Armstrong.

The two news presenters gave each other a knowing wink before one of them turned back to the camera and quipped, "Well, some of us are old enough to *know* that it wasn't a hoax. After all, *we* were around at the time and *saw* it all happen! Goodnight."

For those of you who may need reminding—or who might even need to be informed of the fact—historical record has it that Neil Armstrong and Buzz Aldrin landed upon the surface of the Moon at precisely 4:17 a.m. Eastern Daylight Time, on July 20 1969, in an unlikely-looking spacecraft called a LM, full title Lunar Module (or LEM, short for Lunar Excursion Module) and codenamed *Eagle*. The *Eagle* had earlier detached itself from a larger vehicle called a CM or CSM (Command Module/Command Service Module) codenamed *Columbia*. The mission was the eleventh in an American manned space program called Apollo and the first to attempt putting a human being on the Moon. Piloted by a third astronaut named Michael Collins, *Columbia* acted as an orbiting "mother ship" throughout the period of separation, descent, landing, and extra-vehicular activity (EVA) by Armstrong and Aldrin until, finally, the "ascent stage" of the *Eagle,* with the two Moon explorers on board, eventually blasted off from the Moon's surface to rejoin the mother ship. That feat accomplished, the *Eagle*'s ascent stage was then jettisoned and all three astronauts safely returned to Earth in *Columbia.* Another six

Apollo missions to the Moon were to follow, with five of them repeating the outstanding success of Apollo 11.

But the allegation that had caused such an out of character response from Aldrin was hardly a new one. It is, in fact, a "conspiracy theory" that has been knocking around for quite a few years now, although given short shrift by the vast majority of humanity. The ridiculous, preposterous, even outrageous claim is…

That man did not land upon, walk upon, or even go to the Moon in July 1969! Neither did he do so subsequently a further five times, nor are there any plans for him to do so in the near future!

An ever-growing lobby of informed opinion based chiefly in the very country we all thought had performed this modern set of miracles, now firmly believes that all the so-called "manned" Moon landings were part of a colossal hoax perpetrated by NASA, the American space agency, with the connivance of certain highly-placed members of the United States government. The problem is that adding a certain amount of flesh to the bones of the basic, skeletal outline of this "crazy" notion is the fact that the incumbent American president back in those "heady" days was a certain Richard Milhous Nixon…

"Tricky Dicky" as he is less than fondly remembered by some.

At the time Keith and I began discussing the possibility of writing this book, like him, I possessed a strong conviction that the Moon landings had been entirely genuine. And the truth is, unless I had been the type to sit up watching late night TV shows on Channel 4 (U.K.), or on even more obscure satellite channels, which might occasionally have dared to feature an "oddball" character or two suggesting that the Moon landings of the late sixties and early seventies had all been one colossal hoax from start to finish, there had never been any good reason for me to suppose otherwise. And that probably went, as well, for 99.9 percent of the world's population! By the time I actually commenced this lonely journey, though, having by now absorbed much of the pro-hoax stuff, I think it fair to say that I was about fifty-fifty, which, when one thinks about it, is probably the best way to be if one intends to represent both sides of an ongoing argument. Where am I now, after completing the book? Well, that would be unfair of me to say. What is far more important is where *you* will be after reading it.

As you will soon see, I decided to use a "trial by jury" format (I had already used this in that earlier project I mentioned) to hopefully reach a reasonably fair representation of the truth regarding this affair without the need for anyone to actually go to the Moon in order to prove it all one way or the other. "Jury?" I hear you say. "What jury?"

You, and hopefully many more like you with an interest in this subject, will form my jury. In fact, you have already become a member of it simply by opening this book and having read thus far. Shortly, you and those who hopefully join you will actually be referred to as "members of the jury" as opposed to "my readers." Soon, this whole case will be laid out before you just as it would be in any normal case being dealt with by any normal court of law. But this one will be make-believe *English* court of law, so to begin with there will be what is called "a preliminary hearing." This is, in fact, already in progress by way of this very opening chapter. As I have yet to fully adopt my role as judge—which I will do shortly—this is presently being presided over by me as the author. Normally, a preliminary hearing explains what a case is about, who is being accused of what and by whom, and discusses whether or not such accusations are worth pursuing enough to place anyone in the dock to face a full-blown trial.

Unlike a normal preliminary hearing, though, which has no jury, this one is *not* to decide whether or not there is a case to answer, whether or not any charges should be brought against any of the alleged miscreants, or whether or not there will *be* a trial at all, in effect. I have already decided that there *is* a case to answer and that there *will be* a trial. That is what this book will be—a trial. And charges *will be* brought; not only by me once I don my wig and gown, but by you and anyone else who chances to read this book. Hopefully, in due course of time, you will number at least twelve and become a grand jury representing all the people of this planet. You, my jury, will be presented with the People's case first—effectively the case for the prosecution—and on behalf of the People, and in yet another adopted role as counsel for the prosecution, I will endeavor to act without prejudice, fear, or favor.

It will be the prosecution's task to convince you that the alleged miscreants, who become the defendants, perpetrated a major deception. The defendants will be the National Aeronautics and Space Administration of the United States of America (NASA) and that organization's patron and paymaster, the United States Government.

The case for the prosecution will then be followed by the case for the defense, by which time I will have miraculously enjoyed a complete change of heart and will now be representing the defendants in my new role as counsel for the defense. Again I shall try my utmost to act entirely without prejudice, fear, or favor. My task will become one of persuading you, the jury, that the case being brought on *your* behalf as the People, and earlier presented by the prosecution, is absolute rubbish. That there *was* no hoax. Only genuine scientific endeavor, albeit for ultimate political advantage, and only genuine heroism by the space crews assigned to the Apollo missions. Also, that true brilliance, not a gift for trickery, was shown by the scientists, technicians, and the many other "backroom" boys and girls of NASA who, ostensibly, had to navigate the Apollo astronauts through some very tricky and dangerous situations.

With the prosecution and defense arguments both aired as fairly and squarely as possible, there will then be a "summing up." Again this will be by someone writing in a style uncannily like my own and acting as judge (a role I shall initially adopt shortly in order to formally lay the charges), who will at least attempt to remain as neutral and as unbiased as is humanly possible for someone who, up until quite recently at least, was convinced that mankind's next "giant leap" would be to put a human being on the planet Mars!

My task, as judge during the summation, will be to broadly go—do you know, I almost typed "to *boldly* go!"—over the whole trial again. Here and there, perhaps, I will need to discuss with you, in greater detail, any point raised by either side of the argument that may have lacked clarity or may have confused the jury as to its specific meaning or reason for having been given in evidence.

But there will be *no verdict* delivered by you, as a jury. Not in this "court." And, unlike any normal jury or tribunal, you will *not* be required to reach a majority decision to either indict or convict the defendants. I am already "indicting" the named defendants by virtue of this book and the decision to either "convict" or "acquit" them of the charges (to be formally laid by me at the commencement of the next chapter) will be a purely private and personal one for each of you to arrive at as individual members of the jury.

Although the defendants are Americans, I am an Englishman and this book is, I repeat, already a make believe English court. Under English law you must arrive at one of two verdicts only: Guilty or Not Guilty. There is no in between. If this book manages to convince you, beyond a reasonable doubt, that the Apollo Moon landings were a hoax, then your private verdict will be Guilty. However, if by the time we have reached the end of this trial you are undecided or not totally convinced that NASA perpetrated a hoax with the blessing of the United States Government, then your verdict will be Not Guilty. Hopefully, you will not allow your decision to be influenced one way or the other by any "unshakeable beliefs" you might have previously held.

As I attempted to explain a little earlier, an unshakeable belief is often an obsession rather than a belief. It is a sign of a captive mind rather than a free one. A free mind believes something to be true not because its owner has been told, or has read, that something is true. A free mind believes something to be true because its owner feels there is enough *evidence* that this is so. But a free mind will be prepared to let go a once strongly held conviction if it is presented with enough firm evidence to undermine that earlier belief, however well-ingrained it might have been.

Members of the jury—even in my as yet "unofficial" capacity, I feel it safe to refer to you as such now—the integrity of each of your minds as individuals is a sacred thing. "Listen" to the evidence. Read it *all* thoroughly, not just the side of the argument you might favor at present. Then read it again to make doubly certain that there is no vital point you may have missed or have not fully comprehended. Be prepared to have any profound beliefs swayed or changed completely; but change not a jot of your previously held convictions should you *genuinely* feel there not to be enough concrete evidence for you to do so. Nothing short of these prerequisites would do if you were asked to act as members of a jury in *any* court of law. And the above remarks are fairly aimed at *both* the "non-believers" and the "believers" in the integrity of the Apollo Moon landings.

THE RACE FOR SPACE

Before undertaking in earnest this hopefully thorough examination of the facts that have led many to claim that the manned Moon landings of the late sixties and early seventies were bogus, it may first be necessary to acquaint you—particularly those of you who were not around or were very young at the time in question—with a brief outline history of the "space race" between America and the Soviet Union, which culminated in the Apollo flights and the eventual claimed landings upon the Moon.

Of course, manned space flight, first by the Russians and then by the Americans, was not the start of the space program. The space race between East and West could be said to have been going on in some way, shape or form—particularly research—since the end of the Second World War.

The Nazis had already left both the Allies and the Soviet bloc the legacy of man's first potential space vehicle. This was the V2 rocket. The possibility of using such a vehicle to lift instruments beyond Earth's gravitational pull and into orbit for many peaceful as well as military purposes, was foreseen long before the world's first successful satellite was launched in 1957.

This was not American, it was Soviet, and it was called *Sputnik* (which in Russian means *Fellow Traveler*).It was carried up into Earth orbit by a rocket called *Soyuz,* which had been developed by the Soviet Union from the German V2 rocket.

First blood to the Soviets!

Triumph

The Soviet space program soon notched up another series of spectacular triumphs. It sent the first living, breathing mammal into space, a dog named *Laika*, although the canine cosmonaut perished before the end of the flight. This was soon followed, though, by sending *two* dogs into orbit, *Strelka* and *Belka*, and then returning them to Earth alive! America's response was to send a chimpanzee called *Ham* into space. NASA likewise managed to return the animal to Earth, but only by the skin of the terrified primate's teeth! Then, on April 12 1961, the Russians launched a *Soyuz* rocket carrying a space capsule capable of carrying something considerably larger than a dog or a chimpanzee. But it was not the design or size of the capsule that captured the world's attention and imagination...

It was the *man* inside it!

Cosmonaut Yuri Gagarin became the first human being in space for a total of 168 minutes. By the 169th minute, Gagarin was the world's most famous man. His spacecraft—that is what his vehicle had become—was named *Vostok 1*. Unlike some of his later American counterparts, the young cosmonaut basked in the glory of his tremendous achievement from that moment on until, of course, his sad and ironic death in a plane crash in 1968. Previously unheard of, even by most Russians, no one was ever more deserving of having the spotlight of success trained upon him for the rest of his life than Yuri Gagarin. The handsome young Russian was not just an undisputed Hero of the Soviet Union; he was a hero to all mankind. He quickly became an even bigger celebrity in the West than he was in the East!

The story goes that he was as bright as a button on the morning of the launch. He announced to all that he had slept very well the night before.

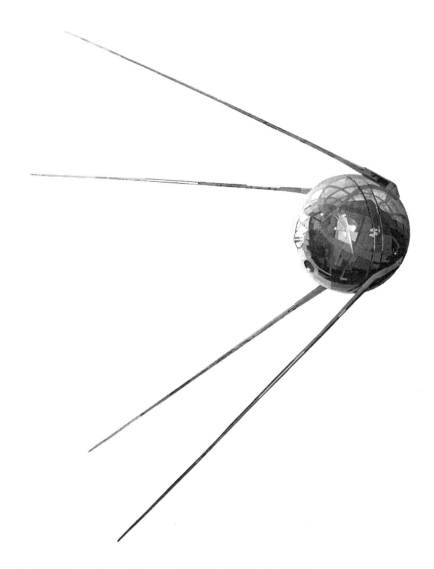

Sputnik: *Beep…Beep…Beep … First blood to the Soviets!*

"Yuri, how could you sleep?" asked a colleague. "We've all been restless and excited all night waiting for the launch—and we're all staying on the ground!"

"How could I function properly on a day like today *without* sleep?" Gagarin is alleged to have replied. "It was my *duty* to sleep—so I slept."

Soviets 2 United States 0

The Americans were not too far behind the Russians, though, especially with their own Mercury manned space flight project. The first American to follow Gagarin into space was Alan Shepard, who later commanded the Apollo 14 mission (allegedly to the Moon). Rear Admiral Shepard was one of the Mercury team named by NASA in April 1959, and he became America's first astronaut and mankind's second. On May 5 1961, in the Freedom 7 spacecraft, he was launched by a Redstone launch vehicle on a ballistic trajectory suborbital flight, which carried him to an altitude of 116 statute miles. True, it was a bold, brave, well-planned mission, even if Shepard never actually went into orbit, and the plucky astronaut was rightly hailed as a great American hero. But it still seemed that everything NASA was achieving was simply following the vapor trails already left by the Soviet Union. To ensure that their rivals could claim no aspect of space travel as "a first," the Soviets then cheekily sent a woman into space. Her name was Valentina Tereshkova.

There remained only one achievable goal as a first for the United States. Only one "killer punch" left, that might put paid to the idea that Communism worked better than Capitalism, which, at the end of the day, was primarily what the space race was all about. It was about a clash of ideologies and attempting to prove that the one had supremacy over the other and, by its very nature, always would have. That goal was, of course, to successfully land a living, breathing human being on the Moon and then bring that man or woman back alive. Ill-fated American President John F. Kennedy, in a speech to Congress in 1962, said, "We must make it our goal to land a man on the Moon by the end of the decade." In these politically correct times I can only presume he meant "or a woman."

The successor to the American Mercury series was a project called Gemini, which involved sending two astronauts, hence the title of the project (*Gemini—The Twins*), into space, for periods of up to two weeks duration. There were ten successful manned Gemini flights between 1965 and 1966 and it was during this project that the Earth was first photographed as a full disc. It was also the project that used computer-assisted re-entry for the first time.

Thus the three-astronaut Apollo series was eventually built upon the foundations already laid by the earlier Mercury and Gemini manned projects. With NASA staff working literally around the clock for months on end and a concentrated program of astronaut training, it was thought that the first three man expedition, Apollo 1, should be ready to go sometime early in 1967.

An Explosion of Pure Oxygen

However, on January 27 1967, with the three crew members strapped in their harnesses awaiting a test run; the countdown proceeding as normal and with all systems apparently "go," an explosion of pure oxygen in the Apollo space capsule caused an intense fire. The three astronauts: Virgil "Gus" Grissom, Walter Chaffee, and Ed White, were burnt to death in seconds, trapped inside their metal coffin as a horrified world looked helplessly on.

It was an unforeseen, unprecedented, total disaster. NASA, though, somehow strove on against all the odds, not the least of which was a shocked and angry American general public and the bloodthirsty media purporting to represent it. Afraid of an even greater hue and cry and the inevitable budget cutbacks—or even total abandonment of the space program!—NASA wisely decided not to risk any more human lives for a time. True, Apollo was intended to be a *manned* space program, but there was an obvious need to test, re-test and re-test again all the hardware and equipment that those astronauts would be utilizing, before placing more precious lives in jeopardy.

The amount of despondency and desperation felt in the offices and workshops of America's National Aeronautics and Space Administration in those days can only be imagined. The Soviets were still managing to put people into space and bring them back alive—no problem. They, the Americans, it seemed, could no longer even get them off the launch pad without first burning them to a crisp! It all

Apollo 1 should have been ready to go in January 1967.

seemed as if they had taken "one small step" forward, only to then take two very large steps back. But press on they must.

Apollos 2 through 6 were all about getting the machinery and technology needed for a manned shot at the Moon *off* the ground and into orbit, first around the Earth and then around the Moon. Of course, there were further mishaps, glitches, gremlins, and technical problems that some of the greatest brains on Earth had to wrestle with and iron out time and time again. Eventually, though, all the hard work and technological wizardry came together. And it came together in a form still bearing the same mythical deity's name, only this time, it seems, the fabulous Greek was also wearing a lucky number…

Mission name: **Apollo 7**

Launch vehicle: Saturn 1B

Mission dates: October 11–22, 1968

Crew: Walter M. Schirra Jr. (Capt.)
       Donn F. Eisele
       R. Walter Cunningham

Mission duration: 10 days 20 hours

Mission accomplishments: 163 Earth orbits. First manned CSM (Command/Service Module) operations in lunar landing program. First live TV from manned spacecraft.

A Greek god and a lucky number

This successful mission was soon followed by...

Mission name: **Apollo 8**

Launch vehicle: Saturn V

Mission dates: December 21–27, 1968

Crew: Frank Borman (Capt.)
     James A. Lovell, Jr.
     William A. Anders

Mission duration: 06 days 03 hours

Mission accomplishments: Flew to Moon and made 10 orbits over a period of 20 hours. First manned lunar orbital mission. Support facilities tested. Photographs taken of Earth and Moon. Live TV broadcasts.

Mission name: **Apollo 9** (Command Module *Gumdrop*, Lunar Module *Spider*).

Launch vehicle: Saturn V

Mission dates: March 03–13, 1969

Crew: James A. McDivitt (Capt.)
     David R. Scott
     Russell L. Schweikart

Mission duration: 10 days 01 hours

Mission accomplishments: First manned flight of all Moon expedition hardware in Earth orbit. Schweikart performed thirty-seven minutes EVA (extra vehicular activity). Human reactions to space and weightlessness tested during 152 orbits. First manned flight of LM (Lunar Module).

Mission name: **Apollo 10** (*Charlie Brown* and *Snoopy*)

Launch vehicle: Saturn V

Mission dates: May 18–26, 1969

Crew: Eugene A. Cernan (Capt.)
     John W. Young
     Thomas P. Stafford

Mission duration: 08 days 03 minutes

Mission accomplishments: Dress rehearsal for Moon landing. First manned CSM/LM operations in cislunar and lunar environment. Simulation of first lunar landing profile. In lunar orbit sixty-one-point-six hours with thirty-one orbits. LM taken to within 15,243m (50,000 ft) of lunar surface. First live color TV from space. LM ascent stage jettisoned in orbit.

Then, at long last, the one we had all been waiting for...

Mission name: **Apollo 11** (*Columbia* and *Eagle*)

Launch vehicle: Saturn V

Mission dates: July 16–24, 1969

Crew: Neil A. Armstrong (Capt.)
      Michael Collins
      Edwin E. Aldrin

Mission duration: 08 days 03 hours 18 minutes

Mission accomplishments: First manned lunar landing followed by first lunar surface EVA. The LM landed on the Moon at 4:17 a.m. EDT (Eastern Daylight Time). Armstrong radioed once the dust had settled. "Houston, Tranquility Base here. The Eagle has landed." After the inevitable time lapse "beep," a strong, clear voice from Earth replied, "Houston to Tranquility Base, we copy on the ground. You have a bunch of guys down here who were about to turn blue. Thanks a lot." Landing site: Sea of Tranquility. Landing co-ordinates: 0.71 degrees North, 23.63 degrees East. Extra Vehicular Activity (EVA): 02 hours 31 minutes. As he stepped onto the surface of the Moon, Armstrong apparently said in error, "That's one small step for man, one giant leap for mankind." He should have said, "...one small step for *a* man." American flag and instruments deployed. Plaque unveiled on the descent stage (to be left behind) which bore the inscription: *Here Men from the planet Earth first set foot upon the Moon July 20, 1969 A.D. We came in Peace for all Mankind*. Lunar surface stay time was twenty-one-pint-six hours. Fifty-nine-point-five hours were spent in lunar orbit, with thirty orbits. After blast off from surface and docking with command module, LM ascent stage left in orbit around Moon. 20kg (44 lbs) of moon rock gathered for analysis back on Earth.

Mission name: **Apollo 12** (*Yankee Clipper* and *Intrepid*)

Launch vehicle: Saturn V

Mission dates: November 14–24, 1969

Crew: Charles Conrad Jr. (Capt.)

The first picture: Apollo 11. The shadow of the LM on the Moon. Oh, it was the LM all right, but…hu-hum…was it on the Moon?

Richard F. Gordon
Alan L. Bean

Mission duration: 10 days, 04 hours, 36 minutes

Mission accomplishments: Intrepid (LM) landed on Moon in Ocean of Storms, 3.04 degrees South, 23.42 degrees West. Retrieved parts of Surveyor 3, an unmanned probe landed on Moon April 1967. Apollo Lunar Surface Experiments Package (ALSEP) deployed. Time on lunar surface thirty-one-point-five hours. Time in lunar orbit eighty-nine hours with forty-five orbits. LM ascent stage deliberately impacted on Moon causing the planet to "ring like a bell," according to instruments back on Earth, as if to suggest it was largely hollow. 34kg (75 lbs) of moon rock gathered for analysis back home.

Then, unlucky for some…

Mission name: **Apollo 13** (*Odyssey* and *Aquarius*)

Launch vehicle: Saturn V

Mission dates: April 11–17, 1970

Crew: James A. Lovell, Jr. (Capt.)
      John L. Swigert, Jr.
      Fred W. Haise, Jr.

Mission duration: 05 days 22.9 hours

Mission accomplishments: Intended third Moon landing and EVA aborted due to rupture of service module oxygen tank. The real success of this mission was the triumph of the human mind and spirit over sheer adversity, and belonged mainly to Mission Control at Houston who—through various trials and tribulations not the least of which was a drastic shortage of oxygen in the spacecraft—successfully guided the three very brave astronauts into Moon orbit and then safely back to splashdown on Earth. Spent stages successfully impacted on Moon.

Mission name: **Apollo 14** (*Kitty Hawk* and *Antares*)

Launch vehicle: Saturn V

Mission dates: January 31–February 09, 1971

Crew: Alan B. Shepard, Jr. (Capt.)
      Stuart A. Roosa
      Edgar G. Mitchell

Mission duration: 09 days

Mission accomplishments: Moon landing at Fra Mauro. Landing coordinates: 3.65 degrees South, 17.48 degrees West. ALSEP and other instruments deployed. Lunar surface stay-time thirty-three-point-five hours. Sixty-seven hours in lunar orbit with thirty-four orbits. Two EVAs amounting to nine hours, twenty-five minutes. Third stage impacted on Moon. 42kg (94 lbs) of moon rock gathered using hand cart for first time to transport material.

Mission name: **Apollo 15** (*Endeavor* and *Falcon*)

Launch vehicle: Saturn V

Mission dates: July 26–August 07, 1971

Crew: David R. Scott (Capt.)
      James B. Irwin
      Alfred M. Worden

Mission duration: 12 days 17 hours 12 minutes

Mission accomplishments: Landed on Moon Hadley-Apennine region near Apennine Mts. Co-ordinates: 26.08 degrees North, 3.66 degrees East. Three EVAs lasting total of ten hours, thirty-six minutes. Worden performed thirty-eight minutes EVA on way back to Earth. First mission to carry orbital sensors in service module of CSM. ALSEP deployed. Scientific payload landed on Moon doubled. Improved spacesuits gave increased mobility and stay-time. Lunar surface stay-time sixty-six-point-nine hours. Lunar Roving Vehicle (LRV) deployed first time. This electric-powered, 4-wheel drive car traversed total 27.9km (17 miles). One hundred forty-five hours spent in lunar orbit with seventy-four orbits. Small sub-satellite left in lunar orbit for first time. 6.6kg (169 lbs) of moon rock gathered.

Mission name: **Apollo 16** (*Casper* and *Orion*)

Launch vehicle: Saturn V

Mission dates: April 16–27 1972

Crew: John W. Young (Capt.)
        Thomas K. Mattingly II
        Charles M. Duke, Jr.

Mission duration: 11 days 01 hours 51 minutes

Mission accomplishments: Moon landing at Descartes Highlands. Co-ordinates: 8.97 degrees South, 15.51 degrees East. First study of highlands area. Selected surface experiments deployed. Ultraviolet camera/spectrograph used for first time on Moon. LRV used second time. Lunar surface stay-time seventy-one hours. In lunar orbit 126 hours with sixty-four orbits. Mattingly performed one-hour in-flight EVA. 95.8kg (213 lbs) of lunar samples collected.

And then, just when everything was going so well that we wondered where it was all going to end, (surely, at this rate, we'll have a man or woman on Mars in a couple of years!) it all did come to an abrupt and seemingly untimely end, with one final mission…

Mission name: **Apollo 17** (*America* and *Challenger)*

Launch vehicle: Saturn V

Mission dates: December 7–19, 1972

Crew: Eugene A. Cernan (Capt.)
        Ronald E. Evans
        Harrison H. Schmitt

Mission duration: 12 days 13 hours 52 minutes

Mission accomplishments: Moon landing Taurus-Litrow highlands and valley area. Co-ordinates: 20.16 degrees North, 30.77 degrees East. Three EVAs of twenty-two hours, four minutes total. Evans performed trans-Earth EVA lasting one hour, six minutes. First scientist (Schmitt) to land on Moon. Sixth automated research station set up. LRV traversed 30.5km (nineteen miles). Lunar surface stay-time seventy-five hours. In lunar orbit seventeen hours. 110.4kg (243 lbs) of material gathered.

And that was it. Not one single manned venture beyond the highly radioactive Van Allen belts (which encompass Earth around 500-700 miles up at closest point) and into outer space proper ever since. And that was more than thirty years ago…

The People versus NASA and the U.S. Government
The Prosecution

2

The Charges Followed by the Testimony of the First Witness for the Prosecution

So it would appear, then, that in a period of little more than two short years after the tragedy of Apollo 1, NASA dramatically reversed its fortunes from abject and dismal failure on January 27, 1969—total disaster for the families of the three brave astronauts who lost their lives that day—by then accomplishing mankind's greatest achievement of all time by July 20, 1969!

Obviously not prepared to rest upon its laurels, it seems, NASA then went on to successfully repeat what can only be described as a technological miracle a further *five* times over the ensuing three years. And their one apparent "failure," Apollo 13, turned out to be the greatest miracle of all...

Three years after the date when it seemed NASA could not actually lift three human beings off the ground without first reducing them to ashes, it was now apparently able to withstand a major malfunction to one of its spacecraft while actually in space! Not only that, but it was able to allow the damaged vehicle to continue upon its journey to the Moon, then orbit the Moon, jettison the redundant LM, return to Earth, successfully separate the space capsule from the damaged Command Module, successfully achieve re-entry, and then successfully deploy the three giant parachutes that would guarantee a nice, safe, gentle landing in the sea for the three incredibly lucky, plucky astronauts—who then emerged from a capsule now bobbing about on the ocean without so much as scratch! In fact, they likely sustained more scratches while later shaving off the beards they had all grown during their classic misadventure!

Perhaps it is just as well that more than thirty years have passed since these truly remarkable events allegedly took place. (Allegedly? But, as that newsreader said, some of us *saw* it all happen, didn't we? No we didn't. We saw a *film* of it all happening!). Perhaps now is the time, with the euphoria and passion of those heady days but a distant memory, to look at it all again in the cool, clear light of day.

Unfortunately—sadly even—the light may be clear; but the air isn't. It isn't all that fresh either. Like those millions of others still shocked by the tragic Shuttle losses in more recent times, I, too, smell at least 50% of a rat. So, let us call them all to task and put some into the dock and some into the witness box, taking the written and published testimony of each as their sworn oath to tell us The Truth, The Whole Truth, and Nothing But The Truth, or So Help Them God. Let us allow them all their day in *our* court, presenting *their* evidence and saying *their* pieces, and let us see if we cannot somehow sort out this bloody mess between us, once and for all!

Members of the jury, I now address you in my role of judge. I find there is definitely a case to answer and very serious allegations indeed that need to be formally brought. The defendants will rise and face the People while the charges are read out:

THE CHARGES

National Aeronautics & Space Administration, who shall hereinafter, for the most part, be referred to acronymically as NASA, as the sole agency working under the direct auspices of the United States Government for the purpose of space exploration, and which government stands jointly indicted beside you in the dock, you are charged *in absentia* (in your absence) that between the inclusive dates of December 21 in the Year of Our Lord Nineteen Hundred and Sixty Eight and December 19 in the Year of Our Lord Nineteen Hundred and Seventy Two, you did knowingly, deliberately and cynically, using taxpayers' money amounting to as much as 40 billion dollars American, an amount allocated by the United States Government on behalf of a largely unsuspecting public, perpetrate a fraud—I shy away from using the too mild a term "hoax"—in the form of a calculated and cunning deception, in that you made a series of allegedly live television broadcasts purporting to show manned expeditions into space culminating in the orbiting of, or actual manned landings upon, the Moon as they were happening.

You are also charged with distributing, at a later date, photographs and film footage purporting to be of these alleged happenings to the world's media. You are also charged with distributing to the world's academic and scientific establishments for detailed study and analysis, samples of material allegedly gathered up from the Moon's surface on these allegedly fraudulent expeditions when, in fact, such samples were either manufactured under laboratory conditions, or were gathered up by the robotic arms of earlier unmanned space appliances, or were parts of retrieved meteorites that had already crashed to Earth, or were a combination of all three alternatives and that this was done with fraudulent intent.

To yet further advance this pretext of reality, it is alleged that you then decided that the general public and the media—sections of whom were already suspicious—would soon sense that a deception was taking place if things progressed *too* well with the Moon landings program. Your problem was to be "seen" to fail but not to lose face by incurring further fatalities. So, the allegation is that you did again deliberately and cynically interrupt the planned program of "successful" missions with yet another purportedly live broadcast. This time for the alleged mission to the Moon bearing the number thirteen, with full awareness that widespread superstitious belief has it that the said number is unlucky, and did allow this mission to "fail," although all too fortuitously, it seems, without loss of life.

Counsel for NASA, how does your client plead?

Defense: Not Guilty.

Judge: Counsel for the United States Government, how does your client plead?

Defense: Not Guilty.

Judge: Very well. Both pleas of Not Guilty to all charges will be entered for the record and we can proceed. But before we begin in earnest, I first need to address the jury again:

Members of the jury, I remind you that from this point onward until after the end of both the People's case and the case for the defense, I shall for a time be forsaking my role as judge. (In fact, the only role that will remain consistent throughout will be as the author of this book). I will begin again by first acting as counsel for the People—in other words, the prosecution. Later on, in like fashion, I shall be acting as counsel for the defense. I shall not be returning fully to my role as judge until my summing up at the latter end of the book. Please be mindful that from here on, until the prosecution case is completed, *my* comments inserted between published sources will either be made purely on behalf of the People or they will be neutral. Likewise during the presentation of the defense case later, my comments inserted between published sources will now favor the defendants (who will, in effect, have become *my clients*), or they will again be neutral. This information will *not* be repeated so it is as well the jury make mental note of it now. Lengthy interruptions to published sources will always be heralded by the word "Paused." My comments will then be followed by the continuation of the published source. The odd aside, observation or explanation inserted both in published sources *and* my comments will often—not always—be prefixed by the italicized initials *A.N.* (short for Author's Note). Interventions inserted into published sources will always be contained between square brackets [ ] and interventions inserted into *my* comments between round brackets ( ). Further note that **round** brackets ( ) appearing in published sources were already part of the original text. As in any normal trial, the judge, or indeed opposing counsel, has the right to intervene at any stage of this trial whenever it is felt something might be amiss or requires clarification. Some editing of published sources has unfortunately been necessary here and there if only to prevent any misunderstanding or ambiguity, which will not do in court. Most of the testimony given to this court will be from American sources containing American spellings, which remain unchanged. For the sake of consistency, therefore, and despite the fact that these proceedings are taking place in an imaginary English court of law, American spellings (*e.g.* "defense" as opposed to "defence") are being used throughout this book. But no more talk of "this book" unless it becomes necessary for me to refer to it as such purely as its author. Let us attempt, as far as possible, to stay with the illusion that we are all sat in court and that you are the jury. Strictly speaking, therefore, you are *hearing* this case, not reading it. You will, though, sometimes be asked to refer to your "case papers," which will of course be this book.

And so, down to business. Is learned counsel for the prosecution ready to present the People's case?

Prosecution: I inform the learned judge that indeed I am.

Judge: Then by all means proceed.

Prosecution: My thanks to the learned judge. I shall call my first witness.

I call engineer and inventor Ralph René.

RALPH RENÉ'S TESTIMONY

Source *www.apfn.org/apfn/moon.htm*—© 2001. Article entitled: *Was The Apollo Moon landing Fake? Did man really walk on the Moon or was it the ultimate camera trick?* asks David Milne.

Begins—

In the early hours of May 16, 1990, after a week spent watching old video footage of man on the Moon, a thought was turning into an obsession in the mind of Ralph René.

"How can the flag be fluttering?" the 47 year old American kept asking himself when there's no wind on the atmosphere free Moon. That moment was the beginning of an incredible Space Odyssey for the self-taught engineer from New Jersey.

He started investigating the Apollo Moon landings, scouring every NASA film, photo and report with a growing sense of wonder, until finally reaching an awesome conclusion: America had never put a man on the Moon. The giant leap for mankind was fake.

It is of course the conspiracy theory to end all conspiracy theories. But René has now put all his findings into a startling book entitled *NASA Mooned America.* Published by himself, it is being sold by mail order—and is a compelling read.

The story lifts off in 1961 with Russia firing Yuri Gagarin into space, leaving a panicked America trailing in the space race. At an emergency meeting of Congress, President Kennedy proposed the ultimate face-saver—putting a man on the Moon. With an impassioned speech he secured the plan an unbelievable 40 billion dollars!

And so, says René (and a growing number of astro-physicists are beginning to agree with him), the great Moon hoax was born. Between 1969 and 1972, seven Apollo ships headed to the Moon. Six claim to have made it, with the ill-fated Apollo 13—whose oxygen tanks apparently exploded halfway—being the only casualty. But for the exception of the known [Moon] rocks, which could have been easily mocked up in a lab, the photographs and film footage are the only proof that the *Eagle* [and all the other lunar modules] ever landed. And René believes they are fake.

For a start, he says, the TV footage was hopeless. The world tuned in to watch what looked like two blurred white ghosts throwing rocks and dust. Part of the reason for the low quality was that, strangely, NASA provided no direct link-up. So networks had to film man's greatest achievement from a TV screen in Houston—a deliberate ploy, says René, so that nobody could properly examine it.

By contrast, the still photos were stunning. Yet that is just the problem. The astronauts took thousands of pictures, each one perfectly exposed and sharply focused. Not one was badly composed or even blurred.

—Paused.

Defense: Objection! That last statement by the witness is purely speculative, as only NASA is in possession of the precise figures for the number of exposures. It is my understanding that the Apollo astronauts did indeed take many thousands of photographs between them, of which, NASA claims, comparatively *few* were deemed worthy of publication. The astronauts were *not* professional photographers, so to suggest that there was not one single poorly taken or badly exposed shot is, in itself, quite ludicrous.

Judge: Objection sustained. Although, I would remind the defense that the witness has an obligation only to tell the truth as he sees it.

Defense: As the learned judge pleases. And I am obliged to him for reminding me of that point.

Judge: The witness may continue giving evidence.

Ralph René continued—

As René points out, that's not all: The cameras had no white meters or view ponders. So the astronauts achieved this feat without being able to see what they were doing. Their film stock was unaffected by the intense peaks and powerful cosmic radiation on the Moon, conditions which should have made it useless. They managed to adjust their cameras, change film and swap filters in pressurized suits. It should have been almost impossible with gloves on their fingers.

—Paused.

Here, the article contains remarks by an impending witness for the prosecution, award-winning photographer David Percy, who is convinced some of the NASA photographs supposedly taken on the Moon are obvious fakes to an expert like himself. But we will hear his opinions in greater detail later. But René puts the question:

Ralph René continued—

"Why would anyone fake pictures of an event that actually happened?" asks René. But the questions do not stop there. Outer space is awash with deadly radiation that emanates from solar flares firing out from the sun. Standard astronauts orbiting Earth in near space, like those who recently fixed the Hubble telescope, are protected by Earth's Van Allen belts. [*A.N.* Mentioned earlier. These belts around the Earth *trap* deadly incoming radiation from space. Hence their own high level of radioactivity!]. But the Moon is 240,000 miles distant, way outside this safe band. And, during the Apollo flights, astronomical data shows there were no less than 1,485 such [solar] flares!

John Mauldin, a physicist who works for NASA, once said shielding of at least two meters thickness would be needed [against such harmful radiation]. Yet the walls of the Lunar Landers, which took the astronauts from the spaceship to the Moon's surface, were, according to NASA, "…about the thickness of heavy duty aluminum foil."

How could that stop this deadly radiation? And if the astronauts were protected by their spacesuits, why didn't rescue workers use such protective gear at the Chernobyl meltdown, which released only a fraction of the dose astronauts would encounter? Not one Apollo astronaut ever contracted cancer—not even the Apollo 16 crew who were on their way to the Moon when a big flare started. "They should have been fried," says René.

Furthermore, every Apollo mission before number 11 (the first to supposedly put a man on the Moon) was plagued with around 20,000 defects apiece. Yet, with the exception of Apollo 13, NASA claims there wasn't one major technical problem on any of their Moon missions. "The odds against this are so unlikely that God must have been the co-pilot," says René.

Several years after NASA claimed its first Moon landing, Buzz Aldrin, "the second man on the Moon," was asked at a banquet what it felt like to step onto the lunar surface. Aldrin staggered to his feet and left the room crying uncontrollably. It would not be the last time he did this. "It strikes me that he's suffering from trying to live out a very big lie," says René. Aldrin may also fear for his life.

Virgil Grissom, a NASA astronaut who baited [made sarcastic comments about] the Apollo program, was due to pilot Apollo 1 as part of the landings build-up. In January 1967, he hung a lemon on his Apollo capsule [in America, unroadworthy automobiles are called "lemons"] and told his wife Betty, "If there is ever a serious accident in the space program, it's likely to be me."

Nobody knows what fuelled his fears, but by the end of the month, he and his two co-pilots were dead, burnt to death during a test run when their capsule, pumped full of high pressure pure oxygen, exploded.

Scientists could not believe NASA's carelessness—even a chemistry student in high school knows high-pressure oxygen is extremely explosive, René says. In fact, before the first manned Apollo flight even cleared the launch pad, a total of 11 would-be astronauts were dead. Apart from the three that were incinerated, seven died in plane crashes and one in a car smash. Now this is a spectacular accident rate.

"One wonders if these 'accidents' weren't NASA's way of correcting mistakes," says René. "Of saying that some of these men didn't have the sort of 'right stuff' they were looking for."

NASA will not respond to any of these claims; their press office will only say that the Moon landings happened and that the pictures were real. But a NASA public affairs officer called Julian Scheer once delighted 200 guests at a private party with footage of astronauts apparently on a [mock-up] landscape. It had been made on a mission film set and was identical to what NASA claimed was the real lunar landscape.

Grissom hung a lemon on his space capsule. Was he suggesting that it wasn't fit to fly?

"The purpose of this film," Scheer told the enthralled group, "is to indicate that you really can fake things on the ground, almost to the point of deception." He then invited his audience to "Come to your own decision about whether or not man actually did walk on the Moon."

A sudden attack of honesty? "You bet," says René, who claims the only real thing about the Apollo missions were the lift offs. "The astronauts simply had to be on board," he says, in case the rocket exploded. "It was the easiest way to ensure NASA wasn't left with three astronauts who ought to be dead," he claims, adding that they came down a day or so later, out of the public eye (global surveillance wasn't what it is now) and into the safe hands of NASA officials, who whisked them off to prepare for the big day a week later.

And now NASA is planning another giant step—Project Outreach, a one trillion dollar manned mission to Mars. "Think what they'll be able to mock up with today's computer graphics," says René, chillingly. "Special effects were in their infancy in the 60s. This time round we will have no way of determining the truth."

—End of this article, not of Ralph René's testimony, which continues:

Begins—

As this issue of *Wired* goes to press, a new book is headed for the stores: *Was It Only a Paper Moon?* by Ralph René, "a scientist and patented inventor." Published by tiny Victoria House Press in New York, in what it has announced will be a first run of 'at least 100,000 copies,' *Paper Moon* supposedly presents the latest scientific findings regarding the Moon landing. René offers data suggesting, among other things, that without an impractical shield about two meters thick, the spacemen 'would have been cooked by radiation' during the journey. Ergo, the lunar endeavors were impossible, and were cynically faked at the expense of gullible people everywhere. [*A.N.* Further extracts from this extensive *Wired* article will appear later on in these proceedings].

—End.

§ § § § §

My thanks to Mr. René and those who have assisted in bringing his testimony to this court. The witness may step down.

3

A Re-Awakened Controversy

On Thursday evening, February 15, 2001, at 9.00 p.m. Eastern Time, in the United States, Fox Television ran a 'special' program. It was called *Conspiracy Theory: DID WE LAND ON THE MOON?*

It could be said to have placed the cat well and truly among the pigeons—if not quite letting it out of the bag!—as far as rekindling the "Were the Moon landings a hoax?" debate, and to have sounded the ringside bell that has brought a somewhat bruised and battered group of contenders back out to fight again.

This hour-long program was visited upon British audiences later on that same year, causing a certain amount of controversy in the Fleet Street press and, no doubt—from the personal experience of the author—around many a dinner table and barbecue fire that summer.

For those members of the jury who at the time missed the program for whatever reason—not discounting total disinterest—here are the main points brought by the "pro-hoax," or "yes, it was all a scam" lobby on the show:

THE PRO HOAX TESTIMONY OF THE FOX TV SPECIAL

Source: Web page entitled: *Comments on the Fox Moonlanding Hoax Special* by Jim Scotti, Planetary Scientist, University of Arizona. © Jim Scotti, May 2001.

(*A.N.* Before proceeding with this source, members of the jury, I have a duty to point out that planetary scientist Jim Scotti, of the University of Arizona, is *not* himself a witness for the prosecution. On the contrary, he will later be called as a witness for the defense, whereupon he will then attempt to answer the questions, suppositions, allegations and charges commonly ranged at NASA, which he lists below).

Begins—

No stars in the images

As usual, this was about the first argument used by the hoax believers to debunk the lunar landings. We see no stars in the images. The Hubble [space] telescope provides crystal clear images showing millions of stars. The Apollo astronauts were armed with some of the best cameras available thirty years ago, which also took crystal clear images of the LM and surrounding area, not to mention of the astronauts themselves taken by each other, but no stars appear in these images.

Likelihood of success too small

Kaysing [Bill Kaysing, Moon hoax investigator] claimed that the chance of a successful landing on the Moon was calculated to be 0.017 percent! Therefore the landings had to be faked, particularly as this success was repeated a further *five* times!

Capricorn One

The Producer of *Capricorn One* [a film about a faked landing on Mars] figures that NASA's huge $40 billion (sic) budget compared to *Capricorn One*'s $4.0 million budget could easily have faked the Moon landings. Comparisons are made between the similarity of the Moon landing scenes and those in the movie.

Area 51

Kaysing claims that Area 51 [a top secret U.S. military 'black project' base at Groom Dry lake, Nevada, where UFO believers are convinced back-engineering of, and research into, captured superior technology to anything yet produced by humanity continues to this day] is where they filmed the Apollo hoax, and similarities to the lunar landscape in the desert surrounding Area 51, as well as craters near the site, is evidence of that.

Astronaut deaths

The show claimed that 10 astronauts died "in mysterious circumstances" during the Apollo program, although these included deaths in service in projects not immediately connected with Apollo. These deaths were:

Ed Givens (car accident)

Ted Freeman (T-38 crash)

C.C. Williams (T-38 accident)

Elliot See (T-38 accident)

Charlie Bassett (T-38 accident)

Gus Grissom (Apollo 1 fire)

Ed White (Apollo 1 fire)

Roger Chaffee (Apollo 1 fire)

Mike Adams (X-15 crash)

Robert Lawrence (air crash)

The show went on to claim that the Apollo 1 fire may have been a murder conspiracy to silence astronaut Gus Grissom's outspoken criticism of the Space Program. A particularly infamous—as far as NASA was concerned!—episode was when Grissom, in apparent disgust, hung a lemon on an Apollo simulator.

The show also claimed that the death of NASA worker Thomas Baron was murder, and a cover-up of a 500-page report [Baron had produced] on the Apollo 1 accident.

No blast crater under the Lunar Module

The hoax believers claim that the LM descent stage used its full thrust of 10,000 pounds at lunar landing and that it should have excavated a large blast crater under the LM.

No dust on the LM footpads

Kaysing cites the lack of dust on the LM footpads as clear evidence of fakery, the suggestion being it was a detail overlooked by the film studio. Unless the astronauts on each mission decided to do some housework before taking their snapshots—which is highly unlikely!—layers of dust should have shown up clearly on the landing pads.

Lack of sound from the LM descent engine

Kaysing claims that you should hear the sound of the descent engine in the audio from the landings, but there is nothing resembling background engine noise in any of the landing audios.

Footprints around the LM

Deep footprints [of the astronauts] are visible in the dust around the immediate area of LM. The hoax believers suggest that the blast of the engine of the descent stage would have dispersed all loose material from the immediate vicinity, making the leaving of such large indentations in the dust impossible that close to the LM.

Pictures of space-suited crewmembers inside buildings

The hoax proponents [in the show] cited pictures of [Apollo] crewmen where [in the background, if one studied the photographs closely enough] walls, overhead lighting, hoses, tiled floors, etc., could be made out.

The "Prototype LM" accident

The program claims that a "Prototype LM" (LLRV) was tested on Earth by Neil Armstrong and that during one pre-mission test flight; Armstrong was unable to control the vehicle and had to eject. The question that the program was asking was this: How was it possible to land the "untested" LLRV so flawlessly *six* times [on an untried lunar landscape] when the "prototype" had so much trouble on Earth?

No rocket plume in video of ascent stage lift-off

Kaysing claims that we should have seen a rocket plume from the engine of the LM ascent stage during lift-off [from the lunar surface] video footage.

Flags waving in the breeze

There is no oxygen, no atmosphere and therefore no wind on the Moon, yet in the video footage shown to viewers of the Fox TV program, American "Stars and Stripes" flags could clearly be seen to be "waving" as if hit by a sudden breeze.

Poor quality of video

Claims were made [in the show] that NASA purposefully provided very poor video footage of the first moonwalks. [*A.N.* Indeed it is true that for such a "no expense spared" project such as Apollo, with its 40 billion dollar budget, taking place as it allegedly was in a nation that has probably forgotten more about the art of film-making than will ever be learned by the rest of the world put together, those first pieces of footage left much to be desired. The only reason the public accepted them so readily and unquestioningly, at the time, was because the overwhelming majority—and that includes myself—were stunned to be viewing *any* film supposedly taken on an *alien world*, let alone query its quality!].

Doubling play-back speed of lunar video/s

The show claimed that if the films taken by the astronauts during their EVAs [Extra Vehicular Activities] are played back at double "normal" speed, the movement of the participating figures is compatible with the way they would normally move on Earth, suggesting that ordinary cinematic "slow-motion" techniques were all that was required to hoodwink the public into believing that the space-suited figures they saw hopping and leaping about near the stationary LM, were "men on the Moon."

Not possible to take "absolutely perfect" photographs wearing spacesuits

The Fox show queried how was it possible for so many "absolutely perfect" photographs of the Moon expeditions to be taken by men wearing spacesuits? The cameras were fixed to the astronauts' chests, so guesswork would of essence have come largely into play as far as pointing and focusing was concerned. Also, frequent changes of film [a fairly tricky operation for most of us at the best of times!] would have been required by hands covered in thick, unwieldy gauntlets. Failing that, any change of film would have necessitated the astronaut climbing back aboard the LM, removing the spacesuit, replacing the film, putting the spacesuit back on again, and then resuming EVA, with all the rigmarole that this would have entailed.

Shadows in wrong directions; not parallel

"The shadows are all wrong, too," said the hoax proponents on the show, indicating shadows of closely situated objects on photographs apparently leading off in opposite directions. Some objects were even giving off more than one shadow. This suggested the use of additional lighting sources, as in a studio environment.

Details in the shadows

Detail is clearly visible in some of the shadows cast by various objects in the photographs, say skeptical observers. This would *not* be possible on a planet with no atmosphere like the Moon, they claim.

Identical backgrounds—different places!

The hoax mongers cite cases of exactly the same mountains appearing in the background in images taken from completely different places around a landing site. In the Fox special, as a particularly good example, they showed a picture of the LM against a backdrop of mountains, and then a picture apparently taken from exactly the same spot *without* the LM in the foreground! So, carrying hypothesis to the utter extreme then, the United States landed *another* spacecraft not too far from this particular landing site prior to the second LM touching down, in order to film or photograph the virgin landscape *before* as well as after the landing! What an extraordinary feat! Why did they not bother to tell us about that one? Was it because it was merely an unmanned probe carrying a camera? No, say the "It was all a hoax" mongers. It was a studio test shot of the "set" as it looked before all the "props" were placed upon it. One that was inadvertently released along with all the other so-called "Moon" photos.

Two different days, two different sites, identical shot

The show's hoax believers show us two virtually identical video clips from the Moon Rover vehicle's camera that, according to NASA, were taken on two different days at two different sites.

Reseau marks disappear behind equipment

"Reseau" marks are the "+" marks [in the U.K. we call them "crosshairs," similar to those that appear in a gunsight] that appear on the lunar photographs. These are caused by marks etched onto the *Reseau* plate, which is a clear glass plate mounted immediately in front of the film on the Hasselblad camera. The pro-hoaxers claim that these crosshairs—always in evidence and irremovable other than by tampering—in some instances, actually *disappear*, either all or in part, *behind* persons and objects!

Radiation, Van Allen and solar storms

Possibly the main argument of the hoax believers. The program details the "impossibility" of any human being passing safely through the ultra-high radiation Van Allen belts, which encompass the earth in a zone roughly between altitudes 1000km to 20,000km, without being encased or cocooned in *fourteen feet of lead!* And then, of course, once through the Van Allen belts, the astronauts would now have extremely intense, further deadly radiation from the sun's frequent solar storms to contend with. To the best of the show's knowledge, though, *none* of the so-called "Moon astronauts" had yet contracted any of the known diseases related to over-exposure to harmful radiation.

Lunar surface temperatures too hot

The show's pro-hoax lobby claim that there is no way the equipment used could have withstood the daytime temperatures (or protect for a prolonged period a human being trying to perform tasks on the Moon's surface), which, with no atmosphere or clouds to shield the surface from the searing sun, can reach 250 degrees Fahrenheit or more! Much of the equipment would soon be ruined by the intense heat, not least the film in the cameras. Lunar nights are no less alien and unfriendly, the temperature often dropping to an awesome minus 270 degrees Fahrenheit!

Apollo sites not discernible through telescopes

The pro-hoaxers ask why none of the supposed lunar landing sites have yet revealed themselves unambiguously to our sophisticated optical telescopes? Oh, we know exactly where they are *alleged* to be all right! But even the Hubble telescope cannot show us anything other than virgin lunarscape where the sites are purported to be! Strange that. We can photograph, with apparent crystal clarity, galaxies of stars countless millions of light years away. We can even take pictures of rocks on the surface of Mars. Yet for some strange reason, it seems we cannot pick out even a single lunar descent stage out of six surrounded by what should be undisturbed signs of many hours of human activity all about, plus all associated paraphernalia abandoned thereof. And this through the cloudless, non-existent atmosphere of the closest planet of all!

No attempt to return by USA or Soviets

The hoax proponents find it peculiar that a country should expend at least $40 billion, resulting in the completion of six successful manned space missions [or so NASA claims] and the successful landing upon the Moon of twelve individuals at a cost of nil casualties, only to then allow the "iron to cool," the moment to pass into history and to not follow through with man's greatest adventure of all time. The Soviet Union of course, as it was then, also suddenly and permanently abandoned all attempts to "follow" the Americans to the Moon. The "space activities" of both great powers, it seems, up to and into the new millennium, being confined to the "user friendly" side of the Van Allen belts where, the pro-hoaxers insist, it *had been* all along anyway!

—End.

§ § § § §

I thank the learned planetary scientist for outlining (with some slight assistance from myself) the alleged inconsistencies pertaining to the Apollo project that were raised in the aforementioned Fox TV program. Although these will form the basis of the People's case, I must again reiterate that Dr. Scotti is *not* a prosecution witness and not a believer in any hoax. I am given to understand that this witness, on behalf of the defendants, is confident of being able to refute *all* the allegations above listed.

4

The Wrong Stuff

The following article was posted on the Internet sometime in the month of September 1994. It contains observations regarding the alleged Moon hoax, which are favorable to both the People *and* the defendants in this case. Some extracts and witness statements, therefore, will be made available to the jury when my learned friend, counsel for the defense, presents his clients' case a little later on.

THE TESTIMONY OF *WIRED* MAGAZINE

Source: *Wired* Magazine Issue 2.09—Sep 1994. Article by Rogier van Bakel entitled: *The Wrong Stuff*. Anagram enthusiasts will find that Rogier van Bakel (*rogiernl@ aol.com*) is Brave Ink Galore. He is a Dutch correspondent in Washington, DC. Copyright © 1993-2002 The Condé Nast Publications Inc. All rights reserved. Copyright © 1994-2002 Wired Digital, Inc. All rights reserved.

(*A.N.* The "wrong stuff" is what those applicants who are either rejected or fail to complete their training, or who show unsuitability *after* they complete their training as astronauts, are reckoned to be possessed of).

Begins—

Are you sure we went to the Moon 25 years ago? Are you positive? Millions of Americans believe the Moon landings may have been a U.S.$25 billion swindle, perpetrated by NASA with the latest in communications technology and the best in special effects. Wired plunges into the combat zone between heated conspiracy believers and exasperated NASA officials.

"Columbia, he has landed Tranquility Base. Eagle is at Tranquility. I read you five by. Over." The voice from Houston betrayed no emotion, although this was anything but business as usual. A human being was about to set foot on the Moon for the first time in history, armed only with the Stars and Stripes, some scientific instruments, and an almost reckless, can-do demeanor that had captivated the world.

The reply from Columbia, the command-and-service module that had released the lunar lander 2 hours and 33 minutes earlier, betrayed only equal professional cool. "Yes, I heard the whole thing," Michael Collins said matter-of-factly.

Houston: "Well, it's a good show."

Columbia: "Fantastic."

That's when Neil Armstrong chimed in. "Yeah, I'll second that," said the 38-year-old astronaut, the moonwalker-to-be, America's own Boy Scout, and the most famous man in the—well, in the universe. And even though the static ate away at the clarity of his consonants, Armstrong's sneering tone came through loud and clear. The mission control man heard it too. And he knew what was coming. Sort of…

"A fantastic show," Armstrong said. "The greatest show on Earth, huh, guys?"

There was a moment's silence. Then a cameraman sniggered. And the director sighed, and did what directors do when actors screw up their lines. "Cut," he groaned. He was a heavy-set man in his 50s, and the combination of the long hours and the hot studio lights had started to get to him.

"Shit, Armstrong, if you're gonna be a smart-ass, do it on your own time, all right? We got 25 tired people on this set. We got a billion people who are going to be watching your every move only a week from now. We're on deadline here. Now, do you suppose you could just stick to the script and get it over with? Thank you."

His assistant stepped forward with the slate. "Apollo Moon landing, scene 769/A22, take three," she announced.

"Action!"

"Columbia, he has landed Tranquility Base," the mission control man began again.

—End of this section from *Wired.*

Here, the *Wired* article goes on to include a section on Bill Kaysing, the Moon landings arch debunker who appeared in the Fox TV program. Although not a rocket scientist himself, Kaysing is knowledgeable enough of the subject to have edited publications concerning rocket science. Utilizing this section from *Wired*, I shall therefore call him as my next witness for the prosecution.

I call Bill Kaysing.

BILL KAYSING'S TESTIMONY

Ditto source.

Begins—

Superfraud

The history books lie. So do the encyclopedias and the commemo-
rative videos and the 25-year-old coffee mugs with the proudly smiling faces of Neil Armstrong, Edwin Aldrin, and Michael Collins.
When Armstrong got down from that ladder, proclaiming that it was
only a small step for him but a giant leap for mankind, he was merely
setting foot on a dust-covered sound stage in a top-secret TV studio in
the Nevada desert. NASA's cold warriors and spin-doctors faked the
whole Moon landing. Come to think of it, they faked all *six* Moon
landings—spending around U.S.$25 billion to prove to the world that
not even the Soviets, especially not the Soviets, could hold a candle to
the U.S. when it came to space exploration.

Well, at least, that's the view of writer Bill Kaysing. It is also the
conviction of millions of Americans who have learned to distrust their
government with a passion. Most of these skeptics don't even appear
to be steamed about the alleged superfraud. They shrug and raise their
palms and go about their business. Not Kaysing. He seems to have
never heard a conspiracy theory he doesn't like, and this one tops
them all! For almost 20 years now, he has been trying to get out "the
most electrifying news story of the entire 20th century and possibly of
all time." He has written a book aptly entitled *We Never Went to the
Moon* and won't give up trying to uncover more evidence.

Kaysing, a white-haired, gentle Californian whose energy level
seems mercifully untouched by his 72 years, worked as head of technical publications for the Rocketdyne Research Department at their
Southern California facility from 1956 to 1963. Rocketdyne was the
engine contractor for Apollo.

"NASA couldn't make it to the Moon, and they knew it," asserts
Kaysing, who, after begging out of the "corporate rat race," became a
freelance author of books and newsletters. "In the late '50s, when I
was at Rocketdyne, they did a feasibility study on astronauts landing
on the Moon. They found that the chance of success was something
like .0017 percent. In other words, it was hopeless." As late as 1967,

Kaysing reminds me, three astronauts died in a horrendous fire on the launch pad. "It's also well documented that NASA was often badly managed and had poor quality control. But as of '69, we could suddenly perform manned flight upon manned [space] flight? With complete success! It's just against all statistical odds."

President John F. Kennedy wasn't convinced at all that the endeavor was next to impossible. In fact, he had publicly announced in May 1961 that "landing a man on the Moon and returning him safely to Earth" would be a Number One priority for the U.S., an accomplishment to instill pride in Americans and awe in the rest of the world. And so, Kaysing believes, NASA faked it, acting in accordance with the old adage that in war, the truth is often the first casualty. (Cold wars, he and his fellow conspiracy believers say, are no exception).

To hear him tell it, NASA had good reason to stage Moon landing after Moon landing, instead of simply admitting that lunar strolls would have to remain the stuff of science fiction novels, at least for a while. "They—both NASA and Rocketdyne—wanted the money to keep pouring in," says Kaysing. "I've worked in aerospace long enough to know that was their goal."

Absent Stars

There is an almost instinctive rejoinder to all of this: *But we saw it.* If television ever had a killer appeal, the Moon landing was it. We bought new [TV] sets in droves, flicked them on as zero hour approached and, miraculously, felt ourselves being locked into an intangible but very real oneness with a billion other people. It was our first taste of a virtual community, of cultures docking. It felt good. And now there's this guy telling us that it was all a lie? C'mon! His rockets are a little loose. What proof does he have anyway?

Kaysing points out numerous anomalies in NASA publications, as well as in the TV and still pictures that came from the Moon. For example, there are no stars in many of the photographs taken on the lunar surface. With no atmosphere to diffuse their light, wouldn't stars have to be clearly visible? And why is there no crater beneath the lunar lander, despite the jet of its 10,000-pound-thrust hypergolic engine? How do NASA's experts explain pictures of astronauts on the Moon in which the astronauts' sides and backs are just as well lit as the fronts of their spacesuits—which is inconsistent with the deep, black shadows the harsh sunlight should have been casting? And why is there a line between a sharp foreground and a blurry background in some of the pictures, almost as if special-effects makers had used a so-called 'matte painting' to simulate the farther reaches of the moonscape? "It all points to an unprecedented swindle," Kaysing concludes confidently.

But just how could NASA possibly have pulled it off? How about the TV pictures that billions of people saw over the course of six successful missions? Wasn't there the rocket lifting off from the Cape Kennedy launch pad under the watchful eye of hundreds of thousands of spectators; the capsule with the crew returning to Earth; the moon rock; the hundreds, perhaps thousands, of space-program employees in the know who would have to be relied upon to take the incredible secret to their graves?

Easy, says Kaysing. The rockets took off all right, and with the astronauts on board, but as soon as they were out of sight, the roaring spacecraft set course for the South Polar Sea, jettisoned its crew, and crashed. Later, the crew and the command module were put into a military plane and dropped in the Pacific for "recovery" by an aircraft carrier. (Kaysing claims that he talked with an airline pilot who, *en route* from San Francisco to Tokyo, saw the Apollo 15 command module sliding out of an unidentified cargo plane!!! But he is unable to provide the captain's name or the name of the airline). The moon rocks were made in a NASA geology lab, right here on Earth, he continues. Not very many people on the Apollo project knew about the hoax, as they were only informed on a need-to-know basis. Cash bonuses, promotions, or veiled threats could have ensured the silence of those who were in on the whole scheme.

"Anything is possible." Kaysing likes to paraphrase Alvin Toffler: "He writes that most people are producer/consumers—he calls them prosumers. They go through life not questioning anything. Not knowing anything. Ninety percent of the American population has no idea what's going on in this country. I'd like to be the one to tell them—tell them at least part of it. I'm either going to share the truth about the Moon with them, or I am going to die trying."

As many as 100 million Americans, says Kaysing, are inclined to disbelieve the whole lunar adventure.

—End of Bill Kaysing's testimony as published by *Wired*.

§ § § § §

I thank the witness for his testimony and those who helped to bring it before this court. Here, the article by van Bakel includes testimony by Bill Brian, the author of *Moongate*, published in 1982. He is my next witness for the prosecution.

I call Bill Brian.

BILL BRIAN'S TESTIMONY

Ditto source.

Begins—

Zero Gravity

Kaysing is not alone in his assertion that NASA has been, hu-hum, "mooning" the public. Bill Brian, a 45-year-old Oregonian who authored the 1982 book *Moongate*, agrees that there is "some sort of cover-up." Although Brian thinks that his fellow investigator may very well be right in saying that we never went to the Moon, he believes there is an entirely different reason for many of the inconsistencies the two have found. Maybe we did go, Brian says, but it is possible that we reached the Moon with the aid of a secret zero gravity device that NASA probably reverse-engineered by copying parts of a captured extra-terrestrial spaceship. Brian, who received BS and MS degrees in nuclear engineering at Oregon State University (although he now holds a job as a policy and procedures analyst at a utility company), uses his "mathematical and conceptual skills" to reason that the Moon's gravity is actually similar to Earth's, and that most likely the Moon has an atmosphere after all. He has crammed the appendices of his book with complex calculations to prove these points, but he trusts his intuition, too: "The NASA transcripts of the communication between the astronauts and mission control read as if they're carefully scripted. The accounts all have a very strange flavor to them, as if the astronauts weren't really there."

But why in the world would NASA feel compelled to cover up knowledge of a high-gravity Moon? "It's a cascading string of events," explains Brian. "You can't let one bit of information out without blowing the whole thing. They'd have to explain the propulsion technique that got them there, so they'd have to divulge their UFO research. And if they could tap this energy, that would imply the oil cartels are at risk, and the very structure of our world economy could collapse. They didn't want to run that risk."

—End of Bill Brian's testimony as published by *Wired*.

§ § § § §

I thank the witness. An interesting argument, I am sure, and perhaps a little more about a possible "UFO" connection later. Hardly,
though, in the way the above witness implies. This next paragraph by
van Bakel might be a little closer to the mark...

Wired article by van Bakel continues—

Other conspiracy buffs don't doubt that men walked on the Moon,
but call the fact irrelevant because extraterrestrials made it there ages
ago—and NASA knows it and has preferred to keep it a secret. In his
recent book called *Extra-Terrestrial Archaeology*, David Childress
points out various unexplained structures on the Moon and argues
that these might be archaeological remnants of intelligent civilizations. Childress, an avid believer in UFOs, also doesn't rule out the
possibility that aliens still use the Moon as a base and a convenient
stepping-stone for their trips to our planet. This might even mean,
enthuses the author, that the Moon is "a spaceship with an inner
metallic-rock shell beneath miles of dirt and dust and rock!" [*A.N.*
Really, Mr. Childress? OK, but I don't think it's a matter we need seriously discuss further in this case—or is it?]

Children and Senators

Although very few Americans subscribe to such grandiose theories
[*A.N.* Neither do I, but as I think I hinted above, there are some startling revelations yet to come from Neil Armstrong himself!], millions
of people doubt the authenticity of the lunar missions, much to
NASA's exasperation. Over the years, the agency's public services
department went through reams of paper answering incredulous
schoolchildren, teachers, librarians—and even U.S. lawmakers like
former Senator Alan Cranston (D-California) and Senator Strom
Thurmond (R-South Carolina), who had written to NASA relaying
the doubts of some of their constituents.

When Knight Newspapers (one of the two groups that later merged
to form Knight-Ridder Inc.) polled 1,721 U.S. residents one year after
the first Moon landing, it found that more than 30 percent of respon-
dents were suspicious of NASA's trips to the Moon. A July 20, 1970,
Newsweek article reporting the results of the poll cited "an elderly
Philadelphia woman who thought the Moon landing had been staged
in an Arizona desert" and a Macon, Georgia, housewife who questioned how a TV set that couldn't pull in New York stations could possibly receive signals from the Moon! The greatest skepticism,
according to *Newsweek*, surfaced in a ghetto in Washington, DC,
where more than half of those interviewed doubted the authenticity of

Neil Armstrong's stroll. "It's all a deliberate effort to mask problems at home," explained one inner-city preacher. "The people are unhappy—and this takes their minds off their problems."

Good Timing

If NASA had really wanted to fake the Moon landings—we're talking purely hypothetically here—the timing was certainly right. The advent of television, having reached worldwide critical mass only years prior to the Moon landing, would prove instrumental to any fraud's success; in this case, seeing really was believing. The magic of satellites, with their ability to enable live global (and interplanetary?) communication, fascinated and awed millions of people, much like anything atomic had caught the public's fancy in the previous decade. Also, space research and rocket science had advanced far enough to make a trip to the Moon likely—or, at the very least, remotely feasible. "The structural nature of technology had changed to make the Moon landing possible, but that also made it possible for people to doubt it," says Gary Fine, a sociology professor at the University of Georgia in Athens specializing in rumor and contemporary legend.

Perhaps more importantly, Watergate hadn't happened yet, and people still trusted their elected officials. "A distrust of authority clearly plays into this whole thing," argues Fred Fedler, who teaches journalism at the University of Central Florida and has written a book on media hoaxes. "With Vietnam and Watergate, people have become less trusting, and to some people it doesn't matter what the government says; their immediate reaction is to disbelieve and to sometimes embrace the opposite view."

The distrust continues to be fed by the mass media, especially in the film and TV business. It is rare to find a movie in which a government agency is actually depicted as a collection of fairly efficient, competent people who serve their country to the best of their ability. Dramatically speaking, an elite of sinister, evil bureaucrats is much more appealing.

007 Uncovers Hoax

The concept of the Moon swindle holds a certain appeal for other filmmakers as well. In *Diamonds Are Forever* (1971), James Bond accidentally stumbles onto a movie set that consists of rocks, a lunar backdrop, and a vehicle that looks like NASA's *Eagle*. Men in spacesuits move about slowly and clumsily, as if simulating low gravity. Bond's pursuers give chase, but 007—stirred, but not shaken—climbs into the lunar lander and makes his escape. The scene is never explained. In the high-tech thriller *Sneakers* (1992), Dan Aykroyd's character, a gadgeteer and conspiracy enthusiast, refers to the Moon landing by casually remarking: "This LTX71 concealable mike is part of the same system NASA used when they faked the Apollo Moon landings." And a small San Francisco Bay area production company with a big name, Independent Film and Video Productions, is working on an as-yet-untitled feature film in which a writer discovers that the Moon landings may have been simulated—and then nearly gets killed in his quest for the truth.

—End of this section of the article, which will continue again shortly.

Do I, as this book's author, need to tread more carefully then? But I see René and Kaysing are still with us, so perhaps I will carry on regardless, continuing to pull no punches in my "quest for the truth" until this trial is over. By which time, hopefully, something at least resembling "the truth" about this whole affair may have become reasonably evident.

Defense: Indeed. Do your worst, because no such "quest" is necessary to begin with.

Judge: That will be for the jury to decide. And I remind any "powers that be" that this is not a "kangaroo court" with its collective mind already made up and out for blood. *Both* sides will get the very fairest of hearings. Counsel for the prosecution should continue "pulling no punches."

Prosecution: The learned judge and my learned friend are informed that indeed I shall.

Again courtesy of *Wired* and Mr. van Bakel, I call upon another non-subscriber to the hoax theory—who will also be giving testimony on behalf of the defendants at a later stage—to give evidence that the prosecution deems of particular relevance to the People's case.

I call Denis Muren.

DENIS MUREN'S TESTIMONY

Ditto source. Muren is a visual effects expert at Industrial Light & Magic, a division of Lucas Digital.

Begins—

Simulating One-Sixth Gravity

Technically speaking, could the Moon landings have been faked? Was the state of special effects advanced enough in the late '60s to fool even the most discriminating eye? Yes, according to Denis Muren. Simulating one-sixth gravity could have been done with the use of hydraulic cranes and thin wires—the *Peter Pan* approach—or even by filming scenes underwater. Muren, an eight-time Oscar winner, is the senior visual effects supervisor at Industrial Light & Magic, a division of Lucas Digital. He was responsible for making the *Jurassic Park* monsters come alive and for key scenes in *Terminator 2, Star Wars*, and *The Abyss*.

"A Moon landing simulation might have looked pretty real to 99.9 percent of the people," he says.

—End of all *Wired* extracts pertinent to the case for the prosecution.

§ § § § §

The prosecution thanks *Wired* Magazine for its much-valued contribution to these proceedings. I must once more point out that the aforementioned witness, Dennis Muren, is *not* himself a believer in the Moon hoax conspiracy theory. He will later be giving evidence for the defense and has actually been accused of being "a NASA agent" by pro-Moon hoax witness Bill Kaysing!

There can surely be no better credentials for placing anyone more firmly in the anti-hoax "camp" than that!

5

"Mankind Cannot Travel in Space" and the Van Allen Belts

My next prosecution witness claimed he once spent part of his career working on a very clandestine project at the Pentagon in Washington, D.C.

I call Philip Corso.

PHILIP CORSO'S TESTIMONY

By the author from various sources.

Colonel Philip J. Corso was a former Pentagon-based army officer who made the startling claim that he once saw the dead body of some kind of "alien being" after the recovery of a mysterious, crashed flying object in the U.S. state of New Mexico in 1947. He also claimed that later on, in 1961, assigned as he then was to the U.S. Army's Research and Development Department (R&D) at the Pentagon, he became involved in a top secret project designed to filter information and material obtained from this "UFO" crash into the industrial and defense programs of the United States.

Not long before his death in 1998, Philip Corso had just completed the co-authoring of a book entitled *The Day After Roswell* (Roswell being the small American town near the area where the UFO allegedly crashed) in which he further claimed that U.S. technological advance in many fields, notwithstanding military and avionics projects, were either a direct or an indirect result of this "alien technology" having fallen into American hands.

However, with all due respect to the earlier witness Bill Brian and indeed to Corso, this present inquiry is not concerned with so-called "alien technology," inadvertently stumbled upon or otherwise, being utilized to enable man to reach the Moon. I, for one, do not believe that such technology—if it ever existed—played any part in our own space program. And by that, of course, I mean the space program of humanity as a whole. (Could anyone conceive of there being any other kind?) However, Colonel Corso's remarks concerning what may have been *piloting* such an ostensibly peculiar flying machine as it is alleged was found are not only interesting, but highly pertinent.

Again addressing the jury more as the author of this book than prosecuting counsel, I first used the following interview with Phil Corso in that previously mentioned earlier work—if I may allude to it yet again—written with the help of my dear departed friend and researcher Keith House. This is provisionally titled *FREAKS* and is thus far only a self-published effort I am afraid, and not yet widely available. This is a similar "trial by jury" look at the above mentioned mysterious air crash: an unsettling and disturbing tale now referred to as "The Roswell Incident." I decided to include in this book also, a segment of the Corso interview because I consider it to be extremely relevant to this case also. Primarily because of what Corso—who claimed he "rubbed shoulders" with some of America's top scientists at the Pentagon and had access to ultra top secret information—had to tell the interviewer about man's lack of ability, or indeed the lack of ability of *any* carbon-based, flesh and blood creature, to become a "space traveler" in the accepted sense. Notwithstanding what I con-sider to be "loaded" remarks by Corso, he never once, to the best of my knowledge, ever spoke publicly about, or made any direct refer-ences to, the Moon landings themselves in any way, shape or form.

The following is the relevant section only of an exclusive interview retired army colonel Philip J. Corso gave to CNI News Editor, Michael Lindeman on July 5, 1997, during the 50th Anniversary cele-brations of the legendary Roswell Incident, which were taking place in the town of Roswell, New Mexico:

Begins—

ML: There has been speculation that the way the Roswell technology was exploited may not be the first time this has happened. Specula-tion that extraordinary Nazi technological developments came from a similar source. What do you think about that?

PC: Yes, true. Russia, Germany—von Braun's team—Canadians and British also, and something in Italy. There were crashes elsewhere, and *they* gathered material too. The Germans were working on it. They didn't solve the propulsion system. They did a lot of experiments on flying saucers. They had one that went up 12,000 feet. But where we all missed out was the guidance system. In R&D we realized that this being [he is referring here to what were seemingly intelligent, non-human but 'strangely man-like' entities of diminutive stature apparently found dead or dying at the scene of the crash] was part of the apparatus himself, or *itself*, because it had no sexual organs.

ML: If we are to develop interstellar travel, will we require a similar relationship with our spacecraft?

PC: Man *can't* travel in space. These 'clones' [as described above] were *created* to travel in space, specifically. They *can* travel in space. Our muscles, bones, brains, can't take space travel. Even today. We can't do it.

ML: We can to some extent.

PC: Well, when they go up in *Mir*, they stay up two or three months. When they come down they have to be carried off. The bones won't hold up. If they stayed longer, they'd never be able to walk [again]. The thing not talked about is the brain is affected up there.

ML: How do you mean?

PC: Gravity, radiation—comes right through the ship. John Glenn saw "fireflies" coming through the capsule. They weren't fireflies, it was some kind of electromagnetic thing came through there. Those beings were created specifically to travel in space.

—End.

§ § § § §

Members of the jury, old soldier Corso was clearly dropping the hint, and not by way of his *own* personal opinions about space travel but those of some of the most brilliant scientific minds around at the time ("poached" from all over the world by the Pentagon, of course, to ensure that the richest and most powerful nation on the planet continued to enjoy such a status) that space travel, that is *true* space travel, in regions well beyond the "apron strings" of Mother Earth, as it were, was impossible. Not only for man, it seems, but also for another intelligence that humanity may have encountered, which is *not* necessarily of extraterrestrial origin even if as "mysterious" as claimed. Corso was basically saying that whatever unknown science lay behind the "clones" allegedly found at Roswell, it had needed to resort to the creation of synthetic beings in order to represent itself in space. In other words, in order to facilitate travel into outer space—and that is where the Moon is—this alternative civilization, or whatever, had needed to create…well, yes…"things." Things that in all probability looked just like them, but were not them. Not really.

But why would they need to do so? This is all beginning to sound rather silly. Can't even bloody aliens live in space then? Read on…

THE VAN ALLEN BELTS

Members of the jury, prior to becoming a much valued part of these proceedings, how many of you had even heard of the Van Allen belts before, let alone be familiar with what they are? Very few of you, I imagine. And those few only lately include myself in their number solely because the Van Allen belts formed an essential part of my research for prosecuting this case. Before explaining what they are in any detail—a task which will actually require far more expert people than myself at various later stages of this trial—let me begin by at least attempting to explain, in relatively simple, general terms, what would happen if these so-called "belts," containing countless trillions of intensely radioactive particles, and which encompass the Earth, were *not* there as part of our planet's protective magnetic field, or *magnetosphere.*

Well, first off, none of us would even *be* here to worry about what the Van Allen belts might have been had they been there. Neither would any other animal or plant for that matter. If we somehow still were, then it would only be because we had evolved from something able to thrive on a world being constantly bombarded, on a second by second basis, by deadly cosmic radiation from space. Radiation that in our present form would kill not only us, but anything and everything else that could be considered alive, in a matter of weeks if not days. And much of it would originate from that great "giver of life" on Earth…

Our very own sun.

In lay terms, the Van Allen belts—named after Dr. James Van Allen of the University of Iowa, who was the first person to discover the existence of these belts in 1958—act like giant filters that "trap" and "store" this deadly cosmic radiation, thereby preventing it, in fact, from ever reaching the Earth.

Although most of this radiation comes from disturbances on the sun's surface (solar flares) some even more deadly stuff would have begun its journey across the galaxy centuries or even thousands of years ago. Some may even have begun heading towards Earth *billions* of years ago! Some might not have originated from our own Milky Way Galaxy at all, but from any or all of the millions of other great galaxies in the universe. Even should much of the original intensity of such radiation diminish with the time taken to reach Earth (simply an obstacle in its path, of course, not a specific target) its potency upon reaching us will depend entirely upon its original intensity at source. If not particularly intense at source and after traveling a hundred years through space, it may well be that some radiation is so harmless by the time it reaches us that even if the magnetosphere and the Van Allen belts were not in place, it would be incapable of harming even a fly. But imagine the intensity of deadly radiation emitted by one exploding—or imploding—star, maybe fifty, 100, or even 1,000 times larger than our own sun! Multiply this by the fact that somewhere or other in this vast universe of ours, such a seemingly cataclysmic event may well be happening on a *daily* basis and then multiply again by the fact that this has almost certainly been happening every day since the dawn of time!

Don't even try to imagine the countless number of stars that have exploded over the millennia, each with a force equivalent to a thousand trillion hydrogen bombs! Don't even try to imagine the intensity of deadly radiation being emitted by that unending procession of white dwarfs (imploding stars), red dwarfs (intensely imploding stars) and supernovas (exploding stars), for the last God-only-knows how many billions of years and even now still heading towards our own galaxy and, of course, li'l ol' Earth!!! Members of the jury, I use the phrase "still heading towards" simply to enable you to somehow picture and comprehend the scenario in some sort of tangible, finite way. Forgive me if that sounds patronizing, but it is not meant to be. I use this same format myself in order to even begin "picturing" the process, which is in fact ongoing, continuous and never-ending. Earlier "radiation blasts" from imploding and exploding stars in other galaxies, as well as our own, have long since hit Earth's atmosphere, been trapped by the magnetosphere and been stored by the Van Allen belts. And there they will stay until their strength eventually diminishes to zero. Later ones are hitting our atmosphere *now*, as I write and you read. Only later ones still are actually "on their way," so to speak. It is a constant and continuous process that has involved the Earth ever since our little planet came into being and will continue to involve it until the very last second of the day our world ends.

Even though the Van Allen belts form a vital part of our protective shield against the "fire and brimstone" of the firmament, without which there would be no life on Earth as we know it, there are some very learned people indeed who claim that they are also our "prison walls." That not only can mankind not pass through these walls without sustaining a fatal dosage of this stockpiled radiation but cannot exist in space for any significant period of time beyond—and I stress *beyond*—these belts, for precisely the same reason that there would be no life on our planet if these belts were not in place.

Not unless man was prepared, of course—even if it were remotely practical—to attempt to perform every tiny little intricate task in outer space with the obvious handicap of being encased in a spacesuit made of lead sheeting far thicker than the human body inside it!

This was of course Phil Corso's point entirely. Mankind *cannot* travel in space, he told us, with all authority of the Pentagon behind him. By that, he also meant mankind cannot *exist* in space. It is also far and away the strongest point being made by the "We never went to the Moon" brigade, now with the agreement of more than a mere smattering of scientists and astronomers. These learned people are saying that they actually *know* we, as living, breathing entities, never set foot upon the Moon—even if most are afraid to come right out and say it publicly—even though man's machines, with markings all too significantly bearing the words "United States" or the initials "CCCP" instead of "Planet Earth," may well have done so.

They say this with absolute, 100 percent conviction and certainty because, they claim, it is quite simply impossible—certainly at present and even more so thirty years ago—that man could have set foot upon the Moon.

Or any other planet for that matter.

And not only does it continue to be impossible even now but, if these people are to be believed, it may *never* be possible. Not ever. And, as I have said, "these people" include some of the world's top brains both in the West and in the old Iron Curtain countries. *Star Trek,* they claim, is but a fantasy that will never become a reality for humankind, even if science fiction, in the past, has in many other instances become today's fact.

I tend to think all that is a bit over pessimistic. Thinking along the lines that today's science fiction, as opposed to yesterday's, might also be a pre-cursor to science fact sometime in the future, let us imagine that prior to the twenty-fifth century and the building of the starship *Enterprise* we are able, as a civilization, to either discover or invent a material or fabric as thin as a sheet of silk that has the insulation value of several feet of lead! We could then use this material to make our spacesuits and also incorporate it into the body shell of our spaceship. Maybe even make the hull of the spacecraft entirely of such material.

In order to gaze out of the portholes of our starship, we would further need to develop a transparent version of this fabric in order to view the universe through it without losing the protection provided by the starship's hull. Similarly, the spacesuits used for EVAs (extra vehicular activities) made of this wonder material would need to be equipped, too, with a helmet visor made of this same transparent derivative. That is if, while undertaking such activities, astronauts would prefer the option of being able to *see* exactly which bottomless pit they are about to tumble into rather than purely relying upon instrumentation to tell them!

But, talking of instrumentation, perhaps a kind of mini personal radar—artificially adding the sense of a bat to those we already possess—might be of definite help in the above precarious scenario. I can envisage such a gadget, but it is doubtful whether it could be relied upon in isolation. In any event, what compensation could there be in being able to "sense" impending hazards compared to being able to actually *see* them?

Another possibility, of course, that could render both us humans and our spaceships immune from the ravages of cosmic radiation for long periods of time has already been a frequent prophecy of countless science fiction novels, films, and television series. It is the force field. Rather than a mini radar built into the space helmet of an astronaut, one type of force field could, perhaps, be some kind of microcosmic "Van Allen belt," thereby trapping, storing and neutralizing anything harmful. The more "traditional" force field, though, takes the form of an invisible, impenetrable cocoon that can be set up both around the spacecraft and—when required—around an individual, too, whenever that person disembarks from the space vehicle.

Such a clever device might even enable astronauts to walk around on another planet without a spacesuit! An astronaut's oxygen "tank" could—for a period at least—be the amount of air from the interior of the spacecraft that would be trapped naturally between the human body and the force field itself. But we will need to come up with a way of temporarily storing, out of harm's way, noxious gases being continually expelled from the human body. Maybe even find a way of filtering them out through the force field without actually breaking it. It would be imperative for the force field to provide a totally impenetrable barrier until safely back on board ship. For it to be "switched off" for even a split second external to the ship might allow in both deadly radiation and a possible poisonous alien atmosphere! By "noxious gases" I mean, in particular, the carbon dioxide that would inevitably build up every time an astronaut breathed out. Methane would not be a major problem, but the atmosphere inside the force field would become rather unpleasant, that's for sure!

With their mini force fields, complete with "exhaust" facilities, in place, there is no reason why astronauts might not one day be able to walk about for a short while on other planets without the encumbrance of otherwise restrictive spacesuits, helmets, and other ancillary life support equipment. Providing there was a means of adjusting the "fit" of the force field to, say, "skin-tight" around the hands and feet especially, this would also allow simple menial tasks to be performed as well. The type of force field envisaged by most science fiction writers also allows light and sound waves to pass through it unhindered. Anything short of this and it would be well and truly back to square one and fourteen feet of lead (which would likely be cheaper, anyway!) because our space "explorers" would neither be able to hear, smell or, most importantly, *see* anything at all beyond the force field!

In the highly unlikely event that such an invention comes to pass far more quickly than I envisage, and some of you young kids out there find yourselves strolling around on Mars one day with one of my mini force fields protecting you, promise me you won't even think about approaching one of the natives and attempting to shake it by the hand saying, "Take me to your leader." Why? Because your force field will probably still be thick enough to knock its "hand" just out of reach every time you try to go for it! Remember, a simple salute or some other kind of friendly acknowledgement will do!

Another invention-to-be, set very firmly somewhere in the far-flung future and again often written about in science fiction, might even allow mankind the chance to set foot upon already colonized planets without the need to endure long, tedious journeys through space beforehand. It has already been named even though it doesn't yet exist.It is called teleportation.

I have no wish—or indeed the necessary knowledge—to get into lengthy discussions as to what teleportation might one day entail. No one knows, as yet, what processes might be involved or even what natural forces might be harnessed. Maybe even so-called "supernatural" ones! In simple language, though, the concept is one of reducing

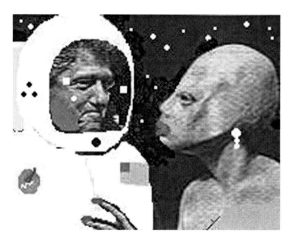

Come on, you handsome Earthman! Take your suit off...I won't mind!

a living being or an inanimate object to its basic atomic-molecular structure and then somehow "transmitting" these disseminated atoms through space, as we do with radio and television. The idea is that once arrived at an intended destination—which could be many millions of miles away from source!—the atoms are then re-assembled to once more form a fully functional being or object, just as happens when your TV aerial unscrambles billions of electrons flying through the air to give you a discernable image on your television screen.

Those *Trekkies* among you, members of the jury, will instantly recall the phrase "Beam me up, Scotty!" Although a well established a part of *Trekkie* folklore, there is some dispute regarding whether or not Captain Kirk ever actually uttered those four famous words as such. William Shatner, the actor who played Captain James T. Kirk, denies ever seeing a script that contained precisely that form of words, let alone that he actually *said* them! But whether he did or not, the point is that he would have been ordering his chief engineer to return him from the surface of an alien planet to the bridge of the *USS Enterprise* by the utilization of a not dissimilar, although equally fictitious, process.

On a rather more serious and far less optimistic note, sadly, it has to be said that should none of the possible discoveries or future inventions foreseen above ever materialize into something practical, then the best that we, mankind, can hope for, if we wish to travel to the stars, is to continue sending machines. Alternatively, we could still "have a go" at building that spaceship with no portholes and with a body shell made entirely of lead, which, despite one estimate of it needing to be fourteen feet thick, might still be viable with significantly less in order to render the crew reasonably safe. But those who volunteer to travel aboard it, as well as being seriously advised to opt for setting down this phenomenally heavy craft on a planet with a low gravity and some good, solid, rocky ground underfoot, should try to ensure before landing that such a world has its own radiation-gobbling version of the Earth's Van Allen belts. Otherwise the crew would have to stay inside the craft until the time came to return to Earth. EVAs in spacesuits made of lead several feet thick are most certainly *not* a viable option under any circumstances!

Even if an expedition were merely "a short hop" to our nearest star neighbor *Alpha Centaurus,* with the ship capable of traveling at the speed of light—or "Warp One," as Captain Kirk used to say—that would still equal a round trip of about ten years.

Imagine spending ten years in a tiny, cramped cocoon with lead walls four or five feet thick! And to what avail? Would we go to *Alpha Centaurus* simply because "it was there"? (This was the explanation given for the need to conquer Mount Everest, of course) One can imagine a future astronaut returning to tell the media, "Well, we got there okay, and the instruments and cameras picked up a whole lot of data, but we couldn't actually *see* anything with our own eyes or *touch* anything with our own hands or *smell* anything with our own noses because five feet of lead kept getting in the way!"

If that is the way it might have to be, then we may as well not go. Not ever, just like the pessimists say. If it comes right down to the wire, we might just as well send machines armed with artificial intelligence, sensitive instruments and cameras in the first place. A lot less risky and a lot less expensive.

And, of course, ever since Apollo, that is precisely what the Americans and Russians have been doing. Sending machines only—not human beings—to the far reaches of space. Robot probes to intercept comets and take pictures of their cores; bleeping drones to investigate and send back information about the rings of Saturn; unmanned spacecraft carrying the finest cameras produced by man going out as far as Pluto to take pictures of the solar system's most recently discovered member, sending them back and then traveling on forever. The space program is not dead simply because there have been no further so-called "manned" expeditions to the Moon for thirty years or more. NASA has not been disenfranchised or dismantled.

"But we already know that," I can almost hear many of you snorting. "And we also know that the manned space program isn't dead either. What about *Mir* and the Shuttle and all that stuff?" Quite right that the ardent space buffs among you should feel affronted if I appear to have conveniently forgotten "all that stuff." But I can assure you that I haven't. Of course NASA has launched many, many manned space missions over the past thirty odd years—sadly, not all of them without a terrible loss of life.

So, too, have the Russians. Yes, in particular the manned *Mir* missions that the Americans not only co-operated in helping to finance, but actually jointly manned on occasions with plucky astronauts joining long-serving and equally brave cosmonauts via the space shuttles blasting off from Launch pad 39A at Cape Canaveral.

No, I am not adding insult to apparent injury, members of the jury. I realize that those of you with any interest in the subject will be well aware of the above facts. But for those of you who, despite your interest in the subject, have not bothered to delve too deeply into the scientific side of things, you may be intrigued to hear about a tiny little unimportant smidgeon of fine detail that neither NASA nor the Russians have ever gone out of their way to publicize.

All the manned space adventures of the past thirty years or so have taken place on the earthward side of the Van Allen belts.

Why? Because that is *only* part of space where we, as living, breathing, carbon-based, flesh, blood and bone organisms, can actually exist and work for any length of time, that's why. Notwithstanding the odd suicidal volunteer with a penchant for dying slowly from leukemia or from some other terrible, debilitating cancer, or more quickly and quite horribly from radiation sickness, anyone with any sense who doesn't walk around with his or her head in the clouds, realizes that this is the *only* part of space that we, in our present form, will *ever* be able to live and work in. Certainly until some of the ideas I discussed with you earlier come to fruition. And, as I said, there is always the possibility that none of them ever will.

And dying from leukemia or radiation sickness—or both!—is not a "consummation devoutly to be wished," as Shakespeare's *Hamlet* would have put it, by either present day American astronauts or Russian cosmonauts. The extreme danger of entering any part of space on the "wrong" side of the Van Allen belts has been known about since 1958. But the very first "manned" mission to venture beyond the Van Allen belts, according to NASA and many of you who thought you

Since Apollo, every manned mission has taken place in "user friendly" space only!

saw it take off—if only on TV—to orbit the Moon and then return without landing, took place in 1968, with Apollo 8.

So the Americans had ten whole years to dwell upon the dangers and to decide whether or not it was worth risking the lives of three family men in their prime. Ten whole years during which—less than two years previously—three other family men in their prime had died horrifically while still on the launch pad, let alone on the nasty side of the Van Allen belts!

So one of the many questions that are being asked these days by some highly-intelligent and inquisitive people is this: Would NASA—indeed the U.S. Government of the day—within two years of already having lost three of its most experienced astronauts, deliberately take the calculated risk of sending three more to an almost certain—if not immediate—death, either by requiring them to pass through or to venture beyond the Van Allen belts?

Can you imagine, members of the jury, the sense of public outrage—not only from their own fellow Americans but from the whole world—if the three Apollo 8 men, Borman, Lovell and Anders, had either perished on the journey or arrived back on Earth terminally ill, especially so soon after the Apollo 1 fire tragedy? Three dead in the space of two years was bad enough, but *six?* Even though much worse tragedies happened later with the space shuttles, of course, for that to have happened at that time would have been catastrophic for NASA and the American administration that had authorized the Apollo 8 expedition, which was that of Richard Nixon. There would have been a justified and overwhelming public reaction that this accident had been clearly avoidable, especially following so closely on the heels of the first.

NASA would have been closed down by Nixon and he would have been extremely lucky for that not to have been his final act as President before resigning from office due, of course, to his impeachment for the Watergate scandal.

But enough of my unqualified ramblings regarding whether or not travel through or beyond the Van Allen belts will ever be an option for human beings determined to live long and stay healthy. I am a writer, *not* a scientist. But the fact that I am a writer also means that, with the aid of a half decent pair of spectacles, I can also read. That means that even though not an expert myself, I have been able to study plenty of material provided by eminent scientists and professional astronomers describing precisely what the Van Allen belts are and what their function is in the great scheme of things. On behalf of the religious, one could say that these barriers were placed there by God to initiate life on Earth and to continue to nurture and protect it. And one could also say that not only were they placed there to keep all this nasty stuff well and truly out, but also to keep *us* well and truly in! We can only hope that this is not the case and that the barriers have become obstacles by default only, because, if it is the case, we have no chance. Again not ever. True, man is capable of overcoming seemingly impossible odds when beset by problems. It is an ability conferred upon him by his Maker. But that ability cannot include defying the very *will* of his Maker. That is a battle mankind cannot win.

I humbly apologize to those more learned members of the jury who may think that I have over-simplified my definition of what the Van Allen belts are all about and indulged in far too sweeping a generalization, but I stand by my definition enough to state that notwithstanding all manner of things we might invent or develop in the future to surmount the dangers posed by space, it suffices to illustrate what is basically mankind's present position as regards space travel. In order to be able to travel into deep space—that is *outer* space—and survive for long periods in these fragile human bodies of ours, my research has led me to the opinion that we would need to take *all* of our environment with us, not just *parts* of it like oxygen, food and water. Our spaceships and spacesuits will essentially need to be microcosms of the Earth itself, replicating as closely as possible, all the succor and protection our planet normally provides, including some form of artificial gravity. Anything less, either in the short or long term, would undoubtedly cause untold damage to our bodies and minds, leading to untimely, and probably agonizing, death.

But just in case some of my over simplifications have resulted in the odd hiccup or two occurring, as a direct result of something less than the bright light of full understanding registering properly in what, for me, sadly, has to pass for a brain, why not hear it from an expert? And they just don't come any more expert than my next witness for the prosecution...

6

Deadly Radiation Beyond the Van Allen Belts: The Testimony of Dr. Henry C. King and David P. Wozney

This next prosecution witness was the former Scientific Director of the London Planetarium.

I call Dr. Henry C. King.

DR. HENRY C. KING'S TESTIMONY

Source: *Book of Astronomy* by Dr. Henry C. King, © 1966. At the time of his authorship of the aforementioned work, Dr. King was Scientific Director of the world renowned London Planetarium.

Extract on Van Allen Belts begins—

In 1741, Olaf Hiorter of Uppsala noticed that compass needles behaved strangely whenever there were vivid displays of aurorae. Instead of pointing to the magnetic poles the needles swung about violently. There the matter rested until 1852 when the English scientist Edward Sabine found that the frequency of these disturbances kept pace with the sunspot cycle. They were most intense and numerous during the times of sunspot maxima and hardly ever occurred at sunspot minima. Disturbances of this kind are, of course, due to disturbances in the Earth's magnetic field. They are known as *magnetic storms* and, as Hiorter had noticed, go hand in hand with the frequency of aurorae.

Do sunspots therefore disturb the Earth's magnetic field and thereby cause aurorae? Not really. Sunspots merely indicate the general trend of change in solar activity. Much more important, as far as the Earth is concerned, are solar flares. They were first noticed in 1859 by Carrington and G. Hodgson as two sudden brightenings near a group of sunspots. They lasted for about five minutes and then disappeared completely. We now know that only the very largest and brightest flares can be seen in this way. Many more are shown up by monochromers adjusted for the light of the red hydrogen line. Small ones are so numerous that as many as a hundred can be seen in a single day. All of them, large and small, occur near faculae and active spots and are therefore closely linked with the sunspot cycle.

The energy sent out by a really big flare is thought to equal the explosion of 1,000 million megaton atom bombs. The parts that affect the Earth arrive in two forms—waves and particles. The waves include an intense burst of ultraviolet and X-ray radiation, which reaches the Earth 8.30 minutes after the explosion. This disturbs the Earth's ionosphere which in turn affects both short-wave and long-wave radio reception. The particles, electrons and the nuclei of atoms, travel more slowly. Those of very low energy and speed take between 20 and 40 hours to reach the Earth, and then only when the Earth is in the line of fire. They then give rise to aurorae and disturb radio reception. Those of high energy, traveling at a speed approaching that of light, take less than an hour. They consist mainly of protons and are known as *cosmic rays.* Fortunately for us they lose their high energy as they travel through the Earth's atmosphere. But in space they could, on hitting the walls of a space vehicle, produce enough X-rays to kill the crew inside. Their accurate prediction will therefore be one of the great problems confronting space travelers of the future.

Thanks to instruments carried by artificial earth satellites and space probes we now know that the Earth's magnetic field acts like a huge trap for electrified particles. Most of them, if not all, are thought to come from the sun. Those of low energy spiral around the lines of magnetic force and are guided towards the magnetic poles. They are therefore very numerous in a zone or ring some 23 degrees from the magnetic poles where they give rise to intense and frequent displays of aurorae. They do this, it is thought, in much the same way as a stream of electrons (that is, an electric current) makes the gas glow in neon and other gas-discharge tubes used in advertising. Parallax observations made from different stations indicate that aurorae occur between 50 and 600 miles above the ground. At these heights the air is so very thin that the passage of streams of electrons and protons could readily make oxygen glow green and nitrogen red.

The region in which electrons and protons are "trapped" in this way is known as the *magnetosphere.* Its outer parts interact violently with the solar wind, that is, with the hot electrified hydrogen gas that streams away from the corona. Owing to the force of the wind it has a shape like that of a tadpole, the "tail" being in the plane of the Earth's magnetic equator and in line with the sun. When the wind is weak the sunlit side of the magnetosphere is about 40,000 miles from the Earth, but when it is strong it can be as close as 25,000 miles. Just how far the magnetosphere extends on the Earth's dark side is not known but it may well be several hundred thousand miles. Inside the magnetosphere are two belts or zones in which the energy of the trapped particles is particularly high. These, the *Van Allen belts*, lie above the Earth's magnetic equator and disappear completely above the poles. For some reason the intensity of the inner belt varies much more than the outer, and at times of high intensity, could be a serious hazard to a spaceman traveling through it. But how these belts are formed and how the particles in them get their high energies is not known. One thing, however, is fairly certain—the particles come mainly from the sun.

—End.

§ § § § §

By including his learned testimony in the People's case, I am not of course inferring that Dr. King—particularly in his reference to at least one of the Van Allen belts constituting a "serious hazard to a space-man traveling through it," was in any way suggesting that to travel in space *beyond* this and the outer radiation belt was fraught with danger—only passing through it. And, to the best of my knowledge, Dr. King has never said or published anything denouncing NASA's claim—merely two years after his book was published—to have navigated three men (Apollo 8) safely through the Van Allen belts, sent them around the Moon and then returned them back through the radiation belts a second time, to Earth. Over the next few years, NASA would send another *twenty-four* men (Apollo 9 tested LM in Earth orbit *beneath* the radiation belts) on both an outward and a return journey through the belts.

Yet, to the best of my knowledge, only one out of a total of *twenty-seven* men who allegedly ventured *through, beyond* and then *back through* the belts a second time (belts of such intense radiation, I remind the jury, that contemporary scientists and astronomers like Dr. King had deep misgivings about were even one astronaut to make such an attempt, let alone getting on for thirty!) suffered any ill-effects either at the time or more significantly since, that can be directly linked to these ventures into outer space!

Of course, "normal" illnesses and tragedies bound to take their toll of twenty-seven men over a period of more than thirty years have applied here and there, and I refer, in particular, to the regrettable deaths of two of the dozen men who allegedly actually walked upon the Moon. James B. Irwin of Apollo 15 died from a heart attack aged sixty-one, in 1991 and Charles "Pete" Conrad of Apollo 12 was killed in a motorcycle accident on July 8, 1999, in California. He was sixty-nine. The only "fly in the ointment" of the point I am attempting to make is Alan B. Shepard, America's first astronaut and skipper of the Apollo 14 mission. Apart from being America's first man in space, he was also allegedly the first man ever to play golf on the Moon. Shepard died of leukemia near his home in Pebble Beach, also in California, on July 21, 1998, at the age of seventy-four, two years after being diagnosed with the disease. His wife of fifty-three years, the former Louise Brewer, died five weeks later.

But, Shepard apart, and his case may simply have been hereditary and largely unavoidable, no trace of leukemia or any alternative form of cancer appears to have manifested in any other case. At least, no other case that I have been able to obtain any information about. And let us be thankful for it, of course. And not just for the sake of those former members of the Apollo crews who are still with us and their families who rightly love them, but because, on the face of it, it means that both as a species and a civilization, our chances of one day traveling to the stars in these frail human bodies will apparently *not* be impaired by so-called "dangerous" or "fatal" dosages of radiation. Despite what those cynics, skeptics and downright pessimists say, it would certainly *appear* that twenty-six of those twenty-seven men have proven beyond any doubt that man *can* travel in space. That man can both survive and live in space. More than likely, or so it seems, women will one day even be able to give *birth* in space as well. Give birth on another planet even!

After all, what on Earth—or off it!—is there to stop us? If, according to the evidence, the large majority of those twenty-seven individuals are still alive and well more than thirty years after the very last three of them returned from that good *Ol' Devil Moon*, it speaks for itself, doesn't it? We, the human race, have obviously got a license to kill out there. One way or another, we appear to have been written a blank check as far as outer space is concerned.

And that is surprising. Let's face it. Even those millions of us who are *not* astronomers and scientists always imagined outer space to be a particularly nasty and dangerous place to be at the best of times…

Airless. A total vacuum.

Cold. So cold, in fact, that as far as the lowest temperatures ever recorded at the South Pole are concerned, "…shall I compare thee to a summer's day?"

And, despite its name, outer "space" is not empty. It is full of "loose cannons" like comets, tumbling asteroids and giant, hurtling meteors. Then there are the much tinier meteors (they only become meteor*ites* after falling to Earth) the size of a marble or a ball bearing, flying hither and thither all the time…in their countless billions! And all traveling faster than speeding bullets! These things will first pierce your spaceship…and then your suit! And let us not forget, either, those even tinier, much more deadly radioactive particles we have already discussed. Again flying about all over the place in their utter *trillions* in never-ending procession. They are, of course, the inevitable debris of millions of suns and billions of planets that have come to grief somewhere or the other in the universe over the centuries and millennia.

And, as if all this isn't enough to be going on with, even deadlier invisible cosmic rays are forever shooting across the galaxy from a billion and one other unearthly sources, let alone from our own sun. Gamma rays. X-rays, God-only-knows-what type of bloody rays.

Outer space. Hmmm…Doesn't the Good Book actually describe it as the Firmament? And rightly so, too. Because "firmament" is exactly what it is. And another word for firmament is…

Hell.

Any suck—I mean takers?

But we're not bothered, are we, members of the jury? *Nah!* Twenty-seven human beings stuck two fingers up at that lot thirty years ago and not only lived to tell the tale, but did so without sustaining as much as a blister! Those before them, and since, that sadly perished, actually died whilst still in their own earthly environment. Even those of the "magnificent twenty-seven" not still around to tell the tale, suc-cumbed, eventually, to the normal hazards of trying to live here on Earth rather than out there in wonderful, cozy, safe, friendly, welcom-ing…

Ermm…outer space?

Yes, indeed. Such a tale to tell. But the trouble is that those twenty-five former astronauts still in a position to be able to do so, more often than not prefer…

Not to tell it.

I can feel myself becoming increasingly angrier at having been taken for such a mug. When I become angry, I tend to become overly sarcastic. Perhaps this is a good time, then, to call my next witness for the prosecution. He will provide us with more testimony from various creditable sources regarding what these sources allege are the inher-ent radiation dangers in outer space.

I call David P. Wozney.

DAVID WOZNEY'S TESTIMONY

> David Wozney has repeatedly challenged NASA over the years regarding various aspects of the Moon landings and has been less than satisfied with the integrity of some of the answers he has received. He recently posted the following article, entitled **Deadly Radiation At and Past the Van Allen Shields** on the internet. © 2002 D.P.Wozney.

Begins—

There is considerable evidence that the Van Allen shields, which start at an altitude of between 250 miles to 750 miles, protect us from deadly solar and cosmic radiation.

Herbert Friedman, in his book *Sun and Earth,* describes Van Allen's global survey of cosmic ray intensity thus:

"The results from Explorer I, launched on January 31, 1958, were so puzzling that instrument malfunction was suspected. High levels of radiation intensity appeared interspersed with dead gaps. Explorer III succeeded fully, and most importantly, it carried a tape recorder. Sim-ulation tests with intense X-rays in the laboratory showed that the dead gaps represented periods when the Geiger counter in space had been choked by radiation of intensities a *thousand times greater* than the instrument was designed to detect! As Van Allen's colleague Ernie Ray exclaimed in disbelief: 'All space must be radioactive!'"

Herbert Friedman later explains that "Of all the energy brought to the magnetosphere by the solar wind, only about 0.1 percent manages to cross the magnetic barrier."

The *April 28, 1997 HST Update: Recommissioning Status Report* states that the Van Allen radiation belts, between 200 and 500 miles high, "act as a thin, protective skin for Earth, trapping charged particles before they bombard our planet and harm us."

The *Space Physics Textbook* by the Space Physics Group of Oulu, lists what high energy particle radiation in the radiation belts does and is:

- "it degrades satellite components, particularly semiconductor and optical devices
- it induces background noise in detectors
- it induces errors in digital circuits
- it induces electro-static charge-up insulators
- it is also a threat to astronauts."

If it is a threat [to astronauts] then why weren't animal experiments beyond the Van Allen belts done first?

From *http://www.ofcm.gov/nswp-sp/text/c-sec1.htm*

"Today, we have far more knowledge of the space environment from the turbulent surface of the sun, with its continuous solar wind and periodic spewing of clouds of energetic ionised particles, to the protective boundary of the Earth's magnetic field. Which provides a partial shield against deadly solar corpuscular radiation. The Earth's magnetic field is highly reactive to the onslaught of energy and pressure originating from the solar particles and fields."

From *http://www.ofcm.gov/nswp-sp/text/d-sec2.htm*

"...Besides being a threat to satellite systems, energetic particles present a hazard to astronauts on space missions. On Earth we are protected from these particles by the atmosphere, which absorbs all but the most energetic cosmic ray particles. During space missions, astronauts performing extra-vehicular activities are relatively unprotected. The fluxes of energetic particles can increase hundreds of times, following an intense solar flare or during a large geomagnetic storm, to dangerous levels. Timely warnings are essential to give astronauts sufficient time prior to the arrival of such energetic particles."

The average orbit altitude of the space shuttle is 185 miles, [way] below where the Van Allen shields begin. How were the Apollo astronauts protected against these deadly energetic particles and solar flares?

From *http://www.geo.nsf.gov/atm/nswp/intro.htm*

"...the sun...occasionally ejects high energy particles that can be deadly to electronic components and biological systems..."

From *http://ess.geology.ufl.edu/ess/Notes/040-Sun/primer.html*

"The area between the sun and the planets has been termed the interplanetary medium. Although sometimes considered a perfect vacuum, this is actually a turbulent area dominated by the solar wind, which flows at velocities of approximately 250-1000 km/s (about 600,000 to 2,000,000 miles per hour). Other characteristics of the solar wind (density, composition, and magnetic field strength, among others) vary with changing conditions on the sun. The effect of the solar wind can be seen in the tails of comets (which always point away from the sun).

"Intense solar flares release very-high-energy particles that can be as injurious to humans as the low-energy radiation from nuclear blasts. Earth's atmosphere and magnetosphere allow adequate protection for us on the ground, but astronauts in space are subject to potentially lethal dosages of radiation. The penetration of high-energy particles into living cells, measured as radiation dose, leads to chromosome damage and, potentially, cancer. Large doses can be fatal immediately. Solar protons with energies greater than 30 MeV are particularly hazardous. In October 1989, the sun produced enough energetic particles that an astronaut on the Moon, wearing only a space suit and caught out in the brunt of the storm, would probably have died."

Why would NASA subject people to this kind of risk? Once you are past the Van Allen shields, for example between the Earth and the Moon, or on the Moon, you would be subject to the full brunt of solar flares. The Van Allen shields protect us here on Earth from this deadly radiation.

However, even in low Earth orbit, below the Van Allen shields, an August 1972 event could have been life-threatening had there been a space walk in low Earth orbit at the time.

From *http://www.hq.nasa.gov/office/olmsa/lifesci/91_92_journals.html*

"…estimates of human exposure in interplanetary space, behind various thicknesses of aluminum shielding, are made for the large solar proton events of August 1972 and October 1989. A comparison of risk assessment in terms of total absorbed dose for each event is made for the skin, ocular lens, and bone marrow. Overall, the doses associated with the August 1972 event were higher than those with the October 1989 event and appear to be more limiting when compared with current guidelines for dose limits for missions in low Earth orbit and more hazardous with regard to potential acute effects on these organs. Both events could be life-threatening if adequate shielding is not provided."

Apollo 16 was in April 1972 and Apollo 17 was in December 1972. Why would NASA proceed with Apollo 17 just after the August 1972 event and risk astronauts' lives? The hulls of the Apollo spacecraft were ultra-thin. They would have been unable to stop any radiation. The same can be said for the spacesuits.

A calculation quantifies the radiation risk associated with solar flares beyond the Van Allen shields:

From *http://pet.jsc.nasa.gov/alssee/demo_dir/se_radia_fs.html:*

"Solar flares…yield very high radiation doses within very short time periods (hours to days)…energies between 0.1 Million Electron Volts (MeV) and 10 MeV during very large solar flares…thousand-fold increase in the radiation dose over a short period of time. Solar flares show a correlation with the 11-year solar cycle."

From *http://www.pha.jhu.edu/~jaw/Seminar/jawsf4.html*

"During a solar maximum, about 15 flares per day emit detectable X-ray energies."

From *http://flick.gsfc.nasa.gov/radhome/papers/seeca3.htm*

"(1964 for solar minimum and 1970 for solar maximum)."

So the Apollo missions, from 1969 to 1972, were occurring during a solar maximum, when there would have been about 15 solar flares per day!

Edward P. Ney estimates the radiation risks in an article entitled *The Sun Under Surveillance* in the *1967 World Book Science Year*: "We have rough estimates of what the moon travelers can expect, based upon a few observations made during the last solar maximum in 1957. The most violent flares probably will produce exposures of 100 roentgens [or rads, explains Wozney. 1 rad = 1 rem/radiation equivalent man] each hour and may hold this level for several hours."

This level of radiation dose is confirmed by Space Biomedical Research Institute in *Humans in Space*: "<u>*Solar Flare*</u>—*Very hazardous and intermittent but may persist for 1 to 2 days*. High energy protons travel at the speed of light so there is no time to get under cover. Protected dose 10-100 REM/hr. Unprotected dose Fatal."

The *Spacecast 2020 Technical Report* puts the space weather radiation hazard to human life in perspective: "…at geostationary orbit, with only 0.1 gm/cm2 of aluminum shielding thickness, the predicted radiation dose (REM) for one year continuous exposure, with minimum-moderate solar activity, is estimated to be about 3,000,000…"

A radiation dose value from a low energy flare is provided from *NASA Mooned America*, p. 134: "On page 256 of *Astronautical Engineering* there is a chart that shows the dosage of four different flares. On August 22, 1958 there was a *low* energy flare that could have been reduced to 25-rem with 2-cm of water shielding."

So, being conservative and using 25 rems per flare, we have 25 rems x 15 flares/day = 375 rems / day for the Apollo astronauts.

These dose limits are from *Radiation Safety* by the International Atomic Energy Agency, Division of Radiation and Waste Safety: "The dose limits for practices are intended to ensure that no individual is committed to unacceptable risk due to radiation exposure. For the public the limit is 1 mSv [0.1 rems—DPW] in a year, or in special circumstances up to 5 mSv [0.5 rems—DPW] in a single year provided that the average dose over five consecutive years does not exceed 1 mSv per year."

How were the Apollo astronauts able to withstand 375 rems per day when the public can only be exposed to 0.5 rems per year?

—End.

§ § § § §

I thank David Wozney for apprising us of such authoritative testimony regarding the inherent dangers to space travelers posed by radiation, difficult to follow though some of it may be for the average member of the jury.

The Prosecution

7

Fakery!

The "camera never lies" according to the now well-established saying. But of course it does—frequently. Not only can it be made to give a distorted image in the first place, but that image can then be tampered with in a variety of ways afterwards, and no more so than in this modern age of advanced personal computer technology.

But the camera, be it a stills or movie camera, does not lie in itself. The machine, on its own, will always faithfully reproduce whatever appears in its viewfinder. It is that which appears in the viewfinder that is not necessarily always what it seems. It could be a totally fictitious setting for a totally fictitious television or film drama, for example. But the camera will still be accurately recording what it "sees," except that what it is seeing is contrived. A fabrication. Or, to put it at its crudest, a lie.

And that, essentially, is what this whole "Did man really land on the Moon?" debate is all about. Could a blatant lie have been filmed and photographed and then presented as if it were the truth? Indeed, *was* something that was a blatant lie—pure Cold War propaganda, in effect—filmed, photographed and presented as fact to an unsuspecting public on both sides of the Iron Curtain?

My next witness for the prosecution, most definitely thinks so.

I call award-winning photographer David Percy.

DAVID PERCY'S TESTIMONY

Source: *www.apfn.org/apfn/moon.htm*—© 2001.

Begins—

Award winning British photographer David Percy is convinced that the [NASA "Moon"] pictures are fake. His astonishing findings are as follows: The shadows [in the pictures] could only have been created with multiple light sources and, in particular, powerful spotlights. But the only light source on the Moon is the sun.

The American flag and the words "United States" are always brightly lit, even when everything [else] around is in shadow. Not one still picture matches the film footage, yet NASA claims both were shot at the same time.

The pictures are so perfect that each one would have taken a slick advertising agency hours to put together. But the astronauts managed it repeatedly. David Percy believes any mistakes were deliberate. Left there by "whistle blowers" who were keen for the truth to one day get out.

If Percy is right and the pictures *are* fake, then we have only NASA's word that man ever went to the Moon.

—End.

§ § § § §

I thank David Percy.

Members of the jury, strongly bear in mind, not only for what follows immediately, but for the future, what the above writer says about photographic expert David Percy's conviction that "deliberate mistakes" were made by "whistle blowers" keen for the truth to one day emerge.

My next witness for the prosecution is a UFO researcher and writer who was originally interested in the Apollo landings only because of the reported sightings by some Apollo astronauts of what appeared to be structured "edifices" and "tracks" on the Moon, not to mention "unidentified flying objects." Indeed, the most significant of these encounters with what were seemingly "UFOs" actually occurred during an EVA by Armstrong and Aldrin! At one point during the Apollo 11 broadcasts "live from the Moon," remarks between the two men clearly mentioned that there were "other spacecraft here." These were clearly heard by myself and countless millions of other listeners and viewers, even if such remarks have long since been edited—or censored—out of official NASA transcripts and footage released for public consumption.

However, the more my next witness delved into the whole Apollo affair, initially to simply pursue his own particular line of inquiry, the more he became convinced that the only other "spacemen" seen by the American astronauts were their similarly attired opposite numbers and the only "UFOs"—standing for Unmistakably Flightless Objects maybe?—the reflections of their own so-called "spacecraft" in their gold-tinted helmet visors!

To the witness it began to appear highly probable that such bizarre references were "jokingly" made during the broadcasting of actual live segments—although obviously not from the Moon—inserted between ordinary black and white movie footage, the latter likely filmed weeks or even months before on the identical studio set from which the live television transmissions were now coming. We may as well face it. Disguised by those spacesuits, the bulky white figures shuffling about in the foreground in the shadow of the LM in the footage need not necessarily have been real astronauts at all! No way did we know for certain that they actually *were* Armstrong, Aldrin, Bean, Conrad and all the other names we came to know so well.

But *why* the need to play a cynical practical joke on a totally captivated and fascinated worldwide audience? Some of these incongruous remarks—the jury will be availed of them in detail at a later stage—were straight out of the *Dan Dare, Pilot of the Future* cartoon strip I read in *The Eagle* comic as a child!

I don't know, but at a guess I would say that the astronauts made them simply to "wind up" (for the benefit of American jurors, that is a slang term for *teasing* here in the U.K.) their frantic NASA bosses who, despite a certain amount of light-hearted "mutiny" in the ranks, were hell bent upon somehow convincing the people of Earth that this was all in deadly earnest; that it was happening for real and wasn't just a pre-launch stunt for the latest Hollywood remake of *The Forbidden Planet*.

A certain amount of editing of the following article has been necessary because those reading it on the Internet were frequently asked to temporarily stop reading in order to view NASA film footage that accompanied the article before resuming. Such a facility, of course, is not available in this printed page format. However, as far as I know, the website named below is still accessible and those members of the jury with Real Player installed on their PCs should be able to view this footage and, because of this, some references to it by the author of the article remain included. Fortunately though, still photographs from the movie clips were also included in the article, and these are reproduced for the more immediate benefit of the jury on the picture plates indicated in the text.

I call David Cosnette.

That's a small step for a man, but under this lot...I'm actually a girl!

DAVID COSNETTE'S TESTIMONY

Source: *http://www.ufos-aliens.co.uk/cosmicapollo.html*—article entitled: *The Faked Apollo Landings* written by David Cosnette © Updated March 2002.

Begins—

This article has been written to prove, once and for all, that we are not being told the truth about the NASA film footage of the Apollo missions. This will astound even the most hardened [anti-hoax] skeptic and convince many people that the whole Apollo Moon project of the late 1960s and early 70s was a complete hoax. Video links are provided, so that you can watch with your own eyes the "official NASA footage" that proves we really haven't been told the whole truth.

Bill Kaysing was head of technical publications and advanced research at Rocketdyne Systems from 1956 to 1963. He states that it was estimated in 1959 that there was a .0014 chance of landing man on the Moon and returning him safely to Earth. This took into account the effects of radiation, solar flares and micro-meteorites. He could not believe in 1959 that man could go to the Moon.

However, only 2 years later, American President John F. Kennedy set a goal in May 1961, when he made the following famous speech:

"I believe that this nation should commit itself. To achieving the goal, before this decade is out, of landing a man on the Moon and returning him safely to the Earth. No single space project in this period will be more impressive to mankind or more important for the long range exploration of space."

It was just eight years later in 1969 that man finally left Earth and set foot on the Moon. Or so we were led to believe...

I would like to show you some astonishing evidence that shows glaring mistakes or anomalies on the "official record" of NASA film footage and still photographs. I have included the actual official Apollo film footage to illustrate and also possibly educate you, the reader, of the anomalies that exist, and to let you see with your own eyes what has become one of the biggest cover-ups in the history of mankind. I will also explain why the U.S. Government has tried to keep the lid on this secret for over 30 years.

I would like to suggest that if man did go to the Moon, first in 1969, then the so-called "Apollo films" that we were told were filmed on the Moon are bogus and not the *real* footage. Evidence suggests that man could not travel to the Moon's surface in 1969, and that instead they had to stay in near Earth orbit within the safety of the Earth's magnetic field, which would have protected them from the radiation [trapped and stored] by the Van Allen radiation belts.

But why would NASA and the United States bother to fake such an event and to what end?—I hear you ask. Well please read on and I will explain. Was man *too* optimistic about what we could actually do in deep space, and was President Kennedy's speech in May 1961 pressure enough to keep the hoax going?

David Percy is an award winning television and film producer, a professional photographer and also a member of the [British] Royal Photographic Society. He is co-author, along with Mary Bennett, of a fascinating book called *Dark Moon: Apollo and the Whistle- Blowers* (ISBN 1-898541-10-8). The majority of the film footage on this [web] page, though, is taken from a film called *What Happened on the Moon?* This film also features Percy and Bennett and I strongly rec-ommend it to you if you have an interest in the Apollo missions. (Details of how to purchase the video are at the foot of this article). Percy firmly believes that the Apollo footage was either faked or is not the original film shot on the Moon. He believes that many anoma-lous features that would alert the eagle-eyed viewer [Oooh! That's a nasty pun, David!] could have been placed in the films by whistle blowers who were deeply dissatisfied at being part of a huge confidence trick. Percy has studied the entire transfer of the original film on videotape, a feat that not many people have done. What nobody realized at the time was that a lot of the footage was actually pre-recorded and not live at all.

Fake Earth?

An example that clearly appears to be faked is the footage of Earth taken from Apollo 11 when it was 130,000 miles away. [See Picture A on Plate 1].

This was the very first view ever taken of the Earth on a space mission and it seems strange that Buzz Aldrin would film the Earth when he was stood so far away from the window. Why would he do that? Surely you would want to get as close to the window as possible to get the best picture, and also to eliminate light reflections that are evident towards the end of this sequence. But no, we see the window frame come into view on the left of the shot. The camera isn't set to infinity either to get the closest shot. The window frame that comes into shot would have been out of focus if it was.

Did the astronauts actually film a transparency of the Earth that was stuck to the window? You may think this odd, but a few minutes after filming the Earth, the cameraman adjusts his lens and focuses on Mike Collins inside the craft. What we see is what appears to be an exposure of the Earth taped to the glass of the window in the background to the right of him. That is the very same window that Aldrin was earlier filming the Earth through.

But the biggest shock is yet to come! The camera pans left past Neil Armstrong towards the left hand side of Apollo 11, and what do we see out of the left window? We see what appears to be...wait for it...*another* Earth!

It must also be noted that Apollo 11 at this point in the mission was supposedly half way to the Moon. The time elapsed was 34 hours and 16 minutes; but from the view of Earth in the right hand window, we can say that in fact they were not in deep space at all, but still in low Earth orbit. Why? Because there is *blue* sky outside! This would also explain why they would be filming an exposure of the Earth that was far away: to give the impression that they were in deep space. The exposure would be clipped to the window and the sun's luminance would light it up, a technique that was also used to read star charts to help with navigation and star reference.

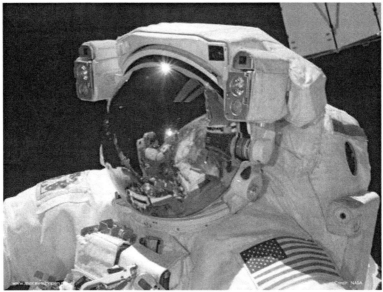

Fake Earth? Oh, well…it's easily done. Can you *Spot The Ball?*

Anomalies with the film footage!

Hasselblad were the manufacturers of the camera that would take all of the photos on the Apollo missions. Jan Lundberg was the Manager Of Space Projects at Hasselblad from 1966 to 1975 and was responsible for the production and building of the Hasselblad 500 EL/70 cameras that were used on the Apollo missions. He says: "Originally, NASA made all the alterations themselves. But then they presented what they had done to us and asked if we could do the same? To which we replied, 'yes, we can, and we can do it better.' We proceeded to make the alterations that were then accepted by NASA." Protective plates were added to the case and film magazine.

Another very important factor to consider is the great variation in temperature that the film would have had to endure whilst on the lunar surface. The temperatures during the Apollo missions were recorded as being between -180F in the shade to an incredible +200F in full sunshine! How could the film emulsion have withstood such temperature differences? The astronauts can be seen to move between the shadows of the rocks and then into full sunlight in some shots. Surely the film would have perished under such conditions?

Crosshairs

On all Apollo footage, there should be cross hairs or reticules present on the film. These crosshairs were, according to NASA, placed on the film to help calculate distances on the Moon. The crosshairs were actually built into the camera, and therefore should be visible on every single picture taken by the astronauts on the surface of the Moon. Incidentally, Jan Lundberg has stated that the only way that you could calculate the distance in the shot using the crosshairs would be if you had two cameras set up to take a stereo picture.

Take a look at the pictures presented here [Pictures B and C on Plate 1] and you will see that parts of the crosshairs have disappeared from the film. This is impossible unless the film has been tampered with. The crosshairs should be completely visible in all shots and not hidden behind some of the objects in the pictures. The only solution must be that NASA has gone to the trouble of either airbrushing out certain objects in the film, or adding them over the crosshairs.

I claim this rock in the name of the United States and will call it "C"

Why does this rock [Picture D on Plate 1] have the letter "C" on it? There is also a "C" on the ground in front of the rock. The use of the letter "C" on film props is well known by the people in Hollywood and is used to show where the center of the scene should be. One skeptic on the *Bad Astronomy* skeptics' web group [the *Bad Astron-*

omy group are staunch believers in the integrity of the Apollo Moon landings] has even said it is a hair! What? On both the rock and the ground? *Now* who's trying to cover things up?

Shadows

One of the biggest anomalies that show up in the Moon photos are the way in which shadows seem to be cast in totally different directions, even when the objects making the shadows are a mere few feet apart. A classic example can be viewed in this picture [Plate 2, Picture A]. If the guy on the left is near a vertical rise of ground [as has been suggested] his shadow would show a definite "crease" where the land begins to rise…it doesn't!

Question: How can an astronaut cast a shadow several feet taller than his colleague who is standing a few feet away from him?

Answer: He is standing farther away from the studio arc light that is illuminating them both! I truly believe that this shot was taken on a film set. You cannot reproduce this strange shadow phenomenon with natural light, and that includes taking into consideration two natural light sources as many [anti-hoax] skeptics would have you believe.

Film footage from the Apollo 11, 12 and 14 missions would suggest that there are many light sources lighting the so-called Moon's surface. In the Apollo 11 film, the shadow cast by Armstrong is strange. The sun angle is estimated at 10 degrees above the horizon, compared with the Apollo 12 footage that shows a longer shadow. The sun is at a 15-degree angle, therefore the Apollo 12 shadow should have been shorter. The Apollo 14 footage shows a shadow performing truly amazing maneuvers, consistent with moving away from a source of light that is much closer to the astronauts than the sun!

Some of the lighting on "official NASA film" is very suspect. This NASA picture [Picture B on Plate 2] should show the astronaut in complete shadow because the sun is behind him, yet the whole of the astronaut is caught in bright light. The shot should actually appear like this one [Picture C on Plate 2], which was simulated by David Percy.

I have had quite a few debates on the web about this picture [Picture B on Plate 2]. I'm told by [anti-hoax] skeptics that the picture appears as it does because you have to remember that two light sources are present on the Moon's surface (the sun and Earth glow). I do not doubt that there could be reflective light from the Earth, but, in my opinion, if a light is bright enough to light up this astronaut's suit, it should also be capable of producing another shadow behind him.

Hoax skeptics say that he is illuminated by light reflecting off the Moon's surface. The reflectivity is only 7% so the theory of the light bouncing from the surface is highly suspect. If this were the case, the rock on the left of the picture would have hardly a shadow because it is closer to the source where the light is reckoned to be reflecting from!

Shadows do not appear to be correct in several so-called "Moon pictures." Take this next picture [Picture D on Plate 2] for example. The shadow on the LM is due East, yet the shadows on the rocks in the foreground are South East.

A simulation by David Percy of how the shadows should truly appear is illustrated. [Picture E on Plate 2]. If two light sources are indeed at work on the Moon's surface, they would combine together and the shadows would fall accordingly, not at random points. Unless, of course, the anti-hoaxers are saying that the sunlight is falling in the middle of the picture and the earthlight at the forefront of the picture.

Stills camera better than movie camera

During the Apollo missions, NASA says movie cameras were fitted with special night lenses to compensate for the lack of light. Due to the atmospheric conditions on the Moon's surface, only 7% of light is reflected from the ground (that is the same reflectivity as asphalt). So, taking this into consideration, how did the Hasselblad stills camera manage to pick up more detail than the movie cameras? NASA have confirmed that no artificial lighting was used on the lunar surface, so how can the stills camera have taken pictures that were brighter and sharper than those shot by movie cameras fitted with special lenses to compensate for the dark conditions? Some pictures, again not necessarily genuine but containing a background clearly intended to show just how dark a place the Moon is, have deep black shadows on the side of the rocks.

It is interesting to note that some still photos seem to have Aldrin [Apollo 11] brightly lit, in comparison to the gloomy images produced by motion picture cameras that had the special night lenses fitted.It appears that artificial lighting was used or has been added to still photos to show better features on Aldrin's suit and the Lunar Lander. But the lack of atmosphere on the surface of the Moon is the fundamental reason why the shadows are normally so intensely black and why the "props artists" would have been instructed to copy this effect as closely as possible.

Why is there such a vast difference in the [natural] light [being uti-
lized] by the two cameras [still and movie], unless the stills shots
were being lit by artificial lighting? NASA has said that no lighting
equipment was ever taken to the Moon. So, if the Moon missions
were genuine, this cannot be true when you view the evidence. The
still pictures seem to show Aldrin being artificially lit as he descends
the ladder [Picture A on Plate 3]. The reflectivity of the lunar surface
is so low that light doesn't even reflect onto the rocks on the ground,
yet the light in these pictures is so intense, even Aldrin's heel protector on his boot is lit up!

Dr. David Groves, who works for Quantech Image Processing, has
done some analysis of these particular shots and has used Quantech
resources to pinpoint the exact point at which the artificial light was
used. Knowing the focal length of the camera's lens and being able to
get hold of an actual boot, he has calculated that the artificial light
source is between 24 and 36 cm to the right of the camera. If the TV
footage is actually real, then I can understand this, as the movie
images are very dark and grainy, but I believe the still photographs to
be definitely faked.

Let us move on now to the famous picture [Pictures B an C on
Plate 3 and both pictures on Plate 23] of Buzz Aldrin, which shows
the LM, Neil Armstrong and the Apollo 11 landing site in the reflection of Aldrin's visor. [*A.N.* According to some, it also shows the
shadow of the studio photographer, in classic pose, who actually took
the picture!]. One of the truly strange things about this picture is that
the reticule [crosshair/Reseau mark] that is supposed to be in the middle of the picture, actually shows up at the *bottom* of Aldrin's right
leg! How can this be when the camera is attached to the picture taker's
chest? A fact that is easily verifiable by the reflection of the presumed
cameraman in the visor.

Many people have speculated that the pictures have been retouched
to bring up the detail of the astronauts. But this cannot be applied to
the Apollo 11 photographs because a duplicate copy of the original
Armstrong film has been analyzed and shows that the pictures are all
on one continuous roll of film containing over 100 images. Even Jan
Lundberg from Hasselblad, the makers of the camera, says that the
pictures seem as though Armstrong is standing in a spotlight. The
only way the reticule could appear in the bottom of the leg is if the
picture has been copied and reframed!

The horizon is about 89 degrees from the true vertical. Dr.
Groves has also worked out, after analyzing the shadows cast by

both the astronaut in the picture and the supposed cameraman in the visor, that Armstrong—who again is only presumably taking the picture—is standing on ground a mere few inches higher than the spot where Aldrin is standing. If this is the case, then that means that whoever took the photograph for real was in fact at least *2 feet* higher than Aldrin, and this therefore means that Armstrong, although visible with the camera in the visor, is not the actual person who took the shot.

Malfunction

[NASA claims] Apollo 12 suffered a camera malfunction apparently after the lens had been pointed towards the sun. Though the viewing public was told at the time that the camera had burnt out, the lens was in fact still working! The camera didn't actually burn out at all. The very same thing supposedly happened on Apollo 15, but again the camera lens didn't burn out. It is interesting to note that during the Apollo 16 mission, one of the astronauts was told by Houston to adjust the camera. He then asks if he should point the camera towards the sun! Even though the manual for the camera specifically points out not to do this!

We have to remember that the cameras supposedly used on the Moon didn't actually have any viewfinders, and the astronauts could not see the whole of the camera strapped to each of their chests. They had to point their chests roughly in the direction of a subject, release the shutter and then hope for the best! The astronauts even had to change lenses whilst standing outside on the lunar surface wearing heavy gloves. A feat quite hard to believe considering the very awkward-looking pressurized gauntlets they were wearing. The precaution of changing film inside the LM was not adhered to either, and so could have ended in minor disaster if the film had actually been dropped into the dust on the ground.

I think it would have been virtually impossible to change the film and adjust the lenses in such apparatus. However, in certain pieces of film, the astronauts seem to be wearing a different type of glove, which doesn't appear to be pressurized!

Detail on Lunar Lander

How come we can see so much detail on the gold portion of the Lunar Lander in this picture? [Picture A on Plate 4]. As is evident by the shadow cast in front of the LM, the sun is in the background and the gold area should be covered with shadow, not sunlight! How does one explain the "sun" having a halo around it if the Moon has no atmosphere?

Multiple light sources

I think this picture [Picture B on Plate 4] beautifully sums up the evidence of several light sources being used. Otherwise, how would you explain the portion on the right of the picture being in what can only be described as a "spotlight"? You only have to look at the shadows cast by the Lunar Lander, flag and the picture taker to see that this shot has artificial lighting centered on the bright area. If this shot was lit by natural sunlight, the shadows would all fall in the *same* direction—not 3 different ones! Even if you take a second light source (the Earth) into consideration, this would not explain why the flag's shadow is not traveling in the same direction as the LM's! As stated earlier, two light sources would combine together and the shadows would still fall in the same direction. Even if the "light spot" is in a crater, the flag shadow would still travel from SW to NE as does the LM's.

Australians say they saw Coca-Cola bottle on the "Moon"

In Western Australia, during the live broadcast of the supposed Apollo 11 Moon landing, several people claim to have seen an astonishing occurrence. One viewer, Una Ronald, stayed up to see the telecast and was astounded by what she saw.

In fact, the residents of Honeysuckle Creek, W. Australia, actually saw a different broadcast to the rest of the world. Shortly before Armstrong purportedly stepped onto the Moon's surface, a change can be seen where the picture goes from a stark black to a brighter picture. Honeysuckle Creek stayed with this picture, and although the voice transmissions were broadcast from Goldstone, California, the actual film footage was broadcast from Australia. As Una watched Armstrong walking on what was supposed to be the surface of the Moon, she claims she spotted a Coca-Cola bottle that had been kicked into the right hand side of the picture!!! This was in the early hours of the morning. Una immediately telephoned some friends to see if they had seen the same thing she was convinced she had seen. Unfortunately, though, it was a detail her friends had missed, but they were going to watch the re-broadcast the next day. But, needless to say, by this time the footage had been edited and the offending "Coke" bottle sequence cut out of the film. However, several other viewers also claimed to have seen the bottle and several articles [about it] appeared in The West Australian newspaper.

Western Australia received its Apollo coverage in a different way to the rest of the World. Australia was, in fact, the only country where there was no delay to the "live" transmission. Bill Kaysing says, "NASA and other connected agencies couldn't get to the Moon and back and so went to ARPA (Advanced Research Projects Agency) in Massachusetts and asked them how they could simulate the actual landing and space walks."

We have to remember that all communications with Apollo were run and monitored by NASA, therefore journalists who thought they were hearing the voices of men on the Moon could easily have been misled. All NASA footage was filmed off TV screens at Houston Mission Control for the TV coverage. No one in the media was ever given the "raw footage."

Bill Wood is a highly qualified scientist and has degrees in mathematics, physics and chemistry. He is also a space rocket and propul-

We've heard of "Moonshine"—but that's ridiculous!

sion engineer. He has been granted high security clearance for a number of top-secret projects and has worked with MacDonell- Douglas and engineers who worked on the Saturn 5 rocket (the Apollo launch vehicle). He worked at Goldstone as a Communications Engineer during the Apollo missions. Goldstone in California, USA, was responsible for receiving and distributing the pictures sent from Apollo to Houston. He says early video machines were used to record the NASA footage here on Earth by the TV networks. They received the FM carrier signal on Earth, ran it through an FM demodulator and processed it in an RCA scan converter that took the slow scan signal and converted it to the U.S. standard black and white TV signal. The film was then sent on to Houston. When they were converting from slow scan to fast scan, RCA used disc and scan recorders as a memory and it played back the same video several times until it got an updated picture. In other words the signal was recorded onto video one then converted to video two. Movie film runs at 30 frames per second, whereas video film runs at 60 frames per second. In other words, then, the "live" footage that most people thought they were seeing actually wasn't, and was actually 50% slower than the original footage!

Danger—Artist at work!

This picture [Picture C on Plate 4] from Apollo 15 is really an amazing feat of camera work, especially when one considers that it was taken without any means of knowing everything was in shot. No viewfinder and no one to tell you that everything's in shot. It is also rather strange that the only thing visible on the dark part of the Lunar Lander is the American Flag! You can't put *that* down to two light sources! Much more likely is a NASA artist and an airbrush! This picture was later used on postcards and NASA advertising.

The only one

Did you know that this picture [Picture A on Plate 5] is the only close-up picture taken of Neil Armstrong on the Moon? This is a little strange don't you think, considering that he was [supposedly] the very first human being ever to step onto the lunar surface?

Why no sound? Why no craters? Why no dust?

Why is there a complete lack of engine noise on any of the films as the Lunar Lander is about to land on the Moon's surface? After all, the astronauts are sitting on an engine capable of producing 10,000lb of thrust and burning at a temperature of 5000 degrees Fahrenheit!

Air inside the module was pressurized to 1/3rd sea level atmosphere, so why no sound and vibration? According to some sources, the astronauts could hear the thrusters charging, yet *we* cannot hear the immense engine in the background in any of the transmissions!

The Lunar Lander actually used two engines stacked on top of one another. The LM's descent engine used hypergolic propellants. That means two different fuels that ignite at the same time [apparently when they come into contact with one another]. The exhaust jet coming out of the LM on descent or ascent should have created an enormous cloud of reddish colored gas. Instead we see the bursting apart of the milar covering as the vehicle leaves the Moon's surface. The fuels used are exactly the same as used on the shuttles today, and we can clearly see the exhaust smoke coming from *them*, so why not from the LM?

Surely there should also have been some type of crater blasted underneath the Apollo landing modules, especially that of Apollo 12 as it slowly moved across the Moon's surface before landing. The 5000 [degrees] Fahrenheit produced by the 10,000lb thrust engine should also have produced at least *some* volcanic rock, [especially] if you compare the molten volcanic rock at Mount Etna, which was boiled at only 1000 [degrees] Celsius. I have heard some [anti-hoax] skeptics state that the engine's force would have been dispersed mainly sideways, but if this is so, what power held up the 2,300lbs of each Lunar Lander as it descended to the lunar surface? Also, why wasn't there any obvious dust to be seen on the landing pads either? There is certainly lots of dust to be seen being scattered about when the LMs are leaving the Moon, so, if the argument is that the *Eagle*'s engine, for example, simply blew all the dust away from around the LM as it landed, how, then, did Neil Armstrong manage to create that famous footprint? [More about footprints later].

Surveyor III

Here's another how? How did the Apollo 12 crew manage to capture footage of the Surveyor III spacecraft [Picture B on Plate 5] that had landed near their landing site in April 1967? Anyone who watches the film [visit the named website] will be astonished to see how skillfully the Apollo 12 crew managed to capture the sequence through such a small window whilst trying to land their own craft at the same time! The Lunar Lander does a 360- degree turn around Surveyor III and not once does the fallen probe go out of focus or move out of shot. Was this "surveyor" simply a mock-up and this sequence shot out of the open door of a helicopter? Readers might be interested to know that one [anti-hoax] skeptic tried to deny that this was NASA footage! Oh? Whose was it then? Beats me too!

Apollo 13

By the time of the Apollo 13 Mission in April 1970, public interest in space travel had begun to wane. This could have been partly due to most of the previous Apollo 12 Mission having to rely mainly on an audio transmission, due to the claimed camera malfunction encountered. Was this "lack of interest" a factor in the alleged near disaster of the Apollo 13 mission? Was NASA simply trying to get back the public's attention and therefore guarantee the continued funding of the U.S. Government? On the 13th hour of the 13th day of the 13th Apollo mission, disaster struck when an oxygen tank exploded.

This picture [Picture C on Plate 5] shows the astronauts of Apollo 13 just before they transferred to the LM. The spacecraft is supposed to be some 200,000 miles from Earth. But if we look out of the window we can see blue sky! How can this be if the mission is supposedly in deep space? Surely the windows should be showing *black*? Unless, of course, it is all happening in *near* Earth orbit!!!

Now take a look at these two pictures [Pictures A and B on Plate 6]. As pointed out by Percy and Bennett in *What happened on the Moon?* The first picture [Picture A] shows the *Odyssey* [the Apollo 13 Command Module] after it was damaged by the oxygen tank explosion. This next picture [Picture B] shows a normal shot of a command and service module with its cover removed from the scientific instrument bay. Do both images look rather similar to you? [*A.N.* In other words, members of the jury, Cosnette is asking if there are any major differences observable between the two photographs? Clearly not, is the answer].

Here's another question. How come Astronaut Fred Haise stated that the crew aboard Apollo 13 could see a place on the Moon called Fra Mauro? At the time of the accident, Fra Mauro, which had been the intended landing site for Apollo 13, was in total darkness, and would have remained so for the entire time that Apollo 13 was near the Moon. In fact, Fra Mauro did not reappear again until 88 hours after Apollo 13 had left Moon orbit and was now on its way back to Earth. By this time Apollo would have been 19,000 miles away from the Moon, making it impossible for any of the crew to have seen Fra Mauro at any time during the mission.

How did NASA recreate the effects of weightlessness?

Some [anti-hoax] skeptics will ask, "If this footage wasn't taken on the Moon, how do you explain the astronauts being able to 'bounce' around on the surface? You couldn't do that here on Earth. How would you reproduce the effects of the 1/6th gravity of the Moon?"

If the same [anti-hoax] skeptics cared to *double the speed* of the film, they would see how. The truth is that the "astronauts" don't actually move any differently to how they would on Earth! [*A.N.* I have tried this for myself, and have to admit that to an extent that causes me great concern, this observation is correct!].

Wires

Some footage seems to show astronauts suspended by a thin wire. In fact, if you look closely, you will see the light reflecting off the wires above the astronaut. Watch [visit the named website] how the astronaut seems to be almost jumping on the spot to turn around in the next sequence; it is all rather reminiscent of the practice rig used in training here on Earth, which this footage also includes. [More about practice rigs later, but please *do* try to view all the clips referred to]. And this last sequence will kill you altogether! In the last sequence of this piece of footage, see for yourselves [visit the website] how the "astronaut" who has fallen over, gets up. He stands up without even attempting to put his hands on the ground or by way of the other supposed astronaut assisting him. It's pure farce! Honestly! He is just suddenly yanked up…just like a puppet on a string! Or, more precisely, like a man attached to a bungee rope!

Gravity experiment

Many [anti-hoax] skeptics will probably say that the hammer and feather experiment, which was achieved during the Apollo 15 mission, could not be recreated in a studio. Well, here is the original NASA footage and a very similar experiment, which is simulated and comes from the *What happened on our Moon?* video. [*A.N.* For anyone who doesn't actually get around to seeing the footage, I can assure you that the end result was exactly the same. The hammer and feather hit the ground at precisely the same moment!]. The simulation is carried out within a 1G atmosphere here on Earth, so that blows *that* one out of the water! It is obvious that NASA correctly presumed that most people would not believe that a feather could hit the ground at the same time as a hammer in Earth's atmosphere and gravity, and that they would assume that the experiment simply *had* to be taking place on the Moon. But even were Apollo genuine, scientists would have known that this was not necessarily so. So, how come they all kept quiet about it at the time?

Radiation

Radiation plays a big part in space travel. Solar flares could have affected the astronauts at any time. And leaving Earth would involve traveling through 2 specific areas of very high intensity radiation called the Van Allen belts. The first shield or "belt" is 272 miles out from Earth. The amount of radiation in the belts actually varies from year to year, but every 11 years it is at its worst when the sunspot cycle reaches its apex. And guess what? 1969 to 1970 was one of the *worst* times to go, as this was the time when the radiation was at its peak. I have had numerous Internet chats with [anti-hoax] skeptics who say that radiation would not have played a part in the missions because the astronauts would not have been in the radiation belts for too long. My answer to that is this: When dentists or doctors take X-ray pictures, they either leave the room or stand behind a sheet of thick lead to shelter from the radiation, or this could build up to dangerous levels in their bodies after a certain number of patients. Why did NASA only use a small sheet of aluminum to protect the astronauts, especially when they knew that the radiation levels in space and on the Moon's surface would be many hundreds of times more deadly?

Did you know that the U.S. Government tried to blast a hole in the [Van Allen] belt 248 miles above Earth in 1962? During Operation Starfish Prime, a one-megaton nuclear bomb was detonated to try and force an unnatural corridor through the lower Van Allen belt. Unfortunately, the radiation levels then got *worse*, not better. What they created was a monster. A *third* belt a 100 times *more intense* than the natural belts, and, as estimated by Mary Bennett in *Dark Moon— Apollo and the Whistle-Blowers*, by 2002 this artificial zone would still equal 25 times more radiation than the other 2 belts put together! There is no agreement as to how wide these radiation belts truly are. Dr. James Van Allen, their discoverer, estimates that they are at least 64,000 miles deep at their widest point. NASA, though, maintains that they are only 24,000 miles deep at their widest point. Each Apollo craft is said to have spent approximately *4 hours* traveling through the belts.

So, to what lengths did NASA go to shield the astronauts against this radiation? It's accepted that a minimum of 10 cm width of aluminum would be needed at the very least to keep out radiation. However, the walls of the Apollo craft and capsule were made as thin and as light as possible and, as a result, the craft initially could not carry enough air inside to withstand the equivalent of sea level air pressure. NASA had to reduce air pressure inside the cabin to cope. Here are the official statistics from a NASA website:

http://www.hq.nasa.gov/office/pao/History/alsj/frame.html: "At sea level, the Earth's atmosphere is a mixture of gases—primarily of nitrogen (78% by volume), oxygen (21%), water vapor (varying amounts depending upon temperature and humidity), and traces of carbon dioxide and other gases. Oxygen is, by far, the most important component of what we breathe and, indeed, the Apollo astronauts breathed almost pure oxygen laced with controlled amounts of water vapor. With the nitrogen eliminated, the cabin pressure could be considerably less than sea-level pressure on Earth—about 4.8 psi (pounds per square inch) versus 14.7 psi—and, consequently, the cabin walls could be relatively thin and, therefore, light in weight."

One of the worst ever flares

One of the worst sun flares ever recorded happened in August 1972, which was between the Apollo 16 and 17 missions. This single flare would have delivered 960 rem of virtually instant death to any astronaut who was up in space, and yet all of the Apollo astronauts were carrying out their missions in what amounts to nothing more than a thick linen suit. These pressurized suits may have helped to protect the astronauts against heat or micro-meteorites, but they certainly wouldn't have provided any radiation protection. By the way, there is no known method of registering when and how strong solar flare activity will be. So, did NASA just struck lucky?

Radiation would also have greatly affected any film shot on the Moon. Physicist Dr. David Groves Ph.D. has carried out radiation tests on similar type film to that supposedly used on the missions and has found that the lowest radiation level (25 rem) applied to a portion of the film after exposure, obliterates the image on the film almost entirely. Why, then, didn't this same thing happen to film used by the Apollo crews? That was dead lucky too, wasn't it? NASA is seemingly capable of defying the laws of physics and chemistry now!

Readers might be interested to note that the biggest solar flare for 25 years was recorded in April, 2001. So [anti-hoax] skeptics who are claiming that NASA knew precisely when the solar flares were going to appear, are talking rubbish as usual. If this were the case, why didn't they bring down the astronauts from the shuttle and ISS if they knew this gigantic solar flare was about to erupt?

HJP Arnold is an astronomer and a keen photographer, an expert on space and astro- photography and was the assistant to the Managing Director at Kodak during the Apollo years. He has also been the author of many books about space photography. He comments that the film that was supplied by Kodak for the missions was essentially the same type as is used here on Earth for snapping the kids frolicking in the backyard. It was exachrome 64 ASA or ISO as it is called today. He commented that you would expect to see small dots on some of the film, where high velocity nuclear particles had hit it. However, nothing whatsoever to indicate that this happened on even a single occasion is evident in any claimed Moon photographs/footage, despite film having been changed whilst [allegedly] outside on the Moon's surface, not in a controlled, protected environment.

The only thing able to protect film from this almost certain damage, ordinarily, is a thick layer of lead affixed around the camera casing, which, according to Hasselblad, was not used.

One with and one without!

In the following sequences [Pictures C and D on Plate 6] the camera pans across the landscape that at one point includes the Lunar Landing Module. In another shot from the same mission, we see the very same mountains, but no Lander!!! Ooops! How can this be when the mountains are exactly the same distance away from the camera in both shots? [*A.N.* The point Cosnette is attempting to make, I think, but unfortunately rather badly, is that Picture D, which shows the LM dead center with the mountains in the distance, was taken from *exactly the same spot* on the so-called lunar surface as the other picture, which now contains no LM. So who—or what—snapped Picture C from precisely the same vantage point as Picture D *before* the LM had actually landed? Hmmm…baffling, that one!].

The hills are alive…

One of the most significant of the many anomalies that now lead me to believe that all the so-called "Moon footage" was fake, and actually taken on a film set is the fact that the *same mountains* seem to have appeared on *different* Apollo missions that were supposed to have landed several hundreds of miles away from each other! I have watched film from what were supposed to be two different Apollo missions in two completely different areas of the Moon, yet footage from both reveals the exact same mountains appearing in the background!

Roving Rocks

I also possess an Apollo film documentary called *Apollo: One Giant Leap For Mankind,* which features all the space missions from before the Apollo project right up until the Soyez-Apollo link up and the Space Shuttle. I would like you now to watch two video clips [visit the named website] from the Apollo 16 Mission. NASA says that this first piece of footage [Picture A on Plate 7] was shot during a trip by Lunar Rover to Stone Mountain. This jaunt was supposedly undertaken on 21st April, 1972 to a point 1 km West of the landing site. The second piece of film [Picture B on Plate 7], was supposedly taken the very next day, this time during a Rover excursion to a point 4 km to the South of the landing site. What you can see in *both* films [shots], though, are *exactly the same rocks* at these two different places, which are supposed to be several kilometers apart!!! We've all heard of the expression "the walls have ears," but these rocks seem to have legs!

Oh, dear, oh, dear, oh, dear, oh, dear! Talk about "Be sure your sins will find you out," huh? Laugh? I nearly cried!

One for the Skeptics

Over the past few months I have been having a debate with several members of the *Bad Astronomy* website. Bad Astronomy is a website which purports to be a general meeting place for people who think they can explain away the hoax theories concerning the Apollo program. This site often goes into detail regarding apparent anomalies that show up on the alleged space footage, which they think can be easily explained. During time spent debating issues with them on their site, I was issued several challenges by its subscribers who said that if I could come up with "official NASA footage" which clearly showed certain anomalies, then that evidence *might* make them think again; that there most definitely could be something amiss with NASA Apollo footage. Needless to say, at time of writing, none of Bad Astronomy's contributors have yet come forward to change their stance.

The three main challenges issued [by Bad Astronomy to Cosnette, the author of this piece] were:

No. 1. Produce pictures showing stars that are taken on the Moon's surface. They say because of the very bright conditions on the Moon, stars would not be visible from its surface!

No. 2. Show an example of Moon movie footage that was taken aboard the Lunar Rover whilst it was in motion. (I asked the pro-Apollo site how the satellite dish at the front of the Rover could relay the video signal to a satellite, or Houston, if it was moving all over the place?). I was even told that such footage didn't exist!

No. 3. Could I provide film footage of the LM producing a flame on the Moon's surface? (This would prove that the movie was *not* taken on the Moon because the Moon's [lack of] atmosphere and the fact that it was a vacuum would prevent such a flame occurring).

Those were the challenges. Here is the evidence. Enjoy:

Challenge No.1—Seeing stars

One of the biggest debates between hoax theorists and [anti-hoax] skeptics concerns the non-appearance of stars from the surface of the Moon. [Here the article alludes to a series of still photographs, which appear to show tiny specks of light in the "Moon's" sky]. If the objects in the sky that appear in the film from the Apollo 15 Mission are *not* stars, what are they? We can rule out marks on the lens of the camera or in the film, because these objects appear on various parts of each shot and not just in one place.

Pictures of Hill 305 and the Hadley Delta, show what appear to be "stars" in the sky above the Moon, [Pictures C and D on Plate 7]. They all show a similar formation from different angles. These pictures are from a set (AS15-9012249 to AS15-90-12269). Most of this set [only two reproduced] show 'stars' in the sky! [*A.N.* I was at a complete loss to understand Cosnette's point here. Surely by producing photos that show stars in the Moon's sky, he was providing ammunition for NASA and the organization's supporters on Bad Astronomy to be able to refute claims that stars did not appear in any of the shots claimed to have been snapped on the Moon. However, I think his point is that those who deny there was a hoax cannot have it both ways. Either no stars at all should appear for the agreed reasons, or, if they do appear, ought to do so in far greater proliferation in zero atmosphere than is apparent from the Hadley pictures, which again makes those pictures suspect. Apart from that, though, I have no further suggestions regarding Cosnette's point, and his article offers no alternative explanation].

Challenge No.2—Clear shots from mobile Rover

I would like to know how the TV signal from the Lunar Rover was relayed to Houston when the satellite dish it was sending the signal through was moving all over the place while the Rover was on the move? Anyone who has set up satellite equipment will know that a dish has to be finely tuned within a few inches to receive a signal. How could the Rover camera have sent a picture when the dish was not pointing in one specific direction?

An [anti-hoax] skeptic on the Bad Astronomy board accused me of lying about there being film footage supposedly shot while the Rover was moving across the surface of the Moon, so for his benefit and yours, here it is! [See Picture A on Plate 8]. I inform Bad Astronomy that I don't have to tell lies. NASA is clearly already telling most of them for me!

Challenge No.3—Show flame from LM

If the LMs were supposedly leaving the Earth and not the Moon [say the anti-hoax skeptics] they should have produced a large bright exhaust flame from the rocket propellant. Instead, zero exhaust. I have turned this one around and have found evidence of a flame on one ascent of the LM, just to prove the [anti-hoax] skeptics wrong! [See Picture B on Plate 8].

A trick of the light?

The [anti-hoax] skeptics reading this article could perhaps try to explain why this movie from the Apollo 11 mission [you will definitely need to view the actual footage here] shows a sudden increase in light when Neil Armstrong reaches the bottom of the *Eagle's* ladder prior to stepping off it onto the lunar surface? It certainly isn't due to the light aperture being changed on the camera because only the light behind the Lander alters and not the actual Lander shadow. It's amazing how Armstrong, who is first in complete darkness on the ladder, is then suddenly lit up when he is halfway down it! Let us remember that there are no clouds on the Moon to first obscure and then pass by to reveal sunlight. The camera obviously doesn't move from the position Armstrong earlier fixed it into either. In this piece of footage, it can also be seen that Armstrong had somehow managed to park the LM in very bright light. How fortunate was that? *That* takes some explaining if artificial lighting wasn't used!

Still Not Convinced?...Here's 32 things that need to be answered!

1. [Anti-hoax] skeptics say there are no stars in the black sky, despite zero atmosphere to obscure the view. The first man in

space, Yuri Gagarin, pronounced the stars to be "astonishingly brilliant." See the official NASA pictures [on Plate 7] that show "stars" in the sky, as viewed from the lunar surface.

2. The pure oxygen atmosphere in the module would have melted the Hasselblad camera covering and produced poisonous gases. Why weren't the astronauts affected?

3. There should have been a substantial crater blasted out under the LM's 10,000 pound thrust rocket. [Anti-hoax] skeptics would have you believe that the engines only had the power to blow the dust from underneath the LM as it landed. If this is true, how did Armstrong create that famous boot print if all the dust had been blown away?

4. The LMs, if truly leaving the Moon, would not have produced any exhaust flame from the rocket propellant. Visible exhaust would be zero. But I have found evidence of a [visible] flame on one ascent of the LM [from the surface of the "Moon"].

5. Footprints are the result of weight displacing air or moisture from between particles of dirt, dust, or sand. The astronauts left distinct footprints all over the place.

—Paused.

Excuse me for interrupting the witness here, members of the jury, but Cosnette has brought up perhaps the most significant point raised on the People's behalf but for the alleged dangers posed by radiation. David Cosnette is correct to mention that footprints, or imprints of any kind, left on the Earth, are a result of displacement of air or moisture, or both, from between the particles that comprise the surface stepped upon. On Earth, such a surface could either be snow, damp soil, dust, sand, or anything covering the ground, which is damp and of a grainy nature. The indentation is caused by a sudden weight being placed where the footprint or imprint occurs. What happens is that air is expelled as the tiny grains of soil or sand are suddenly crushed into closer proximity with each other. Moisture between these particles then provides the necessary adhesion to temporarily "cement" them together in their new positions, thereby enabling the imprint to remain as a perfect reproduction of that part of the object recently placed upon it, and the imprint will then remain for as long as the elements allow. Without moisture, the side "walls" of the imprint will tumble in the moment the weighty object is lifted.

As an example of what happens to areas of fine particles completely devoid of moisture we need only take a trip to our nearest sandy seashore. There will be very damp areas that the sea constantly rolls over as the tide ebbs and flows. Depending upon how tightly compacted the sand already is, you should be able to leave very clearly defined footprints in these areas, especially in bare feet. Now move much further back from the rollers to an area of dunes that the tide seldom reaches, except maybe in a storm. Try as you may, in shoes or bare feet, to make a *clear* footprint, you will simply leave a trail behind you of indefinable troughs. What is happening is that the walls of sand from which the air has been displaced are so fine and dry that unlike those that stayed in place closer to the sea, they are now simply crumbling, or "caving in." That is because the individual grains of sand that comprise these walls lack sufficient moisture to "glue" them together.

This phenomenon might seem all the more surprising when one realizes that this is not the Sahara Desert but a place close to the sea with no shortage of moisture in the air. In a desert one would expect to leave troughs rather than clearly defined footprints when walking about. Even in a scorching desert, though, the air above always contains *some* moisture.

But, consider this…

The Moon has no air to *be* displaced.

Neither, it is generally believed, has the Moon a single molecule of $H2O$ either—that's water to you and me—to enable it to "glue" anything together.

The Moon has no atmosphere whatsoever. It contains no moisture whatsoever (at least, not in liquid form). If it did, then that moisture would invariably contain oxygen because every molecule of water contains two atoms of hydrogen and one atom of oxygen. To the best of anyone's knowledge, *there is no oxygen on the Moon*. So, with no water, no air, no oxygen, no gases of any sort, how could anyone have left in such sharp relief in the dust or topsoil of the Moon…

Bootprints?

Answers on a postcard please!

Cosnette's "32 things" continued—

6. The Apollo 11 TV pictures were lousy, yet the broadcast quality magically became fine on the five subsequent missions.

7. In most Apollo photos, there is a clear line of definition between the rough foreground and the smooth background.

8. Why did so many NASA moonscape photos have non-parallel shadows? [Anti-hoax] skeptics will tell you [it is] because there

are two sources of light on the Moon—the sun and the Earth. That may be the case, but the shadows would all still fall in the same direction, not two or three different angles.

9. Why did one of the pictured rocks have a capital "C" on it and a "C" on the ground in front of it? Were they stage props?

10. How did the fiberglass whip antenna on the Gemini 6A capsule survive the tremendous heat of atmospheric re-entry?

11. In Ron Howard's 1995 movie *Apollo 13,* the astronauts lose electrical power and begin worrying about freezing to death. In reality, of course, the relentless bombardment of the sun's rays would rapidly have overheated the vehicle to lethal temperatures with no atmosphere into which to dump the heat build up.

12. Who would dare risk using the LM on the Moon when it was never ever tested successfully? Would you send a relative to the Moon in a vehicle that had never been driven before?

13. Instead of being able to jump at least ten feet high in "one sixth" gravity, the highest jump was about nineteen inches.

14. Even though slow motion photography was able to give a fairly convincing appearance of low gravity, it could not disguise the fact that the astronauts traveled no further between steps than they would on Earth.

15. If the Rover buggy had actually been moving in one-sixth gravity, then it would have required a twenty-foot width in order not to have flipped over on nearly every turn. The Rover had the same width as an ordinary small car.

16. An astrophysicist who has worked for NASA writes that it takes two meters of [lead] shielding to protect against medium solar flares and that heavy ones give out tens of thousands of rem in a few hours. Why didn't the astronauts on Apollo 14 and 16 die after exposure to this immense amount of radiation?

17. The fabric space suits had a crotch to shoulder zipper. There should have been fast leakage of air since even a pinhole deflates a tire in short order.

18. The astronauts in these "pressurized" suits were easily able to bend their fingers, wrists, elbows, and knees at 5.2 p.s.i. and yet a boxer's 4 p.s.i. speed bag is virtually unbendable. The guys would have looked like balloon men if the suits had actually been pressurized.

19. How did the astronauts leave the LM? In the documentary *Paper Moon* the host measures a replica of the LM at The Space Centre in Houston. What he finds is that the "official" measurements released by NASA are bogus and that the astronauts could not possibly have gotten out of the LM wearing their bulky spacesuits.

20. The water source air conditioner backpacks should have produced frequent explosive vapor discharges. They never did.

21. During the Apollo 14 flag set up ceremony, the flag would not stop fluttering.

22. With a more than two second signal transmission round trip, how did a camera pan upward to track the departure of the Apollo 16 LM?

23. Why did NASA's administrator resign just days before the first Apollo mission?

24. Another overlooked intriguing fact is that NASA launched the TETR-A satellite just months before the first lunar mission. The proclaimed purpose was to simulate transmissions coming from the Moon so that the Houston ground crews (all those employees sitting behind computer screens at Mission Control) could "rehearse" the first Moon landing. In other words, although NASA claimed that the satellite crashed shortly before the first lunar mission (a misinformation lie), its real purpose was to relay voice, fuel consumption, altitude, and telemetry data as if the transmissions were coming from an Apollo spacecraft as it neared the moon. Very few NASA employees knew the truth about the Apollo missions because they believed that the computer and television data they were receiving was the genuine article. Merely a hundred or so knew what was really going on; not tens of thousands as it might first appear.

25. In 1998, the space shuttle flew to one of its highest altitudes ever, three hundred and fifty miles, hundreds of miles below merely the beginning of the Van Allen radiation belts. Inside of their shielding, superior to that which the Apollo astronauts possessed, the shuttle astronauts reported being able to "see" the radiation with their eyes closed penetrating their shielding as well as the retinas of their closed eyes. For a dental X-ray on Earth, which lasts 1/100th of a second, we wear a 1/4 inch lead vest. Imagine what it would be like to endure several hours of radiation that you can see

with your eyes closed from hundreds of miles away with only 1/8 of an inch of aluminum shielding to "protect" you!

26. The Apollo 1 fire of January 27, 1967, killed what would have been the first crew to walk on the Moon. The tragedy occurred just days after the commander, Gus Grissom, held an unapproved press conference in which he complained that man was at least ten years, not two, from reaching the Moon. The dead man's own son, who is a seasoned pilot himself, has in his possession forensic evidence personally retrieved from the charred spacecraft. He claims that the government has tried to destroy this on two or more occasions.

27. CNN issued the following report, "The radiation belts surrounding Earth may be more dangerous for astronauts than previously believed (like when they supposedly went through them thirty years ago to reach the Moon?) The phenomenon known as the 'Van Allen Belts' can spawn newly discovered 'Killer Electrons' that can dramatically affect the astronauts' health."

28. In 1969 computer chips had not been invented. The maximum computer memory was 256k, and this was housed in a large air-conditioned building. In 2002 a top of the range computer requires at least 64 Mb of memory to run a simulated Moon landing, and that does not include the memory required to take off again once landed. The alleged computer on board Apollo 11 had 32k of memory. In today's terms, that's the equivalent of a simple calculator.

29. If debris from the Apollo missions was left on the Moon, then it should be visible today through a powerful telescope. However no such debris can be seen. The Clementine probe that recently mapped the Moon's surface failed to show any Apollo artifacts left behind by man during the missions. Where did all those Moon Buggies and landing stages of the LMs go?

—Paused.

This is another extremely salient point by Cosnette. When the United States and Great Britain invaded Saddam Hussein's Iraq in March 2003, there was much opposition to this course of action from many quarters. One of the main arguments raised against war was that "containment"—that is close monitoring and regular UN weapons inspections—would keep him and his regime under control with no need for bloodshed. In part, the monitoring aspect could, and was indeed already, being carried out by American spy satellites orbiting the Earth hundreds of miles up. Anti-war groups claimed that these contained telescopic cameras sophisticated enough to be able to read the name of the manufacturer stenciled upon the side of an Iraqi missile! Would it not, then, be a relatively simple and comparatively inexpensive task for NASA to place a satellite bearing similar technology in close Moon orbit to take "snaps" of the thirty-year-old alleged landing sites?

The reply from NASA though, would very likely be that no agency of the United States would ask its government for many extra millions of dollars merely to silence a bunch of loony conspiracy theorists! It is a good point and not one that I necessarily disagree with. But I began this interruption of Cosnette's list of questions needing answers with a reference to other "pre-emptive" actions taken by the U.S. Government. So, perhaps a quiet word from NASA into the ear of the person who controls the purse strings in America might remind that person that little good ever came of allowing a snowball to become an avalanche!

Cosnette's "32 things" continued—

30. In the year 2002, NASA does not have the technology to land any man, or woman, on the Moon, and return them safely to Earth.

31. Film evidence has recently been uncovered of a mislabeled, unedited, behind-the-scenes video film, dated by NASA three days after Apollo 11 apparently left for the Moon. It shows the crew of Apollo 11 staging part of their photography. The film evidence is shown in the video *A Funny Thing Happened on the Way to the Moon!*

32. Why did *all* of the blueprints and plans for the Lunar Module and Moon Buggy get destroyed [*A.N.* Yes, that is the mind-boggling claim!] *if* the Moon landings were one of history's greatest—if not the greatest—accomplishments? Normally, there would have been quite literally dozens of copies drawn up and deposited in different places in case of loss by fire, or whatever, if all "eggs" were placed in the same "basket."

So, let's see if we can't try to put all this into perspective…

Could NASA really have faked the Moon landings?

Many people have written to me [Cosnette] and questioned my accusations regarding how NASA could fake the Apollo missions. The majority (most of whom are Americans) cannot believe that NASA would fake the photos. Well, I have news for all [anti-hoax] skeptics and I have reproduced below a little known picture from the Gemini 10 space walk that NASA clearly faked. The astronaut in the pictures is Michael Collins, who was later to be part of the so-called Apollo 11 scam…er, sorry, mission, and the first picture [Picture C on Plate 8] is of him practicing his space walk within a high altitude airplane. When Collins finally achieved the space walk, NASA released several pictures of the event, one of which is the second picture, [Picture D on Plate 8]. If you look closely, you will see that this is in fact the first picture all over again [Picture C on Plate 8] merely reversed to give the third picture [Picture E on Plate 8] with a background of black space added. If NASA has the [Coke?] bottle—or the bare faced cheek!—to release pictures like these, how can anyone think them above faking the Apollo missions?

Let's compare the Apollo cover-up with the USSR launch of the dog called Laika into space. She was launched into space simply to see what the effects of space travel would be on a live creature.

It was later publicly announced that Laika had died painlessly when her oxygen supply ran out, but the truth was finally revealed many years afterwards that the dog had in fact died when the nose cone of the craft carrying her was ripped off after reaching Earth orbit and that the dog probably died from the intense heat of the sun. Further investigation revealed that the nose cone had actually been designed to do this.So, in fact, the makers of the rocket had known that the dog would die even before she was sent into space. This evidence took 30 years to be revealed to the general public.

Thousands of people were employed to work on the Apollo missions, it is true, but very few of them had access to the overall picture. Giving many people very small roles to play in the missions meant that no one except a select, trusted few would get to see the complete picture. [Which, it is being suggested, was possibly only 10% a masterpiece of true courage and endeavor and 90% fakery!].

Eleven Apollo astronauts had non-space related fatal accidents [the Fox TV Special claimed only 10, but not all the deaths were directly connected with Apollo] within a twenty-two month period of one another. The odds of this happening are 1 in 10,000. Coincidence? Maybe.

Further:

• James B. Irwin (Apollo 15) resigned from NASA and the Air Force on July 1, 1972.

- Don F. Eisele (Apollo 7) resigned from NASA and from the Air Force in June 1972.
- Stewart Allen Roosa (Apollo 14) resigned from NASA and retired from the Air Force in February 1976.
- Swigert resigned from NASA in 1977.

Why did all these men resign from the "so successful" Apollo Project?

How come the rock samples if we didn't go to the Moon?

NASA claims that 750 lbs or so of "moon rock" was collected on the Apollo missions. However, according to official NASA records, only a couple of pounds were actually collected by the astronauts. In any case, it would not be impossible to create all the so-called "unearthly" qualities of the "moon rocks" in a laboratory.

Why hasn't anybody spoken out about the fraud and cover-up?

Well, the short answer is that they have. Bill Kaysing got in touch with a friend, a private investigator from San Francisco called Paul

H-o-o-w-l! It's a dog's life up here.

Jacobs, and asked him to help him with his Apollo anomalies investigation. Mr. Jacobs agreed to go and see the head of the U.S. Department of Geology in Washington, as he was traveling there the following week anyway, after his discussions with Kaysing.

He [Jacobs] asked the geologist, "Did you examine the Moon rocks and did they really come from the Moon?" The geologist laughed. Paul Jacobs flew back from Washington and told Kaysing that people in high office in the American Government knew of the cover-up. Well, he probably laughed right along with the geologist at the time, but within the space of 90 days Paul Jacobs and his wife were both dead from cancer!

Lee Gelvani, another friend of Kaysing, says he almost convinced James Irwin [Apollo 15] to turn informant and confess everything. Irwin was even going to ring Kaysing about it. However he [Irwin] died of a heart attack within 3 days of agreeing to speak to Bill Kaysing! Is all the above, then, evidence that something far more sinister than a cover-up is operating to protect the Apollo myth?

—Paused.

Defense: Objection! Not for the first time a prosecution witness appears to be making guarded accusations of murder against my clients. If it pleases the learned judge, would he remind my learned friend that my clients are not on trial for murder, and that testimony and opinions given by this witness should relate to the specific charges laid before this court and nothing else. If this witness has any concrete evidence that my clients possess not only the wherewithal and ability to fake half a dozen Moon landings but to introduce into dead bodies the fake symptoms of illness after having first committed homicide upon them, then he must place this evidence before the proper authorities in the United States.

Judge: Objection sustained, although I am in receipt of more than a hint from the witness that "the proper authorities" are themselves hardly above suspicion in all this. However, I would be grateful if the prosecution could advise this witness to "tread more carefully," and to restrict himself purely to giving evidence pertaining to the charges brought. After which, the witness may continue giving evidence.

Prosecution: If it pleases the learned judge, I of course apologize unreservedly to the defendants for any suggestion, by this witness, that they might be involved in crimes of a far more serious nature than fraud. I feel certain that the witness now fully regrets any misinterpretation that may have been placed upon his words, which might have implied that he thought terrible crimes had been committed under the shabby umbrella of so-called "national security." I seriously advise him now, to heed the words of both my learned friend and the learned judge, although I remain convinced that the witness merely meant that the untimely deaths of those above mentioned were truly remarkable *coincidences*, given the circumstances.

Judge: And you clearly do, too.

Prosecution: Indeed, I do.

Judge: And so do I. The witness may continue.

Cosnette continued—

Why would NASA fake the Apollo Missions?

I think the main reason why the U.S. Government and NASA faked the "official record" is because they could not be seen to be the weak link, especially when you consider that during the 60s, the USA were at the height of the Cold War with Russia. Also their own President had forecast that before the end of the 60s a man would be on the Moon. It would be better to try and fool the public and hoax the footage, rather than let their biggest rival in the world strike a huge moral victory by beating them to the Moon. If man really went to the Moon, why did NASA drop the successful Saturn 5 launch rocket after the last Apollo mission? The shuttle's weight is 3/4ths heavier than the Saturn 5 rocket, puts only 1/6th of cargo weight into orbit and costs 3 times as much to launch. Why scrap a rocket that can outperform its newer model? The shuttle was first flown 2 years behind schedule.

Did you know...?

NASA could have easily launched the shuttle on top of the second stage of the Saturn 5 rocket. The first stage would have dropped into the ocean and the second stage and the fully loaded shuttle orbiter would have traveled into low Earth orbit. The second stages could have then been left in orbit and assembled to make a space station, which would have been well on its way to completion by the time the shuttle was first launched in 1981. They could have had the first launch of the shuttle a whole 5 years before it was finally launched and saved the American taxpayer 20 billion dollars!

Why no Russian follow-up?

Why didn't Russia [Soviet Union] even bother to land a cosmonaut on the Moon after the Americans [apparently] beat them to it? Many people would say that it was because it was too late. But if you want to look at it like that, why didn't this apply to NASA when the Russians beat America by putting the first satellite, animal, man, woman and space station into orbit? Russia would not have thrown in the towel just because America had beaten them at one single thing in space! [Unless, of course, honorably "throwing in the towel," as you put it, was the whole object of the exercise, David!].

Robots

Not one piece of apparatus that appears on the surface of the Moon had to be placed there by man. Be it mirrors to reflect laser beams fired from here on Earth to calculate distances, or seismology equipment. All could have been placed in position by robotic machines. It wouldn't necessarily have required human hands to place them there at all.

Graham Birdsall (Editor of UFO Magazine U.K.) has commented that during the very first Pacific UFO Conference in Hawaii in September 1999, astronaut Brian O'Leary, who worked alongside the likes of Neil Armstrong and Buzz Aldrin on the Apollo 11 mission during 1967-68, commented, 'If some of the films were spoiled, it is remotely possible that they [NASA] may have shot some scenes in a studio environment, to avoid embarrassment!'

During Project Apollo, six highly complex manned spacecraft ostensibly landed on the Moon, took off again and returned to Earth using a relatively low level of technology. An 86% success rate. Since Apollo, though, twenty-five simple, unmanned craft incorporating increasingly higher levels of technology have attempted to fulfill missions to Mars. Only seven of these have succeeded.

Flying the Flag

Watch this great piece of footage! The astronaut has a very hard time trying to keep the flag still as it blows in the wind. [*AN:* Here the article provides the famous—or infamous!—Stars and Stripes flag fluttering movie sequence. For still photos apparently showing this phenomenon, see Pictures A and B on Plate 9].

No Going Back

If we were so successful at landing men on the Moon over 30 years ago, why haven't we been back? In *The Ride Report*, a report headed by Sally Ride, a former astronaut herself, an estimate was made regarding how long it would take to make a similar trip to the Moon. In 1987, if fully funded, NASA estimated that they could land "more" men on the Moon by the year 2010. So they were talking 23 years time from that date!

Since it only took 8 years from President Kennedy's announcement to the Apollo 11 mission, why was it estimated in 1987 that it would take 23 years to send man back to the Moon for the 7th time?

In 1999, this estimate changed. For the worse! Douglas Cook, Director of the Exploration Office at the Johnson Space Centre in Houston, calculated that man could go back to the Moon *within 100 years. I'm* not holding my breath! Are you?

[Here, acknowledging several sources for their assistance in the construction of his article, Cosnette recommends further viewing/reading of the following]:

A great film that addresses the hoax evidence against Apollo called *Paper Moon* is viewable for free over the net at *http://www.apollo-hoax.com/PaperMoon.ram*. Also, for anyone who would like to investigate further into "The Apollo Hoax," I would thoroughly recommend purchasing the video *What happened on the Moon?* and the book *Dark Moon—Apollo and the Whistle-Blowers,* both authored by David Percy and Mary Bennett and available through Aulis Publications. (They can both be ordered by visiting *UFO Magazine* website on the Internet).

—End.

Dasvidania! Have a good trip, Yankees. We won't be tagging along this time. Not if it's all the same to you.

§ § § § §

My thanks to David Cosnette. And, despite the fact that I have had to "tamper" with his highly informative and very interesting article a little here and there, wherever I felt his message might be unclear to the jury, I offer due acknowledgment for all the painstaking research that undoubtedly went into it.

8

If the World Stops, We'll Just Get Off! Er…Won't We?

It is, of course, the inherent right of humanity to question, to accept nothing at face value. If that is cynicism, then so be it. There is nothing wrong with a little bit of healthy cynicism if it is directed at a powerful minority who would lie and cheat and hoodwink an otherwise unsuspecting majority into perpetuity, no matter how sound the original reasons for doing so may once have been.

To lie and to cheat for financial gain—how does $40 billion of taxpayers' money sound?—is a *criminal offence* on the statute books of *all* civilized countries. Although not necessarily always falling into the criminal category, to lie and to cheat to gain *any* kind of advantage over others is most certainly immoral in the extreme. The only time such a practice could even remotely be considered acceptable is during time of war, when a slightly more polite term for it is used…

Propaganda.

The use of propaganda may well be "legitimate" during a war. Maybe even during a cold one. But the Cold War between the "East and the West" is over. The Berlin Wall has come down and the "Iron Curtain" has been well and truly drawn back. Communism has been forsaken in the greater part of Eastern Europe and the Soviet Union is no more. If the Apollo Moon landings were indeed a huge propaganda stunt staged by the United States of America in order to somehow prove to the peoples of the old Soviet empire, China and the Third World that although not ideal, Capitalism will always eventually bear more fruit—even *artificial* fruit!—than Communism, then Uncle Sam undoubtedly succeeded beyond anyone's wildest expectations. And the end result may well have justified the use of such unscrupulous means. Why? Because we now live in a much safer world. Back in the bad old days when the two mighty superpowers were always attempting to shake one another warmly by the throat, we seemed to be forever teetering on the brink of nuclear war. The constant threat of same was already a mushroom cloud hanging over the lives of each and every one of us on *both* sides of the Iron Curtain.

Today there is no such threat, although others have taken its place. Nowadays, though, these are far more likely to be posed by extremists, fanatical breakaway elements or other dissatisfied sections of society. Sometimes made up of the most unlikely bedfellows in any normal circumstances (the Irish Republican Army and Colonel Gaddaffi of Libya were in cahoots at one time, I seem to recall!) these groups are often prepared to employ terror tactics to demand something of common benefit to each of the parties now conjoined, either from a particular government or from the larger world community. Better by far, though, to suffer the occasional bite from a viper nestling in one's own bosom than to have two colossal anacondas thrashing about while attempting to swallow each other whole and, in the process, obliviously laying waste to everything for miles around!

If what history records as the Apollo Moon Landings were in any way, or even in their entirety, fraudulent, with Cold War propaganda their sole motive, then it is time for NASA and their paymasters, the United States Government, to come clean about the whole affair. They should do so not only on behalf of their own citizens, but of all humanity.

And *now* would be an appropriate time to do it. In Britain we have something called the "Thirty Year Rule." This applies to documentation kept on file in government departments and ministries that at one time would have been rubber-stamped "Top Secret" or "Classified" or whatever. These documents are generally released into the public domain after a thirty-year period has elapsed, especially if no longer considered to pose any risk either to national security or to public order. I understand that the same will apply to top-secret documentation pertaining to the Apollo program in the year 2026. At time of writing, that is almost a quarter of a century away. Are we really going to have to wait *that* long before any American administration finds enough balls to admit what more and more people are becoming more and more aware of anyway, with each passing day?

I have for a long time believed that citizens of the United States of America enjoy a far less restrictive, more open, free and equal society than we do here in the United Kingdom. Indeed, that is the very reason why so many of the U.K. and Ireland's citizens long ago forsook the lands of their fathers, set up the Thirteen Colonies and then eventually declared themselves completely independent of Britain.

Despite a post-separation period during which the two newly "divorced" countries enjoyed a somewhat frosty relationship to put it mildly, what with the continued sharing of a common language, a common European heritage and, in the main, the fact that they still shared common ideals and aims, a sense of "kinship" then gradually re-established itself once more between the two countries. Despite its existence sometimes being denied due to the fact that it obviously irritates other European nations whose migrating citizens over the ensuing two centuries or so have also helped to swell the population of the New World, there is almost certainly a "special relationship" between Great Britain and the United States of America. A bond even, that grows seemingly more unbreakable as those decades and centuries continue to pass. A bond of such underlying strength and durability that the two cultures—although "incompatible" still in a few diehard areas, *e.g.* when *will* the Americans stop calling football "soccer" and accept that what they call football is actually a form of rugby—now seem to be virtually interchangeable. The two sovereign states still have their "squabbles" of course, which is to be expected in even the closest knit of families. But the idea that the two countries could ever again be on opposing sides in a war is absolutely unthinkable. Even the very idea that one or the other would go to war against a common enemy without the aid and assistance of the other—or at least without the political backing and logistical support of the other—is again quite unthinkable.

Britannia's relationship with Uncle Sam is now pretty much on a par with the relationships she enjoys with the other great English-speaking nations of the world. America's great neighbor Canada, for example. English-speaking southern Africa, the Indian sub-continent and, of course, Australia and New Zealand. Particularly since the end of the Second World War, the two last mentioned countries have opened up their borders to a wide variety of nationalities, cultures, races, religions and creeds, just as did the United States two centuries earlier. Such multi-culturalism as now exists in those countries, though, has in no way prevented the peoples of the U.K. and Ireland from continuing to look upon Australia and New Zealand as "family." Both great countries in their own right nowadays, being on the other side of the world in no way prevents them from continuing to be considered among our dearest and most trusted friends in an angry world still often fraught with terrible danger. "Kinship" doesn't necessarily have to be anything to do with race or religion. It can be to do with *shared beliefs* about so many other important things. It can be about ideals, goals, aims and aspirations. It can be about different cultures and traditions being tolerant of one another's beliefs and customs and eventually blending into a single society that emerges the wiser and much richer for it. There can truly be no finer examples of this than the Commonwealth, and of course the mighty United States of America herself.

Ah, yes. The Commonwealth. In times of trouble, all these now quite independent countries—and I refer now to our other dear friends in many parts of Asia, the Far East and the Pacific—are at the U.K.'s side without question. Of all the things in our long history and tradition here in these "scepter'd isles," it is the one thing that makes me proudest of all. Not only to be a citizen of that little unique family of nations that includes the English, the Irish, the Scots and the Welsh, but that we have such devoted, loyal and trusted friends of *all* nationalities, colors, creeds and religions scattered about the globe in places often situated many thousands of miles away across the oceans. Friends like the incredible Gurkhas from Nepal in India, who still form a formidable part of the British Army. Friends I sometimes feel we, the British, do not truly deserve. But it tells me that the "experiment" of Empire—much as I personally despise the very thought of empires and their implied "holier than thou" or "we know what's best for you" connotations—was not *all* bad. It tells me that the British—not so much as a race, perhaps more as a civilization, a culture, a way of life—rather like the ancient Romans whose mighty legions comprised every race under the sun—eventually learned well the lesson that in order to keep one's family and friends with one, one sometimes has to let them go, no matter how reluctantly.

Before we properly learned that lesson, the Americans fought long and hard before the mighty Empire was forced to let *them* go too. Free at last of the shackles—"apron strings" is probably more accurate—that had bound her to the "mother" country, America herself then became the mightiest of the mighty. In fact, the United States went on to give the word "mighty" a whole new meaning. It went on to become a country possessed of an arsenal of weapons that all the defenses of the rest of the world put together would fail to equal! It went on to become a country with an economy able to generate more wealth than...yes, you guessed it: again the rest of the world put together! Unfortunately, though, America is now responsible for far more wastage and pollution than the rest of the world put together!

But such things are, of course, the undesirable but inevitable by-products of a highly industrialized society.

I am in danger of digressing. My entire point is that of all the free peoples in this world, I personally consider the American people to be among those least likely to tolerate their *own government* taking them for a bunch of suckers. The least likely to tolerate those they have elected to serve *them*, suddenly turning around and acting instead like their lords and masters by treating the electorate, once in office, like millions of docile, ignorant sheep simply there to be fleeced by taxation and then, with the money thus obtained, going on to fund some secret agenda, for which the blessing of those who had voted them into office was neither sought nor given. A top-secret scam, in effect, designed not only to "pull the wool" (yes, I'm still in sheep mode!) over the eyes of their potential enemies but over the eyes of their own taxpayers as well! That taxpayers should *not* be treated in this arrogant and dismissive way, as if their opinions did not count regarding how their money was spent, was precisely *what* the Americans fought against the British for two and a quarter centuries ago!

Maybe it is time that an American government found the courage to come out and openly admit to its citizens and, by default, the rest of the world, that a previous corrupt—even criminal—administration had funded, through its agency NASA, a huge publicity stunt solely aimed at banging a further nail into the coffin of the detested doctrine of communism. It could argue that although nefarious and underhand in the extreme, and a *modus operandi* (way of operating) which would be unacceptable (and too easily detectable?) in modern times, the means used in those "bad old days" had since been shown to be more than justified by the end result, which had not only rendered the United States and her allies more safe and secure but indeed the whole world. My personal opinion is that once the initial media frenzy dies down, any current U.S. administration will quickly forgiven by the American public. Especially as that earlier administration whose behavior it has suddenly found itself forced to apologize for was that of an impeached president, one Richard Milhous Nixon!

Furthermore, any attempt to indict surviving members of the Nixon administration, or former or currently serving employees of NASA who might have been in on the secret—including the astronauts themselves, of course—would be pointless, not to mention grossly unfair. At the time, Richard "Tricky Dicky" Nixon, for all his sins, was America's commander-in-chief. The Apollo missions to the Moon *all* took place during his term of office and suddenly ceased once he left office at the very height of their so-called success. If these missions *were* bogus, then the buck stopped at Nixon's desk in the Oval Office in the White House. He would have personally sanctioned the whole project, and anyone else involved would simply have been obeying the orders of his or her president. To disobey the president in time of even a cold war is treason.

But Richard Nixon is now dead, God rest him. Were he still alive, far from indicting him too, there is a school of thought that he should instead be awarded a medal for having had the courage and vision to back such an outrageous but undeniably tremendous stunt! Perhaps a posthumous medal is still not out of the question, because there can be little doubt that "beating the Soviets to the Moon" was instrumental in the eventual downfall of communism in that huge country. And Nixon was hardly the sole "villain" of the piece anyway. It is more than likely that the then Soviet leader, Leonid Brezhnev, was also party to the scam. He may even have actually *requested* that it take place due to the fact that the USSR could no longer afford the space race and wanted to see an end to it all.

The rest of the world too, would also quickly forgive any current American government and NASA, providing a clean breast was made of it all. The Russian Federation's top space scientists would long ago have realized it was all pure nonsense anyway, even those not "in the loop" at the time. But no one in officialdom anywhere will ever say anything until the United States decides to "own up" for herself. Be assured that the present British prime minister, like all his predecessors since the seventies, would have been acquainted on his *first day in office* with the truth about Apollo, along with various other "ultra top secrets" presented to him in a locked box to which only he now had the key!

And here's more food for thought. America's reluctance to come clean may again even be at the *behest* of the present day Russian administration! Try to imagine, members of the jury—particularly those of you old enough to recall the period in any detail—being Soviet leader at the time of the Apollo 11 Moon landing, or *alleged* Moon landing. Imagine standing up before a battery of microphones and saying to the gathered world's media the Russian equivalent of, "Man on the Moon? It's all a load of crap. They faked it! Mankind cannot travel in space beyond the radiation belts. Our scientists, academicians and more than a few dead or dying cosmonauts have proven this beyond doubt."

By so saying, you would be admitting that you had likewise wasted billions upon billions of rubles that a population of getting on for 300 million souls had struggled to pay in taxation. Billions of rubles, all literally gone up in smoke, that could have filled the shops with food and made the mile-long daily bread queues unnecessary for all those years. Not only that, but you would also be admitting that you had wasted many brave young lives in the process. All that queuing. All that money. All those valuable young men and women lost. And for what? For the sake of an exercise in utter futility, that's what. Certainly back then, and certainly today too, at this much more advanced stage of man's technological development.

Taken into Nixon's confidence or not, in direct consultation with him or not, Brezhnev would in any case have breathed a huge sigh of relief that the Americans appeared to have won the madcap race for the Moon. It meant that he could at last stop pouring money into a venture he and his government had learned the hard way would provide absolutely no return whatsoever, neither in the short nor the long term. It was well past time that somebody put an end to the sheer lunacy of it all—again literally!—and if it had to be the "enemy" by way of a skill that only he possessed in the art of cinematographic illusion, as well as the unlimited funds at his disposal, then so be it. It meant that the Soviets could now confine their efforts in space to the possible rather than the impossible. It meant that they could go on to win the *real* space race. To be first to conquer *near* space as opposed to the impossible dream of conquering outer space. And they did that, of course, by building man's first viable space station. They called it *Mir*.

My point about the present day Russian administration possibly having put the blocks on any attempt by the present day American administration to admit to the alleged Apollo fraud is a simple one. It is highly likely that the Russian government still has genuine qualms about the way the ordinary Russian in the street might react, even in modern times, to any suggestion that a past Soviet government was not only in on the "Man on the Moon" stunt but, as I say, may even have requested of the Americans that they "stage the show" in ways that only Hollywood knew how. There may be a genuine fear by present day Russian leaders—what with huge country's long, troubled and volatile history—that instead of finding forgiveness in their hearts for past regimes for all those hard times when a fortune was wasted on space exploration, the people might find only anger…

Which would almost certainly be directed *at them!*

If America's reticence to come clean—providing there *is* anything to come clean about, of course—actually has nothing to do with Russian government approval, then what with the Cold War being over the U.S. Government might be deliberately allowing a longer "honeymoon period" with its own electorate to elapse before letting us all in on the secret. That's OK. Providing they don't leave it too long, the administration that ultimately "confesses" should still enjoy our forgiveness worldwide. Of that I am fairly certain. As I think I have already said, perhaps congratulations might even be in order for having pulled off the biggest political *coup de grace* of all time! After all, "all is fair in love and war" as the old adage goes, and the Russians are a gracious and great enough people to hopefully accept this. Especially if it turns out to be true that a former Soviet leader's "down to earth" common sense may have actually been behind the whole thing. Apart from that, the Russians would be among the first to admit, anyway, that the use of propaganda is a recognized and often necessary evil in times of conflict, even if it is only during the course of a tussle for "one-upmanship" between two opposing ideologies.

But, as a civilization—as a species even—humanity, as a whole, has a *right* to know the truth, even if it is something we do not really wish to hear. "Mankind cannot travel in space," said Colonel Philip Corso. The problem is that this was not just *his* personal opinion as an old soldier with views and feet also firmly affixed to the ground. It was almost certainly the agreed view of the *crème de la crème* group of top scientists he worked with at the U.S. Army's Research and Technology Division at the Pentagon in the sixties. Were they right? That is the burning question. The "to be or not to be?" if you like...

Which *must* be answered.

For the past half century or so, the human race has bombarded itself with books, films and television series all telling us that we will one day be able to transport ourselves through space with impunity, by no end of means. These range from Jules Verne's "bouncing ball" coated in anti-gravity paint in *The First Men On The Moon* (turned into a film during the last fifty years, although Verne wrote his story considerably further back than half a century ago!) to Gene Roddenberry's colossal starship "Enterprise" in *Star Trek*. From Arthur C. Clarke's giant space "sailing ships," which utilized solar winds in *The Wind From The Sun*, right through to teleportation, the possible perils of which were all too frighteningly illustrated in the film *The Fly*. These, and countless other stories, have foretold of the possibility of travel through space, not just to the planets of the solar system, but to stars hundreds of millions of light years away.

Before Apollo, these were just stories, purely and simply: Fanciful imaginings that may or may not have been possible in fact. Then, all of a sudden, these stories were lent enormous credibility by what we believed to be actual landings upon another world by fellow human beings, even though that other world was only a mere quarter of a million miles away. This inevitably spawned yet another spate of science fiction stories and movies. The question, though, is this. And I will commence a new paragraph (Ooops! I am speaking in court—not writing a book! Oh well...) to ask it:

Has the human race been lulled into a completely false sense of security by believing that man has already landed and walked upon another planet? Are we now, as a race, living in a Fools' Paradise by believing, possibly wrongly, that it might not matter too much if our home planet, our beloved Earth, one day comes to an end? I do not need to remind most of you that this was the horrendous fate that tragically befell the fictional home planet of an equally fictitious superhero in yet another famous science fiction story. Many of you will recall that this particular story began—in almost Biblical fashion—with a tiny infant boy, wrapped in blue and red blankets, being fired from *Krypton*, his doomed home planet, through space to Earth in a sealed capsule. He was found lying unharmed in the crashed and now open "spaceship" by a childless local couple, the Kents, who rescued him from the wreck and then passed him off as their own son. They named him Clark. Eventually, though, they were shocked to discover that this handsome boy was no normal child. He seemed to possess the strength of a hundred men and developed many strange powers while growing up, which he then went on to use in adult life as the comic-book hero *Superman*, to the benefit of mankind.

But do the vast majority of us now believe that if that terrible *Krypton* day were ever to befall Earth—and one day it will, of course, even if it is debatable whether the human race will still be around by then to witness it!—it will simply be a question of *c'est la vie* (that's life), because by then we will have colonized other planets? At the time of the claimed Moon landings, how many members of the jury will admit to having suddenly felt quite smug about that inevitable calamity to end all calamities, the end of the world? How many of you suddenly felt "safe" in the knowledge that man would survive and continue on elsewhere?

For those countless millions of you—again including me—who, since Apollo, have been inclined to think in this slightly arrogant "Of course, like the *Titanic*, we will be sorry to see the old girl go, and all that, but are we that bothered about the end of *this* world anymore?" fashion, you might be wise to think again. What *appeared* to be the position that happy day in 1969 might *not* be the case. Unlike baby *Kal El* (Clark Kent/Superman), *we* may not be able to escape to another planet. Not for the foreseeable future anyway. Maybe never. As a race, as flesh and blood bio-organisms, we may be condemned to forever living only where we were intended to live. Here on this Earth. If Apollo somehow convinced you otherwise you may have to accept that you have been laboring under an illusion. Like me, in the face of all the evidence—much of it yet to come in this trial, believe me!—you may be forced to accept the very high probability indeed that the best part of thirty-five years ago we were all completely taken in by an elaborate performance from start to finish. But how could we have been so gullible? How come we were all so easily taken in? I will tell you why…

Because we damn well *wanted* to be taken in, that's why!

Some of us, especially those of us who are concerned enough about the dreadful state of the planet to find ourselves wondering where the bloody hell we are all going to go once we finally destroy this beautiful place altogether, might even have been *desperate* to be taken in!

RE-ADJUST YOUR THINKING

It might pay us all to re-adjust our thinking anyway—even if Apollo *was* genuine—along the lines that mankind may never be able to *properly* travel in outer space. Not in these bodies anyway, even if aided by some of the possible future inventions earlier mentioned. You will note that I say, "not in these bodies." This has no doubt aroused your curiosity, members of the jury. I hope so, because that was my intention. However, please bear with me for now. I am confident of the learned judge explaining in detail later on, precisely what I mean by that deliberately controversial remark. For now, though, suffice it to say that you should not, repeat *not*…

Abandon hope all ye who would enter outer space!

Of course, the belief that "we" had already conquered outer space and had actually landed people on the Moon was bound to have given an enormous boost to the broader vision of mankind's future that probably lies within the subconscious minds of us all. A vision undoubtedly further nurtured by the fertile imaginations of a multitude of science fiction writers. Apollo was the basis for a whole new faith. A whole new religion even. Rather than a tantalizing dream about things quite beyond our reach there was now this far more tangible, cast iron belief that man would indeed one day inherit the stars and create his own vast kingdom up there in the heavens.

For those of you who would cling on to this vision, this belief still, then I insist—and have indeed hinted as much above—that you should *continue* to cling on to it. But just be forewarned, that's all. There can be terrible disappointment, let alone danger, in living in a house you think of as being made of bricks when, in fact, it is only made of straw.

Man, in one form or another, *will* one day go out into those far, far reaches of space and build that mighty empire as a seeing, hearing, thinking, sentient being. I, too, still cling to that belief and, as hinted at earlier, the reason why will be explained later. Maybe not so much an empire as a commonwealth. We want no occupation of worlds against the will of any indigenous intelligences, and colonized worlds should always be granted independence, without bloodshed, as soon as such a step becomes the wish of the majority of colonists thereon.

Technology, maybe working hand-in-glove with plain, good, old-fashioned evolution, or the evolutionary process further advanced *by* technology, *will* resolve all the problems in due course of time. Of that I am certain. But I am also becoming more and more certain that this process was *not* begun thirty-five years ago. Mark my words that it will not even have begun thirty-five years *from* now. Start thinking at least a *thousand* years from now and we may be making our first *real* faltering, tentative steps into the cosmos proper. There will be no "giant leaps" into space.

Of course, for a third of a century now, it has been generally understood that the first "giant leap" had already taken place and that the very first footprints on this "stairway to the stars" had already been placed in the dust of the Moon. If this were true—despite the "impossibility" of footprints at all according to the witness Cosnette—then those famous boot prints will *still be there*, just at the base of the *Eagle's* landing stage ladder. Cosnette claims that because there is no air or moisture on the Moon, footprints could not have been made in the first place. But the plus side of the Moon containing no air or moisture is that neither would there be any wind and rain to weather those boot prints away should Cosnette be wrong. Nor would they have been filled in by loose dust showering down after the ascent stage blast-off, because—according to the *anti*-hoax lobby—most of the dust was already blown away during the landing! That is apparently why there was little or no dust to be seen on the LM's lunar lander pads, as I feel certain my learned friend will later claim on behalf of the defendants during this very trial. If the Moon landings *were* genuine, then both Armstrong and Aldrin would have been careful *not* to disturb the area where Armstrong's boots first touched the Moon's surface. Not to do so would almost certainly have been— quite literally—part of their "standing orders."

As I mentioned earlier, there are spy satellites currently orbiting the Earth, which carry cameras of such sophistication that they can sneak a peek over the shoulder of an unsuspecting individual from many miles up and photograph, with a high degree of clarity, the headline of the newspaper that person might be reading. It would surely pay NASA to simply send another unmanned Apollo type command module up to closely orbit the Moon, maybe armed with such a camera in order to photograph those famous first bootprints. It would need to be the bootprints in particular because pictures of the equipment left behind would still not be proof that man was ever there. All that stuff could still have been placed there covertly and robotically a month, a week even, before the publicized return to the Moon by the vehicle carrying the camera. One new picture of those first bootprints would be worth billions of dollars in publishing rights alone for NASA! And, talking of billions of dollars, would it not be worth the U.S. Government and NASA's while—in the long term—to send another *manned* mission anyway, to ensure that the Apollo 11 site, more so than any other, is somehow protected and preserved for all time?

If there ever are any space-traveling generations to come—and I believe there will be, as I have said—can anyone even begin to imagine just how much of a tourist attraction that first Apollo 11 site—if it exists—will become? Not merely an attraction, it will become a shrine, a place of worship. A truly wonderful, sacred place; where God finally agreed to strike away the shackles that had hitherto bound mankind inextricably to the Earth.

If the Apollo landings *were* genuine, not to make some attempt to protect the Apollo 11 site is pure sacrilege. Despite the Moon's natural ability, in normal circumstances, to preserve any imprint—or object—left upon its surface, there is always the chance of a meteor strike happening close enough to smother the whole Apollo 11 site in a thick layer of debris and dust. To lose the footprints at the Apollo 11 site would be quite disastrous because we are not talking just any old set of footprints here! But as far as providing proof of the landings is concerned, *any* set of prints from any mission will suffice because even one clearly defined print left by a booted foot anywhere on the Moon will *prove beyond doubt* that man not only can and *did* travel into outer space, but that for a short time he actually survived beyond the walls of his spacecraft on the surface of another planet!

Although his words were well scripted, when Neil Armstrong supposedly jumped that last couple of feet from the lower rung of the *Eagle*'s ladder onto what history records was the surface of the Moon, the bootprints he made did not merely represent "*one* giant leap for mankind." That phrase, as applied to the event he was describing, should perhaps have gone down as one of the classic understatements of all time. In truth, the action Armstrong was allegedly performing would undeniably have been the single most momentous, most monumental leap *in all the history* of mankind!

Hmmm…if it actually happened, that is…

The Prosecution

9

Standing on a Paper Moon—the Testimony of James Collier

I shall call my next witness for the prosecution. Once a journalist and writer by profession, he too, sadly, is now dead. But he did not go quietly as you will see.

I call James Collier.

JIM COLLIER'S TESTIMONY

Source: *http://www.recrea.19.co.uk/papermoon.htm*. Article— often referred to as 'The Collier Article' that appeared in *Media Bypass* magazine in August, 1997. *Investigator Challenging NASA* by the late James M. Collier © 1997. Collier died of cancer in 1998.

Begins—(slightly edited here and there for the sake of clarity).

In 1994, Victoria House Press in New York received a manuscript titled *A Funny Thing Happened On Our Way to the Moon*. Its author, Ralph René, a brilliant lay physicist who had studied Bill Kaysing's thesis that NASA faked seven Apollo moonshots, wanted it published. Since I had written the investigative report *Votescam: The Stealing of America,* publishers Victoria House Press asked me to investigate René and his manuscript, to determine the credibility of both.

"I read Kaysing's book *We Never Went to the Moon,*" René told me, "and although it was compelling, it lacked technical detail, a grounding in physics that would convince scientists, beyond any doubt, that America never went to the Moon."

René was positive that NASA had pulled off the hoax of the century. "NASA didn't have the technical problems solved by 1969 when they launched the first moonshot," he insisted, "but I believe that they couldn't admit to it or they'd lose thirty billion dollars in taxpayers' money." I read René's manuscript and although I understood basic physics, I couldn't immediately assure the publisher that René's assertions were scientifically accurate. Least of all, I couldn't assure them that we didn't go to the Moon. I needed time.

So what began as simple research turned into months at the New York Public Library, the Library of Congress in Washington and the United States Archives. Surprisingly, precious little had been written about the Apollo missions except standard "puff" pieces in the *New York Times* and the *Washington Post*.My research then turned to Grumman Aircraft in Beth Page, New York. Grumman built the Lunar Module (LM), the unwieldy looking craft that never flew on Earth but supposedly landed safely on the Moon six times. I asked for blueprints detailing the scientific thought behind its design. Did it run by computer? If so, who built the computer? What made Grumman engineers think it would fly? Grumman told me that all the paperwork had been destroyed. I was stunned. The LM's historical paperwork destroyed? Why? They had no answers. I turned to Boeing Aircraft in Seattle. They built the Lunar Rover, the little car that NASA claims traversed the Moon on Apollo missions15, 16 and 17. NASA claims it was transported to the Moon in a five-foot high by six-foot wide, triangular corner section of the LM. (The LM's bottom section was basically a tic-tac-toe design with nine sections. Five sections were squares with the four corners being triangles). But my research indicated that the Rover was at least six feet too long to fit into that corner compartment, thus making it impossible it ever got to the Moon via the LM. Next was the National Air and Space Museum in Washington and the Johnson Space Center in Houston, Texas, where I videotaped an actual LM. Here, research indicated that the crew compartment and hatches were too small for the astronauts to actually enter and exit. After taking the video footage, I challenged NASA to prove that two six-foot astronauts, in ballooned-out pressure suits (4-psi in a vacuum) could either get in or get out of a LM.Trying to understand how the Moon acquired a ten-foot layer of topsoil without wind, rain or water to erode the volcanic-crystalline surface, I spoke to a geologist at the Massachusetts Institute of Technology in Boston.Much of my time was spent just trying to mentally picture the physics of light and shadows, jet propulsion and solar radiation, because most of what NASA was claiming about the moonshots—and what was supposedly discovered on the Moon—appeared to be diametrically opposed to present day text book physics.

The Lunar Rover. Too big a fit for the LM?

Anyway, I was knee-deep in all this research, when René became impatient and decided to self-publish his book. He changed the title to *NASA Mooned America.* I, however, had been hooked. But now, it seemed, there wasn't a book to research. I was left hanging. All kinds of questions were still plaguing my mind. Questions that neither Kaysing nor René had addressed. Their research had led me into a scientific wonderland, filled with possibilities. What was I going to do? I had been thrown out of a great movie and I would never know how it ended. I decided to continue the research. I proposed a book to Victoria House Press titled *Was it Only a Paper Moon?* and I promised it by 1998.

I started with the technical problems NASA still faced in outer space. In fact, I discovered there are two separate zones out there, an inner space and an outer space, and that fact eventually became very significant in my research. [*A.N.* As indeed it did in mine too!]. It appears that human beings are most likely operating in *inner* space (the space lab, shuttles, etc.) but outer space, beyond the Van Allen radiation belts and the whole magnetosphere thing, around 560 miles up, may be too deadly to enter due to solar radiation [and a helluva lot more too!]. If that data proves to be correct, men could not possibly have gone to the Moon and returned without some signs of radiation poisoning, cell damage and DNA alteration and, most likely, death from cancer [Hear! Hear!]

The first concern I faced when I started to write the book was my own public credibility. After all, I was the person who had told the country (in his co-authored book *Votescam*) that their votes were being rigged by a cartel of the powerful elite, including the owners of major media companies in America.Now I found myself investigating the possibility that we didn't actually go to the Moon! "You've got to be nuts," said my friends. "First you told them the vote was rigged and now you question whether we went to the Moon? They'll hang you Times Square!" [*A.N.* Make that Red Square and you'll get an idea what current Russian leaders might be afraid of!]

So I decided to test the water with several talk shows on radio in the Midwest. Most of the callers said they never believed we went to the Moon in the first place. Others protested that I was doing the station and myself a disservice by even bringing up the subject. They argued that I shouldn't malign "those great American heroes, the astronauts." What could I say to these people? I wanted to explain that I not only sympathized with their point of view, but that at one time I had shared it. It wasn't easy being the Cassandra of the airwaves, I can tell you! Telling people things they definitely didn't wish to hear. Half of me wanted to be proven wrong [*A.N.* I know just how he felt!], but the other half of me had both hands well and truly on the tail of something that sure looked like a duck and sure quacked like a duck. The last time *that* had happened, the duck turned out to be an expose of computer vote rigging in the United States. As an investigative reporter, I just couldn't let go of that damned duck.One enraged listener said that the eagle-feather and hammer that the astronauts had simultaneously dropped on the Moon, was an experiment to prove that there was no atmosphere on the Moon's surface. This person was absolutely outraged by what I was suggesting, convinced that I didn't understand basic physics. I explained that the experiment wasn't done to prove the absence of atmosphere but to prove that an eagle feather and a hammer would *both* fall at the same rate of speed because the Moon does have *gravity*, although only 1/6th as strong as Earth's."On Earth," I said, "both the hammer and feather will fall at 32-feet per second because of gravity." The caller then actually started to holler. "No, no, an eagle feather will *float* down on Earth and the hammer will fall faster. On the Moon there is no air so they both fall at the same speed!" I told him to get an eagle feather and try it. [And I suggest that *you* try it too, members of the jury]. It is Galileo's law: no matter what the weight of any two objects, they will both fall at the exactly same speed.

—Paused. The testimony of this witness will continue again shortly.

Here, a note is placed on the site containing the Collier Article, which says, "Jim was called to task on this discussion because of the issue of atmosphere versus no atmosphere and drag versus no drag. He addressed his thinking on this in one of his emails…and others have given clear discussion at the MSN community 'moonomoon' which is also linked from this site. Feel free to join the discussion. Jim is dead now. But the questions he raised linger on, though some were better than others. Get into the discussion *now!*"

The Collier Article continued—

In the final analysis, I had tested the waters by doing radio and found that although they were hot, they wouldn't actually burn me alive. Despite being hollered at much of the time on air, there were still scores of calls from listeners who encouraged me to continue the investigation.

Then, a funny thing happened on my way to writing that book I mentioned. I was trying to use *words* to describe the strange visual phenomena that I saw in NASA photos and videos. Those provocative images being the first evidence that people investigating the supposed NASA Moon landings use to draw you into the fray. "You won't believe *this* NASA picture," they'll say, and then the tantalizing hunt for more clues is on. It was then that I realized you had to *see* some things to believe them. Those NASA pictures were supposedly taken on the Moon's surface, but the lighting from the only available sources, the sun and reflected earthlight, seemed to be all wrong. It was too soft, appearing more like a Disney studio photo, with soft pastels and diffused light. How could there be diffused light on the Moon? Earth's atmosphere takes light and bends it, spreading it around objects. Light reflects off air molecules and lights up the dark sides of objects. It is atmosphere, bending the sun's light that makes the sky appear blue. However, on the Moon there is no prism of atmosphere to diffuse or bend light, so the sky is totally black.On the Moon, the sun's light should be absolutely blinding. In fact, this is the reason the astronauts were wearing, gold tinted faceplates on their helmets to cut down 95-percent of the light from the sun. The dark side of objects in NASA photos should be pitch black, while the lit side should be hellishly bright. Yet, all NASA photos from the Moon are softly lit and appear to have been taken in Earth's atmosphere. Why? If NASA film footage was actually taken on the Moon, then it

would be a tremendous scientific story. One would expect new physics books trumpeting an incredibly new physical reality to have
appeared! That atmosphere has nothing to do with diffusing light!
Therefore, and forever thereafter, a new scientific principle would be
taught in schools: where there is no atmosphere, light will react
exactly the same as light does in an atmosphere. What was wrong in
the world of science? Why were the scientists silent about such an
important discovery? How come the major media companies were
mute on the subject?

The next thing was, I called Kodak, in Rochester, N.Y., the company that supplied the film for the Hasselblad cameras the astronauts
purportedly "used on the Moon." I asked: "At what temperature does
film melt?" The reply was: "One hundred and fifty degrees." But
NASA video and film would appear to show the astronauts on the
Moon's surface with the sun at high noon! [Denied by NASA, which
claims the Moon missions all took place either during a lunar "morning" or "evening," the significance of which will be fully explained
later]. The temperature would have been +250 degrees Fahrenheit!
"Without some method of keeping it cool, the film in the cameras
would melt," Kodak said. So the duck was really quacking now!

When I realized that everything I was trying to describe with words
was strongly visual, I decided to commit the research to videotape
instead of a book. *Was it Only a Paper Moon?* the video, was released
in Spring of this year [1997]. It contains a 90-minute unbroken chain
of circumstantial evidence that, if not refuted by NASA, proves we
could not have landed on the Moon.

I feel there is enough evidence overall to demand a Congressional
hearing to ask the question: did NASA indeed pull off the hoax of the
century?

—End.

§ § § § §

James Collier's video *Was It Only A Paper Moon?* is a product of
Victoria House Press. It is provided by Grade-A Productions. Sadly,
as mentioned above, Jim Collier died in 1998 of cancer—and he
never even went into *near* space, let alone beyond the radiation belts
to the Moon! I apologize unreservedly for occasionally not strictly
adhering to Collier's exact words as originally published in the above
article, but I feel that I stayed with the mood of his article and, if anything, maybe enhanced it a little here and there by removing any unin-
tended ambiguity on his part. This book is possibly the most serious

attempt yet to get to the very bottom of this whole Apollo Moon landings affair and to present all available evidence fairly to a jury made up of its readers. Any lack of clarity will not do in a court of law. And I repeat that this is what this book is to all intents and purposes: an English court of law. Every single word contained between its covers forms a trial. A trial that should be taken as seriously—perhaps even more so—as one heard at London's Old Bailey. I offer my sincere condolences to Jim Collier's surviving relatives and friends at the untimely loss of such a talented and courageous man. I understand exactly what it is like to lose a bright, intelligent colleague and friend to cancer years before his proper time and with so much more still to contribute, and they have my deepest and most heartfelt sympathy.

I thank Jim Collier. I call Gordon Philips.

GORDON PHILIPS' TESTIMONY (MORE ABOUT JIM COLLIER'S TESTIMONY)

Source: *http://www.thewalls.boom.ru/moon.txt.* **Standing On The Moon** by Gordon Phillips, Founder & President of *Inform America,* Vol. I, Issue 3, Aug 18, 1997. © 1997.

Begins—(again some editing for the sake of clarity).

Dear Reader,

Stop your world for a moment, get in touch with your feelings and imagine that you are an Apollo astronaut standing on the Moon. Your fellow astronaut is nearby, inside the Lunar Lander, and you are out there all alone, wearing your spacesuit, standing on the powdery, barren surface of the Moon. It is beyond eerie. It is unearthly. Alien. No amount of training could ever have prepared you for the way you are feeling right now. Your heart is pounding. The landscape is stark and flooded with searing light. With no atmosphere, the stars blaze all around you in a black sky like headlights pointed straight at you. You can see the Earth above, hanging like a light blue tree ornament in the perpetual night. You can almost reach up and touch it. Your family is there. So near, yet so far away too. If the rocket on the LM fails to ignite, you will die here. The quiet is so unearthly you can hear the sound of your own blood flowing through your inner ears. You can almost hear the sound of your own digestion. You are truly and inexorably alone...

Then the NASA movie director yells, "Cut!"

You lift off your helmet and head out through the hangar door. It's time for lunch with the rest of the crew.

Whew! I had you going there for a second, didn't I, eh? Now get in touch with your imagination again, only this time, pry your mind open as wide as it will go. Turn your "truth detector mode" to "high" and consider, if you will, the outrageous, the impossible. That NASA never actually landed a man on the Moon! That they faked it in a simulation filmed entirely inside a floodlit hangar here on good old *terra firma*.

"But that's impossible," did I hear you say? I mean, if you're over about age 40, you can most likely still remember exactly where you were *and* what you were doing when Neil Armstrong spoke the most famous words in all human history. "That's one small step for a man," said Neil, "one giant leap for mankind." *Fake* it? Why on Earth (literally) would anyone want to do that?

Hard-digging investigative journalist Jim Collier of *Votescam* fame [*A.N.* The full title of this book, which is banned in the USA, is *Votescam: The Stealing of America*. It exposes what is allegedly a vote-rigging scam by computer in the United States and was co-written by Collier and his brother and fellow journalist Kenneth] has now hung his credibility and perhaps his entire career on the line with a question regarding whether or not NASA actually "mooned" the entire world with the most colossal hoax of all time.

The premise is that the pressure was on to beat the Soviets to the Moon. Everything was at stake—$30 billion in federal funding, prestige, egos, careers, the whole works. NASA knew how to send three men *to* the Moon and *around* the Moon, [*A.N.* Three single men, tired of life and quite unconcerned about their future good health presumably!] but lacked the technical expertise to land them *on* the Moon, wherein they would need to separate from a "mother ship," descend in a smaller, self-contained "shuttle craft," fire a 10,000-pound-thrust landing rocket and spiral down to a controlled landing, suit up, disembark, romp and play about on the surface, go for a drive, return to the Lander, blast off, rendezvous and dock with the orbiting Command Module and finally return to Earth.

So they [he means NASA] faked it. Considering that this was the same general era as Stanley Kubrick's blockbuster *2001—A Space Odyssey*, the special effects technology indeed existed. And, let's face it, you can make an awful lot of movie with a cool $30 billion at your disposal! In a recently released VHS video titled *It's Only A Paper Moon*, Collier looks straight at the camera and, in a "video letter," asks senior NASA management for proof of their claimed accomplishment—a feat the entire world has long since taken for granted. Collier points out a few "minor" inconsistencies, summarized here:

1. Two fully suited Apollo astronauts could not physically have fitted into the allotted space inside the Lunar Lander and opened the door [hatch], because the door opened *inwards*, not outwards. They could not possibly have exited with their suits on. Collier measures this on film.

2. An Apollo astronaut could not physically have fitted through the tunnel connecting the CM and the LM. It was too narrow. Collier went to NASA's museum and measured it. The ends of the tunnel contained a ring of docking connectors. NASA's "in flight" footage, that we are told was taken *en route* to the Moon, shows astronauts freely tumbling through this connecting tunnel, which would be blatant enough except that there are no connectors visible, plus the tunnel hatch door opens the wrong way. These shots had to be taken on Earth.

3. There is *blue* light coming through the spacecraft windows *en route* to the Moon. With no atmosphere filtering through, space is *black*. That footage must have been taken here on Earth, most likely in a cargo jet placed into a steep dive to simulate the sensation of "zero-G."

4. Photos taken by disembarked astronauts purportedly on the Moon show the LM sitting on a placid, smooth, undisturbed surface. Impossible. If they blasted down to a landing, riding on a 10,000 pound thrust plume of flame [well, plume of *exhaust*, anyway!] the entire area would be severely disturbed. Those shots had to be taken here on Earth.

5. There are no stars in the sky in any of the Apollo photographs claimed to have been taken on the Moon. None. That is impossible. The astronauts were surrounded by blazing white stars, which were at all times unaffected by any atmospheric distortion. These shots had to be taken here on Earth.

6. There are multiple shadows projecting from the astronauts and from other objects supposedly standing on the lunar surface. Shadows of different lengths. That is impossible. There was no other source of illumination other than the sun, the light from which (obviously) came from the exact same direction. Those shots had to be taken here on Earth.

7. Earth never appears in a single NASA photograph allegedly taken by an astronaut standing on the Moon. Not one. [That's my boy!]. If you yourself were some top-notch NASA individual involved in a *genuine* moonshot, wouldn't you want, for publicity purposes if nothing else, *many* shots of the astronauts—maybe of every single astronaut—posing "next to" the Earth? Maybe even appearing to *hold* the Earth in his upturned palm? [*A.N.* A quite brilliant observation is this, although not wholly accurate. On Plate 27, Picture D would appear to show a heavenly body—no not Pamela Anderson's!—hovering in the sky behind the Lunar Rover on an Apollo 17 excursion to Plum Crater. It is far too obvious to be a studio light, unless it is a studio light being passed off as the Earth! But I take this writer's illustration of Jim Collier's point. It seems utterly impossible, notwithstanding ludicrous, that NASA could have overlooked such a golden opportunity to play the song *He's Got The Whole World In His Hands*, doesn't it?].

8. With $1/6^{th}$ the gravity of Earth, the "rooster tail" of dust kicked up behind the wheels of the "dune buggy" (Lunar Rover) should have gone up *six times* as high as it would on Earth traveling at the same speed. But it doesn't. And it also comes back down in sheets—*sheets!* That is impossible in zero atmosphere. The dust should have come back down again in the exact same smooth arc it went up in. [*A.N.* I don't know whether this is true or not. What I do know, though, is that in $1/6^{th}$ of 1G gravity, the dust should not only have been thrown up six times higher from the surface, but should also have fallen back to the ground six times *more slowly* than on Earth. Footage I have studied shows that neither of these things happened. There was what looked to be normal disturbance/displacement of sand, or whatever, that was filmed and seemingly played back at half-speed!].

9. The folded Lunar Rover could not physically have fitted into the allotted storage space on the LM. Collier went and measured it. The space is *several feet* too short. Shots taken "on the Moon" show the astronauts *starting* to unhatch the compartment to remove the Rover, but then it [the film footage] cuts away. When it pans back, the Rover has already been removed from the compartment and unfolded. How convenient!

—Paused.

Defense: Objection! This is hardly a valid observation. The first mission to carry a Lunar Rover on board was Apollo 15. My Client, NASA, claims that the LMs used on missions 15 through 17, were modified to accommodate the Rover vehicles. The witness Collier could not have measured the storage compartments of the *actual* LMs of either Apollo 15, 16 or 17 because like their predecessors of 11, 12 and 14, they were all three abandoned on the Moon! Collier would only have been able to measure the storage compartment of a LM replica on public display. This would neither have necessarily been a replica of the modified versions used to carry the Rover nor would its dimensions in any area necessarily have precisely corresponded with those of the actual LMs used.

Judge: Sustained. I would seriously advise counsel for the prosecution to remind his current witness, although I fully understand that he is only quoting a previous witness, that I will not allow any more time to be wasted in this court discussing the internal dimensions and sizes of doors and windows of museum replicas. One might as well claim that one has measured the dimensions of a mocked-up version of *The Flying Scotsman* (a famous British locomotive) on show at railway exhibition, which is maybe using smaller replicas due to lack of space, and proven that the original was too small to have fitted the gauge of track in general use throughout Great Britain. What nonsense!

Prosecution: I can only apologize to the learned judge, learned counsel for the defense, and the jury. Rest assured that there will be no more talk of museum replicas and their measurements from *any* witness on this side. The witness mentioned that the astronauts plural were about to remove the Rover from the LM when the camera filming the scene apparently "cut away" before "panning" back. As there were only two astronauts supposedly on the Moon at the time, perhaps the witness would have been better off posing the question: who, then, was operating the movie camera?

Judge: Indeed, but I do not preside over these matters in total ignorance. I am given to understand that some element of remote control over certain cameras was available to those at Mission Control in Houston. But it is again hardly a detail around which this entire case revolves, is it? The witness may continue giving evidence.

Prosecution: As well as being very knowledgeable, the learned judge is also most gracious.

Gordon Phillips continued—

10. The LM crashed—*crashed!*—during its only test flight on Earth. So why was its only further test flight the subsequent "successful" attempt to land it *on the Moon*? If you were an astronaut's wife, would you have gone along with that?

11. No Apollo astronaut has ever written an "I Went To The Moon" book or other such memoirs.

12. There is more—much, much more. The location of the directional thrusters, the smoky visibility of the burning rocket fuel, etc., etc.

Now ask yourselves this: if we did *not* land on the Moon and the American public ever discovers that fact, do you think its eyes and minds might be opened just a teeny bit to the *hundreds* of other *confirmed* acts of deception by government? Such as the fact that a Social Security Number is *not* required by [U.S.] law? How about the limited imposition of the income tax? How about Gulf War Syndrome? How about abandoned soldiers in Korea and Vietnam? Gee, how about the entire New World Order?

NASA is stonewalling Collier completely. No response. No rebuttal. No denial. Just silence. Perhaps a few *million* Americans should start turning up the heat by asking NASA and their elected government officials these same questions. First, though, they simply *must* buy and view Collier's video, *It's Only A Paper Moon.* Order by sending a check or money order for F$19.95 plus F$3 S/H to: Victoria Press House, 67 Wall Street, Suite 2411, New York, NY 10005

Oh, by the way, Grumman and Northrop corporations, developers and manufacturers of the LM and the Lunar Rover, told Collier that all of the original blueprints and design records for those two vehicles were destroyed. Not lost. Not misplaced. Or in the Smithsonian proudly displaying the history and development of this monumental feat of human engineering and accomplishment...

Destroyed.

Ready? Sing along, gang! And a one, and a two..."It's only a paper moon, sailing over a cardboard sea, but it wouldn't be make believe if you believed in me."

Yours For Liberty In Our Lifetime, Gordon Phillips, Founder & President: *Inform America!* Mail & Shipping: Nat'l Representative: Save-A-Patriot Fellowship, 12 Carroll Street. E-Mail: gordon@informamerica.com 1787V-Link: (801)715-3890 Voice & Fax Westminster, MD 21157-4831

—End.

§ § § § §

I thank Mr. Phillips for his contribution to the case for the prosecution. Inform America? Between us, indeed we shall.

The Prosecution

10

How Was it All Done?

Of course, there is one question the anti-hoax section of the world's population (at present, the vast majority, I suspect!) would ask—and indeed always *do* ask—either as individuals or vociferous groups often set up solely to counter what they see as the "treachery" of the "outrageous" suggestions being made by the pro-hoax lobby. That question is:

If it was all a colossal fraud, how was it done?

Fairly they ask, how could so many people—including scientists, well-known television presenters of astronomy programs (in many cases qualified scientists and astronomers themselves) as well as "friendly" and "unfriendly" governments the world over, have been so easily fooled?

It is a good question, and no doubt it will be asked again and again by my learned friend for the defense at a later stage in these proceedings. In fact, it is very likely to be the main thrust of the argument he present—

Defense: Objection! I would ask the learned judge to remind my learned friend that he is currently presenting the case for the prosecution and any attempt to pre-empt the case for the defense should be disregarded by the jury at this stage.

Judge: Sustained. Learned counsel for the People will restrict himself to laying before this court all such testimony, evidence and personal opinions pertinent to the People's case only, and make neither this nor any future attempt to lead defense counsel "by the nose," so to speak. The defendants' case will be placed before the jury in due course. But rather than urge the jury to disregard any remarks made by the prosecution that lend credence to the defendants' pleas of Not Guilty to the charges laid, I would, on the contrary, urge them to bear such remarks in mind throughout.

Prosecution: Again my apologies to the learned judge and, of course, to my learned friend, who I am reasonably confident will lead the defense of his clients without being "led by the nose" by anyone, least of all myself!

Judge: Very well. *Should* learned counsel need reminding of the fact, you are currently presenting the case for the prosecution. You may proceed with so doing.

Prosecution: I am indebted to the learned judge whose wisdom, coupled with an ability to immediately grasp a situation would appear to know no bounds.

I was simply going to say that as far as the question: "How were so many fooled?" is concerned, my answer, merely as an individual representing in this court the opinions of those who firmly believe that somehow they were, is that we *are* talking rocket science here: a subject which 99.9999 percent of humanity—including myself—understands very little about. I can only hazard a guess that the tiny percentage who at the time of Apollo *did* understand something about it, were leaned upon from a very great height, easily equivalent to that of the venue where these "adventures" are alleged to have taken place, by both the U.S. and Soviet governments. I will hazard a further guess that all the very top names in the worlds of politics, the military, and science in general, on *both* sides of the East—West divide, were either "in" on the fraud or *knew* that one had been perpetrated!

The *Politburo* of the Soviet Union, of course, would have been "up for it" from the very word go for the economic reasons given earlier and would have quickly persuaded all the top people in Russia and her satellite states not to question something that would bring about an immediate end to the poverty of the previous two decades. Sometime prior to the launch of the Apollo 11 mission, it would appear also that some kind of word went out to every "A-List" personality in the free world, too, which would undoubtedly have included all the big media barons. This was probably along the lines that something truly astonishing was about to happen: Something that would "guarantee" future peace and stability in the world. They were to keep their eyes and ears open, of course, but their mouths firmly shut and simply "go with the flow" of it…

Or else.

If you are a celebrity in any field, and the CIA, MI6 and the KGB tell you to keep your mouth shut about something, then, if you've got any sense, that is what you do. In this particular case, you would have "played along" with NASA and the American Government whenever and in whatever capacity they asked you to do so. This might have entailed you making congratulatory speeches, personal appearances, being photographed with the "returned astronauts," even being required to order subordinates to publish tributes and accolades or to present TV programs about Apollo on hastily constructed sets complete with little models of LMs, Command Modules and samples of "moon rock."

But why would the "Great and the Good" do all this while knowing it was completely bogus? Why would they subscribe to such a scam? I'll answer that too, in a moment, but only in the vaguest of terms. Powerful forces may still be at work to ensure that *everyone* continues to "toe the company line" even now. And I have no desire to attract any of them in *my* direction simply for *presenting* the People's case on behalf of the pro-hoax lobby in particular and humanity in general. Please don't shoot *me*—I'm only the messenger!

Basically, the West's top people would have kept quiet and gone along with it because they *were* top people. Most top people have a lot of common sense and they want to *stay* being top people, able to earn a decent crust for what they do. Apart from which, most may well have accepted—as I do myself—that the peace and security of the world *was* at stake.

Similarly, leading lights behind the Iron Curtain who wished to hold onto their country dachas and privileged positions would have stayed silent too. As I surmised earlier, apart from being perfectly happy to "deliberately, but honorably lose" the space race, the Soviets, had they *not* been in collusion, would in any case have been quite unable to "rat" on the Americans without placing themselves in the invidious position of trying to explain to 286 million frugally-living taxpayers the shocking waste of money that for all those lost years could have put more food and vodka into people's shopping baskets and more medication, staff and beds into Soviet hospitals. An immense fortune wasted on a "mad" venture that was ultimately impossible. Better by far to have the average Russian in the street thinking that he or she could have *won* the space race if only everyone had been taxed *even more* heavily, than to have him or her knowing that all the taxes paid already had been swallowed up by a lost cause from the very beginning.

Furthermore, the Soviets would have had the quiet satisfaction of knowing that in reality they had *not* lost the race for the Moon. That goal, as their scientists were probably even more aware of than were those of the United States, had proven to be unattainable until much further into the future. Or at least far enough into the future for their cosmonauts to be able to return from the other side of the Van Allen belts not just barely alive but well enough for a time to at least be able to *make their reports* before dying in agony!

It wasn't so bad. They had only lost the propaganda battle. And they had lost it to a skill in creating make-believe that they could never have afforded to compete with even had they possessed the know-how. They grudgingly accepted that America possessed not only the cinematographic skills that would be required but also the money needed to pull it off. And talking of money, the Soviets had merely run short of the stuff, that's all. This and only this had forced them to gracefully "bow out" and let a bunch of foreign showmen, entertainers and special effects people take the final "curtain call" that would end the madness. That they had only been beaten by money— or the lack of it—and *not* necessarily by a better ideology or a superior science was something that Brezhnev and the rest of the Soviet leadership could live with. Showmen? Entertainers? Yes indeed. The Americans are the finest in the world, of course, far and away the greatest experts on the planet at the art of celluloid illusion. They were more than capable of staging "the greatest show *off* Earth" as it was to become…

So the inference is that the United States of America won this profoundly significant battle of the Cold War in precisely the same way that it has won many other battles over the years, and solved many of the problems that have beset it as a nation during the 230 odd years since independence. And if, in the final analysis, it has brought stability and peace to the world and maybe saved countless millions of lives into the bargain, then why not?

In short, it would appear that after having received the green light from Brezhnev, Nixon then passed the baton to NASA. This meant that Houston indeed "had a problem." It was a tall order, right enough, but nowhere near as tall as *actually* trying to land men on the Moon. So, in time-honored fashion, the Americans decided to do what they always do whenever there's a tricky problem to be sorted out. They simply…

Threw *dollars* at it.

So many dollars, in fact, that even by today's standards the amount would keep a sizeable African country afloat for a whole year. It is said that they threw *30 billion dollars* just into creating the special effects alone! Make no mistake, members of the jury, when I say that if there was ever a bolder definition or statement of unimaginable wealth, and the unimaginable power that such unimaginable wealth brings, then tell me about it. Echoing the famous words of Sir Winston Churchill as he hailed the success of daring RAF fighter pilots during the Second World War, one could say that:

Never in the field of human deception was so much money, time and effort expended by so many, to secure the mere illusion that so few had landed upon the Moon!

Anyway, that's enough from me for the time being about how NASA "got away with it" once the dastardly deed—multiplied by six times for maximum effect—had been done. Let us now take a look at another web-posted article that gets hands (again not *mine,* thankfully!) well and truly dirty, by delving deep into the nitty-gritty of how some of the practicalities and technical detail of the deed itself, may have contributed towards making it all look so utterly convincing at the time. I will call my next witness for the prosecution.

I call John Doe.

HOW ILLUSION BECAME REALITY

Source: *http://www.geocities.com*. Article entitled *Apollo Reality*. The true identity of this witness is to be deliberately withheld by the prosecution. Some editing of this testimony has again been necessary because this article, in its original form, was highly illustrated by some startling photographs, frequently alluded to, but which are now to be found on the picture plates provided. Certain words and phrases have also been changed, omitted or added, again purely for the sake of clarity and to hopefully prevent any misunderstanding by the jury regarding what is being said. The flavor of the original article, though, hopefully remains intact.

Begins—(some editing).

This web page will show you how, and where, NASA faked the lunar approach, the lunar landing, and lunar take off, for all 6 supposedly successful Apollo missions.

Contrary to what many believe, the sequences were not shot in a desert, or a Hollywood studio, or Area 51. Of course, the odd picture may have taken at Area 51, and a few Apollo 17 pictures, in particular, were maybe taken in some remote desert area. The majority of stills and videos, however, were actually taken at Langley Research Center, Hampton, Virginia.

NASA scientists already knew in the early 60s that a manned mission to the Moon was impossible, especially within 8 years, so a plan to fake the Moon landings was put into operation. NASA's fake Moon pictures were taken at various locations, such as the Kennedy Space Center, the Johnson Space Center and Langley Research Center, and of course, there were the odd one or two desert locations. I would point out to all the pro-Apollos [the anti-hoax lobby] out there, that the art of manipulating photos and moving film is as old as photography and moving film itself. The 1930s film *King Kong* has a huge gorilla scaling up the Empire State Building. But just because it's on film, it doesn't make it real. In fact, the opposite is true. But that the Apollo Moon landings, which they have only ever seen film and photos of, *were* real, is precisely *what* the pro-Apollo groups are claiming!

Langley Research Center is a top-secret NASA research facility, and staff members there are sworn to absolute secrecy. All files pertaining to the Apollo missions are stored there, along with other artifacts, which include the burnt out Apollo 1 capsule which killed Grissom, Chaffe and White. At Langley, they have the facilities to simulate anything. Fake backgrounds, simulated orbiters…you name it. The first piece of crucial evidence is this large 250-foot traverse crane. [See Picture A on Plate10].

This huge traverse crane was *purpose built* in 1963-64, specifically to imitate a lunar landing as closely as possible to the real thing. It was used to suspend both the astronauts and the LM itself. It enabled movement of the astronauts and the LM in all directions, *i.e.* up down, left right, forward and reverse. The trial runs were that good, and what with NASA being fully aware that a Moon landing was impossible, they opted to use this set-up to fake the cine film of the lunar landing and take off, even to the extent of the flag being blown over by the "blast." According to Bobby Braun and other NASA officials, when challenged as to the specific purpose of the crane, the idea was to teach the astronauts how to land the rocket propelled LM on the surface of the Moon. However, no "rocket powered" LM was ever suspended from this crane.

In any case, anyone who has the slightest knowledge of rocket-science knows that it is *impossible* to control a rocket engine. If the pro-

Apollos disagree, and are all such experts at rocket science, then maybe they can direct me to an explanatory video or piece of film footage that clearly demonstrates how controlling a rocket engine can be accomplished. It's pure nonsense. The so-called "Moon" landings were controlled purely by the traversing and lowering of the LM in the normal way, just as would happen if it were suspended from a conventional crane.

These pictures [Picture B on Plate10 and Picture A on Plate 11] show a mock LM suspended from the crane. Note [in Picture A on Plate 11] the circular objects on the ground. The vast expanse of ground area beneath the crane was ideal for creating a mock lunar landscape, so this area was first covered with a mixture of gray ash, (possibly from some coal-fired power station or boiler house) and plain cement. The circular objects were then raised by crane to create authentic looking Moon craters, as you can plainly see in these photographs. [See Pictures A and B on Plate 12].

This picture [Picture B on Plate 11] is a time-lapse sequence taken at night.Spotlights on the crane gantry illuminated the ground surface. The mock *Eagle* was then traversed full length of the crane, and simultaneously lowered at the same time in order to create what looked to be an authentic lunar landing filmed from within the mock LM itself. Power supply to the mock LM was by cable from the crane tower. This enabled a large fan, (fitted beneath the LM), to create the dust scatter effect of a rocket engine as it descended to the fake Moon surface.

The films shown to the public of the *Eagle* and other LMs apparently blasting off from the Moon's surface, were all created beneath this crane at LRC. A mock LM would simply be attached to the crane and then hoisted up very rapidly, at the same time as a pathetic-looking "blast off" would be enacted beneath it. Each film was then speeded up for showing to the public. It is interesting to note, though, that filming was always cut short once a LM had reached a certain height or, in truth, had reached the crane's maximum height! If you have ever wondered *why* no camera ever continued to film a LM until it became so small as to disappear from view in the Moon's sky, then now you know. Quite simply, it was because it was not possible to do so under the circumstances in which the "lift offs" were faked.

These next pictures [Pictures A and B on Plate 12] were taken by Bob Nye on June 20, 1969. This was one month before Armstrong supposedly stepped onto the Moon. The first picture [Picture A on

Plate12] shows the mock lander on the fake Moon crater surface beneath the crane.

Believe me, this is how it was done, even if the pro-Apollos will say, "no way." This next shot [Picture B on Plate 12], taken at night, looks like a realistic Moon setting, although I have no doubt that some people will say this photo was taken on the Moon. I have heard so much bullshit from the pro-Apollo groups that nothing would surprise me anymore! The light source seen in the picture [Picture B on Plate 12] is the *same* light source that highlighted Buzz Aldrin in the controversial picture of him allegedly standing on the Moon [Pictures B and C on Plate 3]. Those lights are fixed at the top of the crane gantry.

This monochrome photo [Picture A on Plate 13] shows Neil Armstrong in January 1970, visiting the site where they faked it all at LRC. This was 6 months after he supposedly landed on the Moon, and, by then, Apollo 12 had apparently done the same thing. Armstrong no doubt returned to the simulation site to try to figure out how he could possibly have gotten away with conning the world into believing he had actually landed on the Moon and returned with no ill effects, 6 months before this picture was taken!

—Paused.

I think the writer of this piece is actually suggesting that apart from what he may have simply pre-recorded in a sound studio somewhere for the benefit of an actor in a studio prop spacesuit who would later pretend to be him, Armstrong had little or no knowledge of how the rest of the "production" he had starred in had been put together and wished to see for himself how all the special effects were created!

Apollo Reality continued—

The astronauts were suspended from the crane in order to simulate low gravity. They eventually settled for an upright position, with the astronaut suspended by a strong elastic bungee cord, so that his feet were only just touching the ground, the same way as happens with a baby bouncer. As the astronauts walked in a given direction, the overhead crane moved in the same direction. This enabled the astronauts to literally float along in a crude "moonwalk" fashion.

There is a classic piece of film that I have only ever seen once. It shows two astronauts supposedly on the Moon, but one astronaut is following behind the other in a dead straight line and at a fixed distance. Would two partners in a strange, desolate place be walking so far apart? It's obvious that both were following a given line/route, *i.e.*

the line or route that the overhead crane's bungee cables are forcing them to follow!

In these pictures it can be seen how astronauts were suspended from this crane. [See Picture B on Plate 13 and Pictures A and B on Plate 14].Note the high backward leg swing in this picture [Picture B on Plate 14]. That high backward leg swing is *identical* to the back leg swing in the Apollo 17 photo of Harrison Schmitt allegedly tripping up on the Moon!

In another video sequence of Apollo 17 astronauts supposedly cavorting on the Moon, one of them is actually *suspended by 2 feet* off, and at right angles to, the ground. This sequence lasts for a couple of seconds, so how do NASA officials explain that, and why is it that nobody of note has passed comment on this totally absurd picture? It is clear evidence that the person wearing the space suit is suspended by a wire, rope, or some other line. In this picture [Picture B on Plate 13] notice the rail moving device in the background for dispensing the ash/charcoal over the fake lunar surface.

Still not convinced? Then maybe this NASA archive, dated 26 August, 1969, and copied word for word, will change your mind. It relates to Donald Hewes, who oversaw operations/filming with the fake landing and take off. Read it first, then think hard about it. Why were NASA still phaffing around with fake lunar landscapes one month *after* Armstrong supposedly pulled it off for real? The answer? To make the fake film look even more realistic when future, higher quality images were broadcast to an already gullible audience.

Looking down from the top of the gantry onto the simulated lunar surface, James Hansen writes: "To make the simulated landings more authentic, Donald Hewes and his men filled the base of the huge eight-legged, red-and-white structure with dirt and modeled it to resemble the Moon's surface. They erected floodlights at the proper angles to simulate lunar light and installed a black screen at the far end of the gantry to mimic the airless lunar 'sky.' Hewes personally climbed into the fake craters with cans of everyday black enamel paint to spray them. This was said to be so that the astronauts could experience the shadows they would see during the actual Moon landing."

And this from A.W. Vigil, *Piloted Space-Flight Simulation at Langley Research Center*, Paper presented at the American Society of Mechanical Engineers, 1966 Winter Meeting, New York, NY, November 27–December 1, 1966: "Ground-based simulators are not very satisfactory for studying the problems associated with the final phases of landing on the Moon. This is due primarily to the fact that the visual scene cannot be simulated with sufficient realism. For this reason it is preferable to go to some sort of flight-test simulator which can provide real-life visual cues. One research facility designed to study the final phases of lunar landing is in operation at Langley. The facility is an overhead crane structure about 250 feet tall and 400 feet long. The crane system supports five-sixths of the vehicle's weight through servo-driven vertical cables. The remaining one-sixth of the vehicle weight pulls the vehicle downward, simulating the lunar gravitational force. During actual flights the overhead crane system is slaved to keep the cable near vertical at all times. A gimbal system on the vehicle permits angular freedom for pitch, roll, and yaw. The facility is capable of testing vehicles up to 20,000 pounds. A research vehicle, weighing 10,500 pounds fully loaded, is being used and is shown. [See Picture A on Plate11]. This vehicle is provided with a large degree of flexibility in cockpit positions, instrumentation, and control parameters. It has main engines of 6,000 pounds thrust, able to throttle down to 600 pounds, and attitude jets. This facility is studying the problems of the final 200 feet of lunar landing and the problems of maneuvering about in close proximity to the lunar surface." The latter was published in James R. Hansen's, *Spaceflight Revolution: NASA Langley Research Center—From Sputnik to Apollo*, (Washington: NASA, 1995).

We will now go inside the Langley Research Center complex itself to see how they faked the lunar approach and close orbit of the Moon's surface. We've all seen the film taken from a spacecraft as it supposedly approached the Moon, and then began to orbit. The speed at which it changes from approach to lunar orbit is utterly ridiculous, as any craft traveling at that speed would crash straight into the Moon. *No one* could control a speeding craft in such as way as shown on the film, and in reality *no one* did. There are pictures in existence that show *exactly* how it was done.

Ladies, size *does* matter believe me, especially when NASA want to create a fake lunar surface as shown [Picture C on Plate 14]. It literally dwarfs the two men standing in front of it! It is quite unbelievable the amount of time, trouble and expense that NASA went to in order to fake the lunar missions. It was of course done purely to convince the world that the Americans were the undisputed leaders in space. [*A.N.* Hu-hum…"convince the world" of American supremacy in space? I don't think so after all their dismal failures and all of Russia's successes. It only convinced the

Soviets of something they already knew: that the Americans were merely the undisputed leaders of the movie special effects industry! But Leonid Brezhnev and his cronies would have been quite happy for this false impression to have been given providing it ended the nonsensical space race for good]. This very large fake lunar surface, and others, were used in conjunction with a rail mounted camera, which was also used to focus on a large rotating *plaster of paris* model of the Moon which can be seen in this picture [Picture A on Plate 15]. NASA knew (after Kennedy's speech in 1961) that a lunar landing before 1970 was totally impossible. Realizing this, there was no other option but to try to fake a Moon landing. A program was then launched at LRC to design props/ backgrounds, etc., that would hopefully convince the media that they had indeed achieved the goal set by President John F. Kennedy. Pictures B and C on Plate 15 were scanned from a book, hence the poor quality. However, there are more and better pictures actually from NASA.

This [Picture A on Plate 16] is a high-resolution photo showing a 20-feet diameter sphere, which can be rotated from below. In the left of the picture can be seen a huge blank placard. This is the scene *before* LRC staff began work on converting the sphere to an authentic looking "Moon" complete with craters. Notice the rail track around placard, (there were 3 placards in all). Note moving trolley on that track. The camera was mounted on that trolley. It first began to film the rotating sphere for the fake lunar approach. It then swung around and began scanning the fake lunar surface on the placard for the fake lunar orbit.

This picture [Picture B on Plate 16] shows the sphere *after* the modeling work was completed…Pretty impressive, huh? Notice how the background is in the dark. Remove that bloke from the picture and you could *easily* pass this photo off as having been taken by the Apollo Command Module as it circled the Moon. It is evident that there were many people involved in the faking of Apollo, and NASA claims that if it had been faked, someone was almost bound to have spoken out by now. Not necessarily. You see, all the LRC staff were sworn to absolute secrecy. If anyone were to ever tell they would face prosecution, possibly jail and, of course, a complete loss of pension rights. In any case, those involved were 100% patriotic to the USA and, because of a "pride in America" that had been thoroughly instilled into them; they would not in any way have condemned what they were doing.

These pictures [Pictures A and B on Plate 17] show how LRC made *plaster of paris* copies of the Moon craters on the placard and sphere. They are checking that the craters are exactly to the same scale and layout as craters shown on the lunar photographs previously taken by high magnification telescopes. This one [Picture A on Plate 18] is again of the sphere after modeling work. The sphere had a light inside it, which was translucent on the outside, hence the strange, ghostly, "other world" appearance. The large placard with Moon craters was also backlit. Turn off all your lighting and you end up with the picture shown. This is how the Moon would look in the void of space *if* you could get close enough to it. However no one, not even Armstrong, did actually get anywhere near the real thing.

NASA claims this picture [Picture A on Plate 18] is the far side of Moon, taken by the Apollo 8 crew. The picture is, of course, a fake, because I do not believe Apollo 8 ever left Earth orbit, and the far side of the Moon would be in total darkness anyway. Compare this sphere with the one shown in this picture [Picture A on Plate 17]. It speaks for itself does it not? In all of these pictures notice the "blackness" of the backgrounds. This, of course, made it easier for touching up photos to ensure that background space was indeed black. However, as everyone knows, they 'forgot' to put any stars in the final pictures! [*A.N.* Well, faking the Moon is one thing—faking the entire Milky Way galaxy as a backdrop, which would be perpetually *moving* anyway, would have been an entirely new ball game involving many millions more "new balls"!]

This one [Picture B on Plate 18] was taken by an unmanned space probe orbiting the Moon in 1966, or so we are told. However it bears a striking resemblance to pictures supposedly taken 3 years later by [the Apollo] manned spacecraft orbiting the Moon. What I am saying— and this is directed purely at the pro-Apollo people—is that this picture shows it is perfectly possible to get a photograph of an Earth rise above the Moon's surface without a *manned* spacecraft orbiting the Moon—unless this picture is also a fake!—when compared with the photos above mentioned.

There are some other photographs, taken inside the LRC complex, which show astronauts cavorting around in spacesuits beside the LM shown in the alleged Apollo 11 Moon pictures, and watched over by NASA officials. However those pictures are fairly common and can be seen on the Apollo 11 website, hence I have not included them here. It may be that the *very first* fake pictures were indeed taken inside some large complex; if that be the case, then it was most certainly LRC and *not* some studio situation put together in an aircraft hangar at a top-secret base in the Nevada desert, as many anti-Apollo people proclaim. Incidentally *all records* pertaining to the faking of Apollo are stored under 24-hour guard at LRC, and are not due for declassification until the year 2026AD. [*A.N.* I think I mentioned that, didn't I? Hmmm...not good.].

We now move away from LRC to KSC—the Kennedy Space Center—for some genuine simulation pictures taken here on planet Earth, but which were then doctored for use as Moon photographs. I do not believe for one moment that they were training for a Moon mission. Moreover, the simulations were done purposely to create photographs, which, after alteration by Michael Tuttle [Michael J. Tuttle was allegedly NASA's chief photo "doctor"] could be passed off as Moon photographs.

The sole purpose of these pictures [on Plate 19] is to show you how simulation photos could easily have been later doctored and classed as Moon pictures. In this one [Picture A on Plate 19], note the sandy surface between the boulders. Imagine this surface covered with Moon boot footprints similar to the Moon boot footprints allegedly photographed on the Moon. [*A.N.* See Picture A on Plate 22, which I have included for the benefit of the jury].

Now look at this next picture [Picture A on Plate 19] of the astronauts—clearly on Earth—riding around in the Moon buggy. But the main point to observe is the relative flatness of the foreground and the *abrupt straight line* where that foreground meets the background, which is a bushy, grassy area just a few yards away. Notice the obvious straight line that divides the foreground, which contains the LM ascent stage supposedly taking off from the Moon's surface, from a plainly superimposed background of hills! Applying the same tactics as Tuttle, we could easily paste our mountain scene in the background, black out the sky, and there you have it [as in Picture B on Plate 19]. One photo apparently taken on the Moon. Clever, eh? But—tut, tut—not clever enough.

Both these photographs [on Plate 19] were taken here on planet Earth [claims this writer]. This one [Picture A on Plate 19] was, of course, quite obviously taken here on Earth. Actually at KSC. Nobody's denying that. However, those randomly spaced boulders that you can see would appear to have a *very similar* layout to the boulders shown in this second picture [Picture B on Plate 19]. NASA claims that this picture (Picture B on Plate 19] is a picture taken on the Moon. But how could NASA create a boulder-strewn landscape *before* visiting that area of the Moon and having any knowledge of the detail of the landscape there? [*A.N.* Well, all I can suggest is that 35 years ago, NASA must have had telescopes so powerful that cameras attached to them were able to snap both large and small boulders in that particular area of the Moon. However, they must have "destroyed" the blueprints for *those* too, because they don't appear to have them anymore. If they did, they would be able to quite easily snap the abandoned landing stages of the LMs from all six "success-ful" missions and bring this whole unseemly debate to an abrupt and immediate end. Even if a few of us might end up with egg on our faces, I for one would actually be ecstatic at having been proven wrong!].

This picture [Picture A on Plate 20] was taken at Houston. It shows how NASA created a fake Moon crater. That *same* crater, however, appears in quite a number of "Moon" photos from supposedly *differ-ent* missions! In the Apollo 16 picture of Charlie Duke standing by a crater, *this* is the crater. The rough edges, gaps and pockets between boulders were smoothed out with ash and charcoal. Overall, this fake crater appeared in rough format for the Apollo 14 and 15 fake pic-tures, and smooth format for the Apollo 16 and 17 fake pictures. This shot [Picture B on Plate 20] is how the crater appears in NASA's fake Moon pictures.

During the mid sixties, astronauts went on training missions to Ice-land. One of the reasons for this was due to the belief that the remote Icelandic landscape closely resembled the Moon itself in places. [So remote, in fact, that in future it would likely be visited by nosey tour-ists *even less often* than the Moon itself maybe?]. This picture [Pic-ture A on Plate 21] shows astronauts on one such a mission to Iceland. However, the background mountain scene in the picture is *identical* to a mountain scene from one of the Apollo 15 pictures! Evidently NASA decided to use some of these remote scenes, (in monochrome of course), for pasting onto the simulation pictures taken at KSC and LRC, and then pass them off as genuine photos taken on the Moon.

This, my last picture [Picture B on Plate 21], shows Buzz Aldrin in his Apollo 11 capsule, "returning from the Moon" in the full black-ness of space. But hold on a minute. We can see something very strange through the capsule window? No. Not a UFO. It's something even more startling. It's the light blue haze of Earth's atmosphere! Proof positive that the spacecraft was only in Earth orbit. In fact, a similar light blue haze streaming through the capsule window can be seen in video coverage from *every* Apollo mission. From Apollos 8 through 17. Hardly surprising, though, as *all* were simply orbiting the Earth for the duration.

So there you have it. You've seen how what appeared to be film of the lunar approach, lunar orbit, Moon landing and take off, were all accomplished. So, next time you see a film on TV of any Apollo craft supposedly approaching, orbiting, landing on, or taking off again from the Moon, you will know exactly how it was all done. I have also shown you how the fake Moon pictures were created. Those under age 35 can be excused for being somewhat gullible and for maybe having believed that film and photos from supposed Apollo missions to the Moon were real. Why would they have occasion to disbelieve recorded history that happened before they were born? However, all those over age 50, who can recall how shocked and surprised we all were at how quickly and successfully it all seemed to happen after the terrible disaster of Apollo 1 only a couple of years earlier, should perhaps have "wised up" many years ago—34 to be exact!—that the so-called manned Moon landing missions were all well and truly faked.

—End.

§ § § § §

Well, how's about *that* then? Phew! Excuse me one moment, members of the jury, but I definitely need a drink of water after that lot! Mainly to wash a very nasty taste from my mouth. Can't quite make out what it is…Seems to be a mixture of ash, charcoal and cement. Well, well, well. I can only say that the prosecution is extremely grateful to that last witness for providing such lucid and highly detailed explanations regarding how the special effects—if there *were* any in reality—of the alleged Apollo hoax might have been put together. It suddenly becomes all the more apparent to me that, under the direction of NASA space experts and cleverly manipulated by the very best hired help from Hollywood, such props, tricks, and illusions would undoubtedly have fooled an unsuspecting viewing public, including me, most of us watching only on black and white TV screens, into believing that what we were watching…

Was the real thing.

Indeed, when it happened, even though it all seemed utterly incredible and hundreds of years ahead of its time, why would we have *not* thought it to be the real thing? I was watching Moon coverage broadcast by the BBC, probably the most respectable and respected broadcasting company in the world. The epitome of what it means to have a free, unshackled media, the BBC always bends over backwards not to tell lies or give false impressions. To tell it exactly like it is, no matter what political shade the government in power at Westminster happens to be. Why should I, and millions of others just like me, have suspected for a moment that the BBC could be in on the biggest scam the world has ever known. But they *must* have been in on it. And at the very highest level.

And what of the roles played by the "astronauts" themselves in all this? And I have deliberately raised the job description in quotes. Not, however, to cast doubt upon whether or not the likes of Schirra, Lovell and Borman ever went into space at all, because it is accepted that they did. Even those who for one reason or another didn't make it into space were most certainly trained to a degree of excellence in order to be *able* do so. But it has to be said that apart from the inevitable pre-launch interviews and the physical presence of three *bone fide* astronauts each time the hatch of a returned Apollo capsule was opened after a spashdown, there would have been no other pressing need for *genuine* trained astronauts to have physically participated in the hoax at all, apart from possibly lending their voices to pre-recorded material. Some claim—to use the most important mission of all, Apollo 11 of course, as an example—that Armstrong, Aldrin, and Collins were *not even on board* Apollo 11 when it took off! These claimants maintain that the three men, having been ordered not to shave for a week, were then flown out on July 24, 196—along with a "pre-roasted" space capsule!—in a high flying B52, and then, over a specified drop zone, were simply dropped, inside the space capsule, by parachute into the Pacific.

Others, like Bill Kaysing, suggest that the genuine astronauts took off all right, but as the orbiting Command Module passed over some remote region like the South Polar Sea, the crew, after having climbed aboard the re-entry stage capsule, then either "self-jettisoned" from the CM or were separated from it remotely by Mission Control. That much done then, the parachutes were deployed and they were deposited safely into the sea to be picked up by an aircraft carrier already expectantly standing by. A dangerous enough exercise in itself, you will agree. Not only that, but it was one that, on the face of it, the same three men would have to face all over again little more than a week later!

If Kaysing's guess is correct, and I don't happen to think it is, then that sounds to me like subjecting three genuine astronauts to a totally unnecessary risk at least *once* too often. After all, these "big names" were the stars of the show. I cannot stress enough that the *only* time throughout the whole of a bogus mission when it would have been absolutely essential for the *real* astronauts to appear *in person*, so to speak, would be when the divers opened the hatch on the second re-entry capsule (possibly even the same one recycled!) as it, too, now bobbed about in the middle of the ocean at the correct minute, hour and date of its "official" return to Earth.

If a fatal accident had befallen any one of the three key men involved in any mission during Bill Kaysing's alleged earlier return, it would have jeopardized the whole charade there and then. We might as well face it, it would not have been necessary in the slightest, even for the sake of the watching media, for the three real astronauts to have boarded the Command Module of Apollo 11 before launch, fixed as it was atop the giant Saturn V rocket. Three reasonably close body doubles in those head-phoned, Balaclava helmets and white coveralls, bedecked with all those omnipresent Stars and Stripes, NASA and mission badges would have sufficed just as well. And, prior to the actual launch of course, even the body doubles would have been smuggled off and whisked away to a safe distance or to a thick-walled, concrete bunker.

What I am suggesting is that if the Apollo Moon landing missions were indeed all part of a thirty or forty billion dollar film production, then the accepted wisdom is that body doubles and stunt men—and *women* of course, where applicable—take *all* the unnecessary risks.

MOONGATE?

If I were to believe in "Moongate" at all, I have a somewhat different theory. I say "if" because although I am presenting, as earnestly as I can, the People's case at this moment in time, that does not necessarily mean that I, personally, believe that any part of Apollo was a hoax. Whether I personally do or not is irrelevant. But, to use that same little word again, it is my duty to present this case as *if* I do. It is as well to bear in mind that I shall later on be presenting the case for the defense, whereupon I shall not only fight tooth and nail to deny absolutely that anything about Apollo was fraudulent, but shall do so with a vengeance. But I am *not* presenting the case for the defense at this current time, so it continues to be incumbent upon me, in this part of the trial, to think and speak exactly like a believer in the hoax theory. Thus, my own ideas regarding what might have happened after the giant Saturn V rockets carrying the Apollo crews disappeared from view at the Cape, differ from anything mentioned above.

As I mentioned earlier, members of the jury, I am no rocket scientist. In fact, I am not a scientist of any kind. To be perfectly honest, science was one of my *least* favorite subjects when I attended school all those years ago, although I found some of it reasonably interesting here and there, of course.

But, as I also mentioned earlier, I *can* read. And so can all of you. So let us together indulge this ability by taking another look at the irrefutably *genuine* achievements of a certain Apollo mission that, so far at least, no one has even dared to suggest was part of a hoax. Not even the most cynical of the pro-hoax lobby…

Mission name: **Apollo 9** (Command Module *Gumdrop*, Lunar Module *Spider*).

Launch vehicle: Saturn V

Mission dates: March 03–13, 1969

Crew: James A. McDivitt (Capt.)
        David R. Scott
        Russell L. Schweikart

Buzz, what do you mean, "Is that you, Neil?" What other darn fools apart from you and me would wanna be seen dead in getups like these?

Mission duration: 10 days 01 hours

Mission accomplishments: First manned flight of *all Moon expedition hardware in Earth orbit.* Schweikart performed thirty-seven minutes EVA (Extra Vehicular Activity). Human reactions to space and weightlessness tested during *152 orbits. First manned flight of LM* (Lunar Module) in space.

Members of the jury, are you beginning, even if ever so slightly, to get my drift? If not, then read the above summary of the Apollo 9 flight again, making a special note of the sections I have underlined. Do you get it now? I feel certain that you must. Fundamentally, I am suggesting that NASA had no reason to fake the Apollo 9 mission. Three genuine astronauts blasted off on a genuine date atop a genuine Saturn V rocket carrying a genuine CM, a genuine LM, and a genuine re-entry stage or Space Capsule. Three human beings, and all necessary machinery for a genuine Moon mission, genuinely flew to a great height above the Earth and genuinely completed 152 orbits whilst genuinely test flying the LM to see how it handled in a genuine vacuum. Apollo 9 also genuinely practiced docking procedures…

But in *phony* space. Not *outer* space.

Yes. Phony space. That is precisely how it is sometimes referred to. It is that part of so called "space" that closely surrounds our planet. An area beyond Earth's gravitational pull and atmosphere, true, but which is still protected from deadly cosmic radiation by the Van Allen belts.

But let us leave presumably genuine Apollo 9 aside for a moment and get back to allegedly fraudulent Apollo 11, which we may as well continue using as our example. There is absolutely no reason whatsoever why the Armstrong, Aldrin and Collins who strode confidently over the "bridge" that crossed from the towering gantry to the tiny spacecraft affixed to the nosecone of the enormous Saturn V rocket—and it *is* enormous, believe me, I have seen it!—on Launch pad 39A, at Cape Canaveral on July 16, 1969, would not have been anything other than the genuine article. No reason, either, for them to have been smuggled off again before the launch, or indeed any reason why the three genuine, famous names themselves should not have been strapped into their seats at lift off.

Why?

Because there is absolutely no reason why Apollo 11 couldn't have been a carbon copy of Apollo 9 over a period of eight days instead of ten…

Unlike Apollo 9, though, there would have been no need for more EVAs in near Earth orbit or for any further testing of the equipment. Oh, the equipment, namely the LM, would be deployed all right, *but not boarded*. Not this time…

On July 16 or 17, 1969, perfectly genuine signals from an "object" now gone way beyond the Van Allen belts and heading for the Moon, might easily have been coming from the LM, unmanned of course, having already separated from the Command Module while in Earth orbit, just as an earlier one did during Apollo 9, although this wasn't supposed to happen during Apollo 11 until the mission had reached Moon orbit. Meanwhile, the CM, containing the crew, might simply have been maintaining its orbit around the Earth, which it would then have continued to do for the duration. All that was necessary then was for the LM to be directed by remote control from Houston to within a close proximity of the Moon and for it to then orbit as many times as the mission called for, all the while still "pretending" to be the Command Module. The signals and beeps from the LM, all perfectly genuine, of course, could have been backed up by a sophisticated tape recorder containing the pre-taped voices of the astronauts and connected to an in-board computer, which would have allowed a member of Mission Control to remotely operate the recorder as if it were sitting on his desk in front of him! This would have resulted in "programmed responses" to well-rehearsed questions from Mission Control by various members of what was in fact an absent crew. Likewise, pre-recorded questions from the astronauts would invite equally well-practiced answers from specially selected Houston operators only. In other words, human voices would undeniably be coming from inside a machine circling the Moon, but those voices would not be attached to bodies!

By far the vast majority of the technicians on duty at any given time at MC were simply watching their own little monitors and doing their own individual little tasks; they would have had absolutely no idea that they were taking part in anything other than a *genuine* operation like Apollo 9. And, of course, to a certain extent, they were. There would still have been three living, breathing, fragile human beings "up there" somewhere, whose minds and bodies needed to be very closely monitored and protected. The only difference being that they would, in fact, still have been orbiting the Earth as opposed to by now orbiting the Moon, as most would have believed!

By this time, though, the real Command Module would have switched to "radio darkness." Notwithstanding any gadgetry attached to the crew for monitoring purposes, it would have been emitting no signals or communications whatsoever that might be detectable by any agency other than Houston. Complete shutdown, but for life support systems. Any detection of the CM's continued presence above the Earth by any other means—e.g. by powerful amateur telescopes— resulting in a "large, cylindrical object" being observed still orbiting the Earth when the only one most people knew about was supposed to be on its way to the Moon, could have easily been explained away as a "dead" Saturn V stage, which had become trapped in orbit after burning out and being jettisoned.

Apart from any videotapes made weeks or months earlier, the astronauts would still have had time, before firing off the LM to the Moon and complete shutdown, to film lots of inboard action, banter and exchanges of repartee with each other and with Mission Control at Houston. NASA could then have released edited portions of such exchanges at "appropriate" stages of the supposed journey to the Moon. Remember, NASA had *total control* of what was seen and *not* seen on television, as well as *when* it would be seen. It must be possible, ultimately, to fake anything, providing the end result is only seen on a TV or motion picture screen.

Other than the above—which would have included the LM discarding its descent stage (lander section) to crash onto the Moon—all that needed to happen then was for the lunar ascent stage to return towards Earth, still bleeping away, with remotely operated tape machine continuing to do its job perfectly. Even if it hadn't, there would have been a back-up system, because there would have been plenty of spare capacity on board! Pretty much all else seen on television, *i.e.* the "Men on the Moon" sequences of the drama, would have been filmed in a studio weeks or even months previously.

But we mustn't forget that before Apollo 11, there were two other missions that allegedly went to the Moon and back. Apollo 8 and Apollo 10. So that should mean that the three relatively unsung Apollo 8 astronauts, namely Frank Borman, Jim Lovell and Bill Anders were, according to the *official* record of history...

The very first human beings ever to travel through and beyond the Van Allen radiation belts and into...

Outer space.

So, what happened there then? Of course there was a certain amount of rejoicing and a little razzamatazz after the "success" of the Apollo 8 mission. But taking nothing away from Armstrong, Aldrin and Collins if the hoax theorists are wrong, surely the Apollo 8 crew were…

So unbelievably brave and so utterly heroic as to defy suitable epithet?

Yet *why* is it that I had to go back and check my notes before being reminded of exactly *who* they were?

LYING LOW?

The allegation being made on behalf of the People is that man has neither gone to, landed upon, nor walked upon the Moon. Let us suppose, then, that my own personal theory of what might have happened in the event of a hoax having been perpetrated, is basically correct, and that all Armstrong, Aldrin and Collins had to do was *copy* what their colleagues of Apollo 9 had already accomplished and then simply keep their heads down or "lie low" for five or six days. Let us suppose that during this period, NASA Film Productions Inc. took over, and more or less played back the previous Apollo 10 mission, whatever that mission truly entailed. Apollo 10's journey to the Moon and back need only have been bogus insofar as being actually *manned* is concerned, with the CM and LM *genuinely* reaching Moon orbit, where all the necessary footage of craft separating and docking by remote control could have been obtained by in-board cameras triggered from Mission Control. There is absolutely no reason to suppose that Apollo 10 did not genuinely accomplish everything that was later required of Apollo 11, including landing the LM, although NASA described a somewhat "heavy" landing on that occasion. The Moon apparently "rang like a bell" after the LM, which was simply jettisoned because there was no intention of trying to land it properly, struck it! We could actually say that Apollo 10 did everything that Apollo 11 did, *except* provide us with images of two completely unrecognizable, space-suited, bulky figures cavorting about on the surface of the Moon.

Essentially, I am suggesting—on the People's behalf—that not only was Apollo 9 and probably Apollo 7 too, *totally* genuine in every respect, but that all the Apollo missions may have been *technically* genuine. The problem is, though, that they may not always have been genuinely *manned*. There is absolutely no reason, especially with a $30 billion budget to play around with, why we shouldn't suppose that everything that was possible and achievable *without* endangering human life was perfectly genuine and that only the far too dangerous, or downright impossible, was "handed over" to the cinematographic tricksters and special effects merchants.

So, maybe we are talking only a partial hoax here, not a complete hoax. If that is the case, then there is no doubting that those magnificent men—and women—at Mission Control and their remotely controlled "flying machines" played an absolute blinder throughout! Only the impossible was "colored in," so to speak. In fact, there was only one thing that was fundamentally wrong about it all. The *impossible* is precisely what the Apollo program was supposed to be about. It was about sending three living, breathing human beings through the Van Allen belts into outer space as far as the Moon, landing two of them *on* the Moon, and then bringing all three back through the radiation belts for a second time, not only alive, but well.

Part of the impossible then—all of the above except landing on the Moon—was first required of Apollo 8. So if there was a scam, it must have *begun* with Apollo 8, not Apollo 11.

But in case I am confusing you, members of the jury, let us move quickly on, still using Apollo 11 as our example, and I will attempt to clarify my point: If the three named astronauts for each mission lifted off and encircled the Earth in radio silence until it was time to return, this undoubtedly flies in the face of any suggestion that a certain amount of truly live TV broadcasts may have been made directly from a faked Moon studio set by those same three named astronauts. Obviously, they could not have been in *two* places at the same time. This is why some pro-hoaxers subscribe to the philosophy that the astronauts were picked up from the sea a few hours after taking off and then flown to the studio. This might well be true, of course. We are, after all, merely discussing *hypothetical* situations and scenarios here, none of which are cast in stone. Neither too, remember, is the hypothesis that there was actually a hoax to begin with.

But I have already suggested that the figures wearing those bulky spacesuits and helmets with mirror-like gold-tinted visors, whether boarding the spacecraft at the Cape or hopping about in a studio, might easily have been body doubles (roughly same height, weight and looks). Many astronauts were about the same height, age and build anyway (which could have been why they were selected for the training program in the first place!) so it still plausible that the earlier and later stages of all the Apollo missions in question were perfectly genuine, but with other trained astronauts—definitely not actors on these occasions—standing in for their colleagues, thereby leaving *them* free to appear live on the studio set whenever it was deemed necessary. There is even a possibility that certain ad-lib jokes were inserted in scripts during this pre-recording stage by the now famous names, which were somehow overlooked and not "edited out" by editors. Maybe even *deliberately* overlooked by an editor turned whistle-blower. Who knows?

As for the apparently live, in-board banter between the three named members of the crew and Mission Control, there is again absolutely no reason why all such stuff could not have been pre-recorded weeks earlier. Like almost everything else associated with the Apollo missions in question, perhaps we only *thought* it was all happening while we watched!

However, I would still go with the plausible option that Armstrong, Aldrin and Collins *did* actually take off from Cape Canaveral on July 16, 1969 in Apollo 11, and carried out more or less everything that was required of them prior to an equally genuine splashdown, *except* go to, land upon and then walk upon the surface of the Moon. In other words, they merely repeated what other Apollo missions had done before them. In so doing, they would have participated, and I repeat, only in a partial fraud, not a complete fraud. And this applies, too, to those who followed. I believe it to be highly unlikely that the astronauts themselves were involved in anything fraudulent—unless genuine gut-heaving blast-offs, real Earth orbits, dangerous re-entries, hazardous splashdowns and accurate technical conversations to do with the flight can be construed as fraud rather than sheer heroism—and that film industry professionals, under technical guidance from NASA employees, were used to fabricate *everything* else, including all the pre-recordings if pre-recordings there were. When a person from Hollywood is asked by his government to make a film about man landing on the Moon that is to be part truth, part fiction, and that he will have a 30,000 million dollar kitty to dip into in order to take care of the parts that will need to be acted out, then he hires *actors* for those parts, *not* astronauts!

VOICE DOUBLES?

We have already discussed the possible use of body doubles, but there are, of course, voice doubles, too! Expert mimics. And these could well have been employed as separate entities from the body double actors by the studio. With the lip and facial movements of the body double actors obscured by their space helmets, all that would be required, then, would be "voiceovers" by professional impersonators.

We might as well face it. Hollywood literally teems with such people; the vast majority forever seeking work. And, as far as this particular job was concerned, they wouldn't have needed to provide precise matches either. The most distinctive of voices can suffer distortion and become almost unrecognizable when conveyed via an electronic medium, to which we can add the practiced art of *deliberate* distortion by expert sound engineers. This would have made it seem as if at least one crackling voice in the frequent exchanges between Houston and Apollo, separated by realistic pregnant pauses and *beeps* to imitate the time delay factor, was truly originating from another planet a quarter of a million miles away!

By accepting, though, the probability that the real astronauts may not have been personally involved in any aspect of Apollo that could be deemed fraudulent, it becomes quite reasonable to suppose, therefore, that what should have counted as the most famous utterance of all time was not actually uttered by Neil Armstrong at all...

But by a voiceover actor impersonating his laconic drawl!

Hence it becomes even more theoretically plausible that actors—unlike much more disciplined people with military backgrounds who are trained to obey orders implicitly and without question—in other words, men like Armstrong and Aldrin—might mischievously deviate from an established script while playing just "another part" here on Earth. Even more especially so, if they considered what they were doing to be a complete farce and totally hilarious from start to finish. With massive fees from the $30 billion budget most likely already deposited with their agents before they had even agreed to take part, and with no pension rights to lose anyway, one could hardly blame the voice doubles for deciding here and there to "ham" out a few lines straight from science fiction that would have caught the producers totally unawares during any "live" slots. Especially when one considers that such people want nothing more than to be "spotted."

"Hey! There are other spaceships here! They're lined up on the other side of that crater. Wow! Those babies are *huge!*"

And if you think I made that up, then for those of you who would insist upon retaining your "unshakeable belief" in Apollo no matter what, you will need to prepare yourselves for a nasty shock! But more about that later.

Were anyone of the acting profession engaged on the project, it is not difficult to imagine the amount of larking about that would have taken place and the kind of banter that would have passed between them, either between takes or as they jumped about—or fell about laughing!—in those bulky, white Michelin Man suits. After all, people were seriously gonna try and pass all this shit off as science fact when it was pure science fiction…

Wasn't it?

The Prosecution

11

Look, I May Be the Most Famous Man in History, but I Don't Do Autographs— Okay?

Dearie me! The number eleven at the top of your case papers already! How appropriate is *that* for introducing my next witness? And it happens to be pure coincidence too! It may be a sign. Anyway, as opposed to being merely surprised, members of the jury, prepare now to be stunned, for I am about to do something that would be considered quite extraordinary in any normal court of law. I am about to call a person whom one would have expected to be the star witness for the defense, to actually give testimony on behalf of the *prosecution*. And, in effect, he becomes my final and most important witness to testify for the People. Once you have become fully acquainted with his testimony, I feel quite confident, members of the jury, that you will understand why I consider his testimony to be so vital to the People's case.

I call former astronaut Neil Armstrong.

NEIL ARMSTRONG'S TESTIMONY

Source: *www.lerc.nasa.gov/WWW/PAO/html/neilabio.htm.* February 1994.

Begins—

National Aeronautics and Space Administration

John H. Glenn Research Center

Lewis Field

Cleveland, Ohio 44135

Biographical Data

Neil A. Armstrong

Neil A. Armstrong is the Chairman of the Board of AIL Systems, Inc., Deer Park, N.Y., an electronic systems company. He was born in Wapakoneta, Ohio, on August 5, 1930. He received a Bachelor of Science degree in Aeronautical Engineering from Purdue University in 1955.

After serving as a naval aviator from 1949 to 1952 and completing his studies at Purdue, Armstrong joined the National Advisory Committee for Aeronautics (NACA) in 1955. His first assignment was with the NACA Lewis Research Center in Cleveland, Ohio. For the next 17 years, he was an engineer, test pilot, astronaut and administrator for NACA and its successor agency, the National Aeronautics and Space Administration (NASA).

As a research pilot at NASA's Flight Research Center, Edwards, Calif., he was a project pilot on many pioneering high-speed aircraft, including the well-known, 4000-mph X-15. He has flown over 200 different models of aircraft, including jets, rockets, helicopters and gliders.

Armstrong transferred to astronaut status in 1962. He was assigned as command pilot for the Gemini 8 mission. Gemini 8 was launched on March 16, 1966, and Armstrong performed the first successful docking of two vehicles in space.

As spacecraft commander for Apollo 11, the first manned lunar landing mission, Armstrong gained the distinction of being the first man to land a craft on the Moon and step onto its surface.

Armstrong subsequently held the position of Deputy Associate Administrator for Aeronautics, NASA Headquarters, Washington, D.C. In this position, he was responsible for the coordination and management of overall NASA research and technology work related to aeronautics.

He was Professor of Aerospace Engineering at the University of Cincinnati between 1971–1979. During the years 1982–1992, Armstrong was chairman of Computing Technologies for Aviation, Inc., Charlottesville, Va.

He received a Bachelor of Science Degree in Aeronautical Engineering from Purdue University and a Master of Science in Aerospace Engineering from the University of Southern California. He holds honorary doctorates from a number of universities.

Armstrong is a Fellow of the Society of Experimental Test Pilots and the Royal Aeronautical Society; Honorary Fellow of the American Institute of Aeronautics and Astronautics, and the International Astronautics Federation.

He is a member of the National Academy of Engineering and the Academy of the Kingdom of Morocco. He served as a member of the National Commission on Space (1985–1986), as Vice-Chairman of the Presidential Commission on the Space Shuttle Challenger Accident (1986), and as Chairman of the Presidential Advisory Committee for the Peace Corps (1971–1973).

Armstrong has been decorated by 17 countries. He is the recipient of many special honors, including the Presidential Medal of Freedom; the Congressional Space Medal of Honor; the Explorers Club Medal; the Robert H. Goddard Memorial Trophy; the NASA Distinguished Service Medal; the Harmon International Aviation Trophy; the Royal Geographical Society's Gold Medal; the Federation Aeronautique Internationale's Gold Space Medal; the American Astronautical Society Flight Achievement Award; the Robert J. Collier Trophy; the AIAA Astronautics Award; the Octave Chanute Award; and the John J. Montgomery Award.

—End.

§ § § § §

An impressive potted biography to say the least, eh? A *curriculum vitae* to be proud of, without doubt. But Armstrong's list of medals, in particular, likewise impressive as it is, is interesting more because of the medals that are missing, rather than listed. It is my personal opinion, given as a humble Englishman, that what Armstrong allegedly achieved was in the name of humanity. Why, then, was he only honored by seventeen countries out of a total of *191* member states of the United Nations? Why is that the only medal of any note awarded by the United Kingdom was merely one from the Royal Geographical Society? Especially when one considers that the U.K., quite apart from being America's closest ally and greatest friend, actually *participated* in numerous aspects of the Apollo program!

As august and revered a body as the Royal Geographical Society undoubtedly is, for that one single extraordinary act of unsurpassable bravery in having been the very first human being to step onto the surface of the Moon, no less a dignitary than the Queen of England herself should have been asked to reward Neil Armstrong with the highest accolade for bravery that can be awarded by Great Britain in peacetime, which is the George Medal.

And certainly, at sometime during the past thirty years if not almost immediately afterwards, he should also have received an Honorary Knighthood, which is the highest honor Britain can bestow upon a person not born of the Commonwealth at any time. Why, we've even given one to a fumbling, bumbling, faltering, former American president and an Irish pop star! Nor did we forget a certain Jewish American movie director who once made a film about flying saucers! I am not suggesting for a moment that the awards to the above mentioned were undeserved. Far from it. But whatever happened to priorities? What the hell is going on here, for fuck's sake?

Judge: Tut-tut! Temper, temper. Remember where you are, Mr. Frank.

Prosecution: My apologies to the learned judge and to you all. But I am beginning to think I don't *know* where I am with all this.

Judge: Indeed. I understand exactly how you feel. Have a glass of water. Then pull yourself together and continue. Your apology is accepted.

Prosecution: My thanks to the learned judge, who is most generous.

Likewise, Armstrong's two colleagues on that momentous occasion—as we have been led to believe it was—Edwin "Buzz" Aldrin and Michael Collins, should have been similarly honored. It matters not that Armstrong supposedly took that first, heroic step. If these three men actually did achieve what NASA and the American Government continue to claim they achieved, then all three were equally as stout-hearted, equally the three bravest men who have ever lived. Although protocol demands that as non-Commonwealth recipients of the title they should not refer to *themselves* as such, for the last thirty years or so the newspapers, whenever appropriate, should have been referring to *Sir* Neil Armstrong, *Sir* Edwin Aldrin and *Sir* Michael Collins. If there is any sane, rational person out there who might hesitate and deliberate over that for a moment and say, "Hmmm...but honorary knighthoods for all three might be overdoing things a little," try to imagine being Collins, orbiting alone in a glorified "tin can" around a strange new world a quarter of a million miles from home, not knowing whether his two crewmates would ever return...

As an earlier witness reflected, also try to imagine being one of the first two human beings to be genuinely standing on the surface of the Moon. There you are, looking up at your home in utter disbelief. A beautiful, lonely orb of blue light set in a clear, star-studded sky, even though the photographs you would take later might not show this! The Earth is now where you would normally expect the Moon to be! Everyone and everything you have ever loved and cherished is now a *quarter of a million* miles away! On that good old Earth, too, are your beloved partners and children, your relatives and friends, all praying for your safe return. You too are praying that your next faltering step isn't straight into an abyss, concealed as it might well be by a wafer-thin crust of heaven-only-knows-what covered by a deceptive layer of dust…

Through your helmet visor, you allow your eyes to wander back and forth over this indescribably alien landscape. You have seen all this happen before in movies and TV shows you watched as a child. But now it's happening for real! Not only that, but it's happening in front of an audience that dwarfs the biggest box office success ever known! And little old shy, self-effacing, always modest *you,* are the *star* of the show! You are the *real* Flash Gordon! You are the *real* Buck Rogers—in the flesh! If it wasn't for those thick pressurized gloves and your spacesuit, you would be pinching yourself in order to awaken from what is obviously a dream…maybe a nightmare! That's it…any moment now you *will* wake up…not screaming exactly…but certainly in a cold sweat. But not to worry. You can later join your pals in a bar somewhere and have a jolly good old laugh about it, can't you?

No, you can't, because there's something wrong. As much as you strive to mentally wrench yourself back to reality, hoping to find yourself in your perspiration-dampened, but still highly preferable bed, you *don't* wake up. The nightmare continues. And now, as you turn, your gaze reluctantly falls upon a macabre, unearthly-looking machine. You had hoped against hope that it would no longer be there when you turned to look, because that, at least, would have confirmed that this was all just a horrible dream. According to the "dream," this *thing* is your spacecraft. It not only *looks* funny, but it also has a funny name. It is called a LM. What…as in the first syllable of *lemon?* Oh, no! Oh, God! There it stands, forming a stark, ugly, black silhouette against the blinding, never before experienced intensity of the sun. It is just sitting there…like some loathsome, leggy, gigantic spider waiting to pounce. The nightmare is all but complete. And as you stare at it, you cannot help wondering…Okay, so it landed. No problem. But what if any single *one* of the many hundreds of thousands of parts and intricate mechanisms and miles and miles of wiring malfunctions…

And it doesn't take off?

If it isn't a detail that would have crossed *your* minds, members of the jury, I can assure you that it would most certainly have crossed mine! Not once, but over and over and over again…

And what about that huge crater over there, much of which is in shadow? *Black* shadow. You have never seen the color black looking quite so black before. All right, so all the signs from previous unmanned missions and probes have suggested, with 99 percent accuracy, that the Moon could not possibly support indigenous life. But that means life as we know and understand it, doesn't it? This is an *alien* world, not another country! This is *not* like being the first European to set foot on the shores of Botany Bay in Australia. This is *not* like sailing around the Earth waiting to "spill" over the edge if it turns out to be flat. This is alien. *Utterly* alien. None of the rules that apply on Earth apply here. You *have* not and *could* not have been given any positive assurances that some monstrous entity, looking like something straight out of hell, wouldn't suddenly emerge from that crater and begin slithering towards you like the giant serpent it is, gobble *you* up, then your comrade, and then the spider-like machine—more than likely what attracted it in the first place—just for dessert! In fact, as you will eventually see, it seems that "something" *did* emerge from that crater, or was to materialize in the most bizarre fashion close to it! Something far more awesome and potentially threatening than any imagined monster. If true, and Armstrong and Aldrin were not hallucinating, then rather than merely *suspecting* that the Moon landings were bogus we might all need to get down on our knees and start *praying* that they were! Not going to fall asleep in court now are you, members of the jury? No, thought not.

Much worse, the alien life form might be tiny. So tiny, in fact, that the two of you wouldn't see him—or *them*, rather—creeping up on you both. Microscopic. Smaller even. Working their diabolical way through the very atomic structure of your spacesuit until…there they are…on your skin. Then *under* it! Then eating you up…from the *inside out!* Billions upon billions of tiny worms, so small that many thousands would fit on a pinhead still with room to spare! And, all the time they're eating out your guts, they are multiplying. It's all happening in the space of minutes. Six billion become twelve billion. Twelve billion become twenty-four billion…twenty-four billion become…

Yes, my imagination *is* working overtime, members of the jury. Exactly as it would be if I happened to be standing on the Moon right this minute instead of addressing all of you. And that is precisely my point. Armstrong and Aldrin were highly intelligent men: Sensitive human beings with vivid imaginations. Not a couple of dim-witted fools. Multiply those feelings and fears that we have merely been *imagining* by at least a hundredfold and we might be close to envisaging what it must truly be like to step onto the surface of an alien planet for the very first time in all history. But Armstrong and Aldrin didn't panic. Neither did Collins somewhere overhead. They just got on with the job. *Now* tell me that all three didn't deserve to be knighted!

In all likelihood, if Apollo *was* for real, Armstrong and Aldrin would have been the very first and *only* living creatures *ever* to have walked upon the surface of the Moon since the very dawn of creation.Actually, I think I'll put that another way. If I *had* been any one of the two of them, with a corner of my eye still firmly refusing to budge from the deep, dark, utter blackness of that nearby crater, then *that* is what I would most sincerely and earnestly have been *hoping* was the case!

There is, in fact, only one other scenario I can imagine that fills me with even greater horror than the one described above. That is one where I have to spend the next thirty, forty, fifty years, forced by duty, what is expected of me, and a desire to continue receiving my full pension entitlement until the day I die, into situations where I am surrounded, wherever I go, by adoring crowds of people slapping me on the back and congratulating me and demanding my autograph for having achieved the above, when…

I hadn't actually done it!

I say, jolly well done, old bean! Top-notch stuff, eh? What?

How does one cope with hero worship when one is not a hero? How does one handle it when everywhere one goes thereafter, there are hundreds of people jostling to shake one's hand because they, too, can imagine—as I did above—the pure terror of having undergone such an awesome, unearthly experience. Just how does one deal with an adoring public telling you that they would never in a million years find the kind of courage to do what you have done? At least, not without sustaining a complete nervous breakdown in the process! And especially how does one deal with it when the last thing one can do is to be honest about it all and tell one's flocking admirers that you didn't sustain a nervous breakdown simply because you, like them…

Weren't actually there?

For anyone otherwise of great integrity and possessed of a high sense of self-esteem, the scenario is actually beyond terrifying. Imagine all that adulation, all that acclaim, all those honors and medals and the inevitable rounds of "Thank you, but I feel I don't really deserve this" type speeches when the nearest you ever came to being in any real danger was when one of the heavy studio light stands was accidentally knocked over and cracked your dummy space helmet, which was then immediately replaced by another studio prop. But this applies only if the astronauts were involved in the actual fraud, of course.

How does an otherwise sincere, true, trustworthy, thoroughly honest and good man—whose record proves his undoubted heroism in other areas and circumstances—live a lie of such magnitude for getting on for forty years? Especially when the sole reason for him having ended up in this invidious position is purely down to him having patriotically decided, at one time, to become a genuine astronaut, only to then have the terrible misfortune of being selected and ordered to perform a certain duty for his country that he not only had very deep misgivings about at the time, but deeply regretted having been involved in ever since.

Members of the jury, would such a situation not be enough to turn he who could be described as being the happiest, friendliest and most outgoing of men normally—although this is not a description applicable to Armstrong at the best of times—into a tetchy recluse? One who would only emerge, all too grumpily, from his "hideout" in the hills

Pardon me. A bit tough, those little white ones. Quite delicious though!

whenever necessity in the form of the continuation of his pension rights and other material benefits from the American Government and NASA so decreed?

Defense: Objection! This is all pure speculation. There is no evidence whatsoever that Neil Armstrong has ever been forced or pressurized into taking part in anything against his will since he returned from the Moon, or that the loss of his pension has ever been threatened. We are talking here about the United States of America, the land of the free, in the twenty-first century. Not Soviet Russia under Stalin or Nazi Germany under Hitler!

Judge: Overruled. Many people on both sides of the Atlantic, no doubt including the ordinary citizens of modern day Russia and Germany, have been puzzled for many years by Neil Armstrong's lack-luster attitude and demeanor regarding the moment that should have been the proudest not only of his life, but that of all mankind. The prosecution has a right, indeed a duty, to avail the jury of very plausible reasons why "the first man on the Moon" has for some reason become the last man one would approach to talk about such an incredibly brave venture, according to many reports seen by me.

Defense: As the learned judge pleases.

Judge: The prosecution may continue giving evidence on behalf of the witness Armstrong.

Prosecution: My thanks to the learned judge.

Neil Armstrong's Testimony (continued)

Source: http://history1900s.about.com/cs/armstrongneil/ **Neil Armstrong, 30 Years Later—Still Reticent After All These Years**. By Marcia Dunn. © 1999. The Associated Press. All rights reserved.

Begins—

CAPE CANAVERAL, Fla., July 14—Neil Armstrong was standing at the pad where he blasted off on July 16, 1969, watching the tower roll away from the soon-to-be-launched space shuttle Columbia, when a technician approached him.

Would he kindly sign a photo? After all, the shuttle worker explained, "We're all following your dream, Neil." The first man on the Moon replied, "Sorry, I don't sign autographs." The worker, irritated, walked away.

Ten or 15 minutes later, Armstrong went over, asked to see the photo again, and scrawled his name on it. The technician thanked him. But more than two years later—the encounter was in April 1997—it still irks him [the technician].

"I could understand it if you were outside with a big crowd and everybody was bombarding you," the worker, who did not want to be identified, said when recalling the story last week. "But I don't know why he's got so many hostilities when he's around the launch pad."

Three Decades Later

Just days away from the 30th anniversary of his first step on the Moon, when he proclaimed, "That's one small step for a man, one giant leap for mankind," Armstrong is as reclusive and reticent as ever.

While the 68-year-old commander of Apollo 11 has agreed to attend a banquet at Kennedy Space Center on Friday, he will take part in no interviews or news conferences, and give no autographs.

He will be joined by Buzz Aldrin, who followed him down the ladder onto the Sea of Tranquility. Aldrin, 69, is pushing space tourism these days as president of a couple of Los Angeles companies.

Michael Collins, 68, who circled the Moon in the Apollo 11 command module, is skipping the banquet. He's retired in Marco Island, Fla.

As usual, Aldrin is the only one of the three publicly reminiscing on this 30th anniversary of man's first Moon landing.

"As time has passed," Aldrin told the National Press Club in Washington last month, "I've come to understand that the true value of Apollo wasn't the rocks, wasn't the data that we brought back. It was the worldwide sense of participation, of people everywhere recalling where they were at that moment, and how they shared in a human adventure that brought out the best in all of us."

His Silence is Understood

Other fellow astronauts wish Armstrong would speak out also, but they respect his decades of silence.

"I think there's a reason to tell the story. Neil has the capability of doing that very, very well. He chooses not to," said Apollo 12's Dick Gordon. "You can imagine what would happen if he started something like that. I mean, the poor guy would never have any peace of mind." Instead, that burden is borne by the Neil Armstrong Air & Space Museum, in his hometown of Wapakoneta, Ohio.

—Paused.

Excuse a brief interruption of this testimony, members of the jury, but there is an urgent need to consider what has just been said for a moment or two. What would happen if Armstrong said something like what? Why "no peace of mind"? And why does this journalist use the term "speak out" rather than simply "speak"? Is there, perhaps, a vital point being made in former astronaut Richard Gordon's somewhat guarded and ambiguous statement here?

Associated Press article on Armstrong continues—

"We [the museum] are just bombarded weekly with requests for him [Armstrong]. Everything from 'Will you sign this for me?' to 'Will you come to my son's Boy Scout or Eagle Scout program?'" said museum manager John Zwez.

Zwez wishes Armstrong, who has an office in Lebanon, Ohio, would drop by now and then and mingle with the visitors. Armstrong has no ties to the 27-year-old museum and has been there only five or six times in all.

"On the other hand," Zwez noted, "the fact that Neil Armstrong is quiet and reserved about the whole thing, is perhaps the better approach, rather than going out there and selling this and selling that, promoting this and promoting that." [Maybe. But it's hardly natural, though, is it?].

Surprise Guest Appearances

Armstrong surprised many when he threw out the ceremonial first pitch at the Houston Astrodome in April. Two months earlier, he introduced a singer at Italy's San Remo music festival. Aldrin was at the songfest too.

In response to the common question regarding whether he would have made a better first man on the Moon from a PR perspective, Apollo 12's boisterous commander, Pete Conrad, refused to comment.

"Come on. I mean, Neil's entitled to do his thing," third man on the Moon Conrad said in a recent interview. Conrad was fatally injured in a motorcycle accident last week.

But Armstrong's loner status goes way back. Conrad noted that during Armstrong's test-pilot days at Edwards Air Force Base in California, he lived miles away in the mountains. His piloting skills, though, were legendary. Unmatched.

"So was his cool," said Conrad. [Was that "cool," or did he mean frostiness?]

A Proven History

As a fighter pilot in Korea, Armstrong lost part of a wing over enemy territory, but still managed to return to safety. He struggled to regain control of his tumbling Gemini 8 spacecraft in 1966 and brought it down early. He ejected from a Lunar Lander trainer in 1968 just moments before it crashed in flames. And he was down to about 15 seconds of fuel, after dodging boulders on the Moon, when the Eagle landed on July 20, 1969.

"I can't, offhand, think of a better choice for first man on the Moon," [Michael] Collins wrote in his 1974 book, Carrying the Fire.

Dick Gordon would have preferred to see his Apollo 12 buddies Pete Conrad and Alan Bean as the first two men on the Moon, instead of third and fourth. But he acknowledges Armstrong "has to command a lot of respect for his capabilities and for what he did."

As for the 30 years since, Gordon said: "I make no judgment—Neil did what he wanted to do."

—End.

§ § § § §

Far be it from me, members of the jury, even while currently acting for the People and calling the purported manned Apollo Moon missions into question, to cast any slight on the integrity and character of a man held in such high regard by so many. And that many still includes myself, regardless of whether or not Neil Armstrong actually set foot upon the Moon. Whether he did or did not, he was a military man seconded by choice into the space program. As such, he was at all times bound by a solemn oath that all servicemen swear, to do his duty without question. And his duty would have been to do whatever his country, or those running it at the time, decreed that duty should be. But, to be fair, it has to be remembered that everyone—first human being on the Moon or not—is entitled to be a little grumpy now and then, and indeed also to some measure of privacy.

Defense: Hear! Hear!

Sound of gavel banging: Rata-tat-tat-tat!

Judge: Order in court! Learned defense counsel will be silent unless raising a legitimate objection. This is a court of law, not a debating chamber where all and sundry are entitled to yell their support, or otherwise, for a particular comment or motion. Learned counsel for the People may continue.

Prosecution: My thanks to the learned judge. And indeed to my learned friend, for at long last finding himself in agreement with at least *something* I have said. I feel he will like this next piece of evidence even more.

Although it is not part of my brief to have to do so in this part of these proceedings, the following article may serve to redress the balance just a little. It reveals Neil Armstrong in a far more affable and entertaining light. However, there is still that subtle evasiveness regarding his own alleged past efforts and the combined efforts of those who allegedly put him on the Moon.

Neil Armstrong's Testimony (continued)

> Source: *http:/www.space.com/peopleinterviews/neil_armstrong_000222.html*. Article entitled: ***Neil Armstrong: Self-Proclaimed "Nerdy Engineer"*** By Paul Hoversten, Washington Bureau Chief. Posted: 06:22 am ET, 23 February 2000. © 2002 SPACE.com, Inc. All Rights Reserved.

Begins—

WASHINGTON—The world knows him as the first man to walk on the Moon, but Neil Armstrong takes a somewhat different view of himself.

"I am, and ever will be, a white socks, pocket protector nerdy engineer," he told an overflow luncheon crowd on Tuesday at the National Press Club. "And I take a substantial amount of pride in the accomplishments of my profession."

In a light-hearted speech on behalf of the National Academy of Engineering, Armstrong announced what engineers voted as the 20 top engineering achievements with the greatest impact on quality of life in the 20th century. Heading the list: electrification, followed by the automobile, the airplane and clean water.

Space exploration came in twelfth.

"Spaceflight was one of, and perhaps the, greatest engineering achievement," said Armstrong, who walked on the Moon 30 years ago on July 20, 1969 with Apollo 11 colleague Buzz Aldrin. Their fellow crewmate Mike Collins circled the Moon in the Command Module.

—Paused.

Now hold on just a minute, Neil. Surely landing an incredible machine and two living, breathing human beings on the Moon and then returning them safe and well to Earth was far and away man's greatest *claimed* engineering achievement, not spaceflight itself? Russian cosmonaut Yuri Gagarin was the first man to achieve spaceflight. As well deserved as his fame was and still is, are you saying that *his* accomplishment was greater than yours and Aldrin's? If he were alive today, I feel certain Yuri himself would profoundly disagree with you. In all honesty, I cannot even begin to imagine any scientific endeavor, coupled with extreme heroism, that could even *remotely compare* to a first successful manned Moon landing combined with the subsequent return to Earth of all concerned with no ill-effects. And that still applies whether or not such a thing has already taken place, or whether or not such a thing might take place in future.

Nerdy Engineer article continues—

But Armstrong said, "I do not disagree" with space exploration's ranking when held up to the yardstick of quality of life. "Others [on the awards list] were judged to have a greater impact in that regard."

Clean water, for example, is largely credited with holding back the spread of disease and boosting the average life span from 46 years of age in 1900 to 76 today, while agricultural mechanization, ranked seventh, made possible the growth of the world's population to 6 billion people today.

Armstrong, 69, schooled as an engineer and trained as a test pilot, went on to teach college-level engineering after he left the astronaut program. He is chairman of AIL Technologies, an aerospace electronics manufacturer in Long Island, New York and serves on several corporate boards, including that of SPACE.com.

"Science is about what is; engineering is about what can be," he said, noting that this was National Engineers Week.

Years from now, he said, the year 2000 "may be viewed as a primitive period in human history. It's something to hope for."

As for what he thought the greatest engineering achievement of this [new] century might be, Armstrong deadpanned: "Getting rid of the credit card."

Most of the audience, though, wanted to hear about space and Armstrong—one of the most private of the astronauts—charmed listeners with self-effacing humor.

What did he think about sending humans to Mars? "Well, uh, I'd volunteer," he said to laughter and applause.

Did he suffer any lingering physical problems from spaceflight? "You mean like this?" he said, holding up a wobbling hand. No, he assured the laughing, clapping crowd; people can safely go into space for extended periods of time and return to Earth with no problem.

And does he ever dream about that famous "stroll" on the Moon? "I can honestly say, and it's a big surprise to me, that I have never had a dream about being on the Moon. And that has been a great disappointment to me."

—End.

§ § § § §

"Disappointment?" Come on, Neil. Why? Leave the dreams—or *nightmares* in my case!—about it, to the rest of us. You don't *need* to dream about it. And you certainly shouldn't feel any disappointment about not doing so. After all, *you* did it for real…

Didn't you?

That concludes the case for the prosecution.

Picture A—*Shot allegedly of Earth taken on Apollo 11 mission.*

Pictures B and C—*Indelible crossheairs playing hide and seek?*

Picture D—*"We claim this rock as US territory and name it Rock-C."*

Plate 1

Picture A—*He's a man who casts a long shadow—unlike his friend!*

Picture B—*Life in the spotlight?*

Picture C—*Not me, I stick to the shadows...*

Pictures D and E—*...OOOPS! Even when they lose direction...for a while. All OK now!*

Plate 2

Picture A—*Aldrin raises a leg in public... Oooh! What a relief!
I reckon he's on his way down, that man!*

Picture B

Picture C—*You smiling for the camera, Buzz? Only it's hard to tell!*

Plate 3

Picture A—*All that glisters is not gold?*

Picture B—*Shadows of suspicion?*

Picture C—*Artistic license?*

Plate 4

Picture A—*I might be the most famous man in history—but please, no facial photos, huh?*

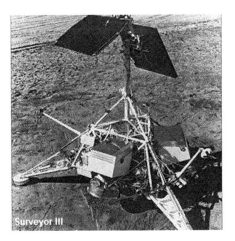

Picture B—*Steady as she goes—nice shot!*

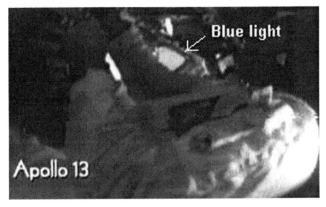

Picture C—*Blue skies—nothin' but blue skies!*

Plate 5

Pictures A and B—*Can you spot the difference?*

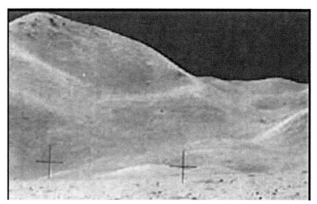

Picture C

Picture D—*And again?*

Plate 6

Pictures A and B—*Same mission. Same rocks. Only one thing wrong. Different place!*

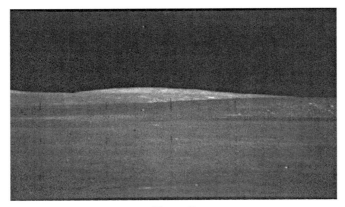

Picture C

Picture D—*Shooting stars!*

Plate 7

Picture A—*"I'll be a wild rover, nay never..."*

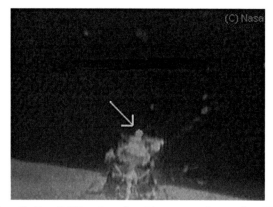

Picture B—*Blast off—literally! There is no fire without oxygen, but a yellow "flame" is visible!*

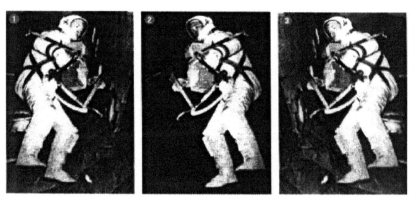

Pictures C, D and E—*"How d'you get this darn thing in reverse? Ah, that's it!"*

Plate 8

Picture A—*Blowing in the wind?*

Picture B—*How proudly she furls…"For fuck's sake, shut that hangar door, somebody!"*

Plate 9

Picture A—*It is claimed that this monstrosity was purpose-built to fake the Moon landings. Maybe—but it hardly looks very top secret. At least one doesn't have to crane one's neck to see it!*

Picture B—*Ready to rig a Moon landing? Or simply a rig ready for astronaut training?*

Plate 10

Picture A—*It might have been hush-hush and about to play a starring role in the scam of the century, according to some, but it seems NASA wasn't afraid to slap its logo on the side of it!*

Picture B—*Steady...steady...picking up some dust...*

Plate 11

Picture A—*Craters? How creative!*

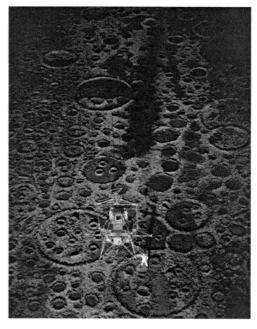

Picture B—*Supertrouper?*

Plate 12

Picture A—*Super Trooper?*

Picture B—*Slippy Tripper?*

Plate 13

Pictures A and B—*Moonwalk? Michael Jackson, eat your heart out...not.*

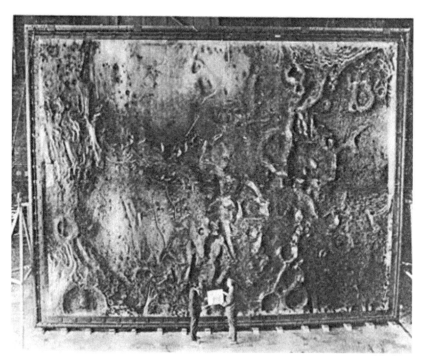

Picture C—*But is it art?*

Plate 14

Picture A—*Off to the Moon? OK. We'll let the train take the strain!*

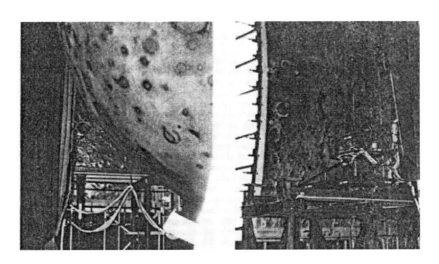

Picture B—*It's a whole world of responsibility being a "props" man for NASA!*

Plate 15

Picture A—*To hell with the handicap! Let's tee off for the Moon!*

Picture B—*Moonlight becomes me—it goes with my chair!*

Plate 16

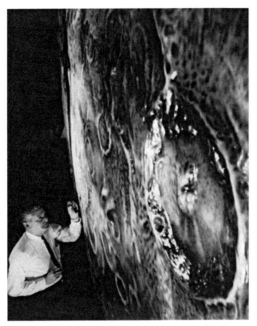

Picture A—*I'll shut my eyes and where this pin goes will do for the next landing site!*

Picture B—*Why don't we just go film the darn thing for real? It's gotta be a lot easier than this!*

Plate 17

Picture A—*"Tall" Tales from the Dark Side?*

Photograph courtesy NASA/Lunar Orbiter 1

Picture B—*Is somebody taking the Earthrise out of us?*

Plate 18

Picture A—*The rocks are good, but all this greenery sorta spoils the illusion, huh, buddy?*

Picture B—*Ah! Now that's a bit more like it!*

Plate 19

Picture A—*I think we're digging ourselves into a hole here...*

Picture B—*...effectively!*

Plate 20

Picture A—*Why waste $30 billion going to the Moon when we've got all the craters and mountains we need right here in Iceland!*

Pale blue light

Picture B—*Well, here I am in the wide blue yonder...er... Cut that! Hey! Can somebody do me black? Black would be nice!*

Plate 21

Picture A—*Super Slipper? That's how to make a lasting impression!*

Picture B—*The LM, sitting there like a giant spider waiting to pounce... the first "legs" home?*

Plate 22

The famous shot of Aldrin, allegedly taken by Armstrong on the Moon.

Plate 23

But when the reflection in Aldrin's visor is enlarged using modern day computer enhancement techniques, it appears to show the long shadow (arrowed) of someone not wearing a spacesuit apparently taking the photograph in a normal viewfinder to eye fashion and not from a camera affixed to his chest! In the background, between a leg of the LM and a patch of reflected light, the space-suited figure of "Armstrong" can be seen, but seemingly walking away from shot—his backpack is towards us—not actually taking it, as claimed! The long shadow, which NASA claims is Aldrin's (minus his arms, apparently!) and that of the LM must have been cast by a light source (NASA insists it was the sun, which would have been blinding!) shining almost directly into the face of Armstrong. Being fair, however, if one studies NASA movie footage, there are many examples of these long shadows, seemingly the result of the low positioning of a very bright light, arguably the sun. To give it the benefit of the doubt on this one, though, I do not personally believe NASA would have been so careless as to not pre-empt the possibility of visor reflections being enhanced in future years. No, the shadow, despite the almost uncanny message it appears to convey, likely is Aldrin's. But the question remains: Was he actually on the Moon when he was casting that long, slender shadow?

Plate 23A

*The Case of the People versus NASA and the
United States Government*

Continues with The Case for the Defense

The Defense

12

Radiation? What Radiation?

Members of the jury, it would be perfectly reasonable for me to expect that one or two of you, after having been acquainted with the so-called "evidence" presented by my learned friend regarding an alleged hoax perpetrated by NASA, the American space agency, who may previously have been staunch believers in the Apollo Moon landings of the late sixties and early seventies of the last century, might now be wavering ever so slightly. I say "one or two of you," because I seriously doubt whether the number would be much greater after so much drivel. However, I accept that the many of you of who may already have harbored certain doubts about this business will have had such opinions further strengthened. That's okay. It means that what my learned friend has succeeded in doing, largely, is to preach to the converted.

But participation in this trial as a jury member was never intended or expected to appeal to closed minds either way. People with closed minds not only do themselves a grave disservice, but others as well. We are human beings. We are the only intelligence we know about anywhere in the universe at this time that is capable of thinking deeply about something, weighing up the pros and cons, and then making a judgment. In order to make that judgment fairly, we need to keep our eyes, ears and minds wide open, especially when listening to experts talking, and often *disagreeing*, about subjects that *we* know very little about technologically and scientifically. We must be prepared to look at *all* the available evidence, to listen very closely to *all* the arguments, and only then decide who is likely telling the truth whenever such differences of expert opinion occur with regard to a particular subject.

As the learned judge pointed out earlier, but which cannot be reiterated enough, a closed mind is not a *free* mind. A closed mind is a prisoner of its own fixed position. Still worse, as the learned judge further pointed out, such a mind is often the victim of an obsession. But, to be perfectly fair, it is just as possible for one to be the victim of an obsessive belief that there was nothing fraudulent about the Apollo landings as it is to have a fixed opposite view. On behalf of my clients, NASA and the United States Government, I accept that. But, it became all too painfully obvious that some of the "witnesses" my learned friend resorted to calling to give "evidence" to this court on behalf of the People, *were* victims of such an obsession.

A free and intelligent mind is one prepared to be persuaded by evidence, reasoned argument and the application of logic and common sense, to an opposing point of view. One that is prepared to accept— however reluctantly—that this opposing point of view *might* be the correct one, or, at the very least, contain an *element* of truth.

That much said, I assure you, members of the jury, that my clients, if only by virtue of the sacred oaths that were sworn by their highest representatives when they took up office, will be expected to tell the truth, the whole truth, and nothing but the truth.So too, will others prepared to speak in support of my clients. And I trust that they will do so on behalf of every citizen of the United States of America or so help them God.

If it should please the learned judge, I would like to call my first witness for the defense.

Judge: Then by all means do so.

This witness is a research scientist attached to NASA who has given an authoritative reply to a question NASA is frequently asked. The following response to this FAQ was posted on the *Ask an Astronomer* website.

I call Dr. David Stern.

DR. DAVID STERN'S TESTIMONY

Source*: http://imagine.gsfc.nasa.gov/docs/ask_astro/ask_*

an_astronomer.html

Begins—

FAQ—I wonder if you could tell me exactly what the VAN ALLEN BELT is and how much radiation does it contain, *i.e.* how many rems of radiation are there out there? Plus, what protection would organic life need to be protected from this radiation?

David Stern, a researcher in one of our (NASA) labs here at Goddard [Spaceflight Center], has graciously supplied an answer to the question, and it is given below:

Answer: The radiation belts are regions of high-energy particles, mainly protons and electrons, held captive by the magnetic influence of the Earth. They have two main sources. A small but very intense 'inner belt' (some call it "The Van Allen Belt" because it was discovered in 1958 by James Van Allen of the University of Iowa) is trapped within 4000 miles or so of the Earth's surface. It consists mainly of high-energy protons (10-50 MeV) and is a by-product of the cosmic radiation, a thin drizzle of very fast protons and nuclei, which apparently fill all our galaxy.

In addition there exist electrons and protons (and also oxygen particles from the upper atmosphere) given moderate energies (say 1-100 keV; 1 MeV = 1000 keV) by processes inside the domain of the Earth's magnetic field. Some of these electrons produce the polar aurora ("northern lights") when they hit the upper atmosphere, but many get trapped and, among those, protons and positive particles have most of the energy.

I looked up a typical satellite passing the radiation belts (elliptic orbit, 200 miles to 20000 miles) and the radiation dosage per year is about 2500 rem, assuming one is shielded by 1 gr/cm-square of aluminum (about 1/8" thick plate) almost all of it while passing the inner belt. But there is no danger. The way the particles move in the magnetic field prevents them from hitting the atmosphere, and even if they are scattered so their orbit does intersect the ground, the atmosphere absorbs them long before they get very far. Even the space station would be safe, because the orbits usually stop above it—any particles dipping deeper down are lost much faster than they can be replenished.

If all this sounds too technical but you still want to find out—what ions and magnetic fields and cosmic rays are, etc.—you will find a long detailed exposition (both without math) on the World Wide Web at: *http://www.phy6.org/Education/Intro.html*

Another point of particular interest to us in high-energy astro-physics is the South Atlantic Anomaly. This is a region of very high particle flux about 250 km above the Atlantic Ocean off the coast of Brazil and is a result of the fact that the Earth's rotational and magnetic axes are not aligned (see *http://www.oulu.fi/~spaceweb/ textbook/radbelts. html*).

The particle flux is so high in this region that often the detectors on our satellites must be shut off (or at least placed in a "safe" mode) to protect them from the radiation.

—End.

§ § § § §

I thank Dr. Stern. If both he and my next witness will forgive the inference, I shall now call another "mad scientist."

TESTIMONY FROM A *MAD SCIENTIST*

Source: *Mad Scientist* website. *http://spider.ipac.caltech.edu/staff/ waw/mad/mad19. html*. Article entitled: *The Van Allen Belts and Travel to the Moon* © 23 May, 2000. *Mad Scientist* website includes NASA among its sponsors.

Begins—

FAQ: [by pro-hoax theorists]. Is it impossible to travel to the Moon, and is this because of the radiation contained in the Van Allen belts and beyond?

Answer: This is an especially interesting question, though maybe more about psychology and epistemology than about astronomy or physics. Nevertheless, the same question comes up again and again, in one form or another, so it really is very important. It has a number of possible answers: The question—

Prosecution: Objection! I must interject here. Should it please the learned judge, the question is quite straightforward. We suggest it requires a simple, straightforward answer in response. Yes or no?

Judge: I agree. I would remind learned counsel for the defense that his current witness will do his case no favors by using words like "epistemology" when what is being inferred is "ignorance," or by giving indirect answers to direct questions, even if they be ones that on this occasion the witness has asked of himself.

Defense: Should it please the learned judge, I am assured that the witness will go on to give straightforward answers to all questions raised.

Judge: Then the witness is advised not to be too long-winded about it, and may continue.

Mad Scientist continued—

The question implies that it is not possible to pass through the Van Allen belts without sustaining a fatal dose of radiation. The Apollo spacecraft passed through the Van Allen belts quite quickly, so that in the short time they were exposed, the astronauts did not receive a dose of radiation considered dangerous, at least not compared to the inevitable other risks in the mission.

This is the straightforward, scientific answer. It is correct, to the best of my knowledge and belief.

The question also implies that it is *not* possible to go to the Moon. It has to be possible to go to the Moon, because we who are old enough all saw them on TV; a million of us (me included, for Apollo 11) saw the actual launch; a few of us (me included, for Apollo 8) saw the Trans-Lunar Injection burn, from low-Earth orbit to trans-lunar trajectory in the dark sky over Hawaii; so how could anyone fake all of that?

This is a simple common-sense answer. Also correct, I think.

FAQ: Could it all have been a conspiracy that you yourself were part of?

Answer: The question implies there was a monstrous government conspiracy, and the whole thing was faked and, as you think I may have been a *part* of that conspiracy, you cannot trust my answer. I know for a fact this one is false—but how can *you* know that?

FAQ: Is it possible that even scientists, like yourself, were somehow deceived too, and that you defend Apollo in good faith not realizing that you were duped?

Answer: The question implies that if there was a monstrous conspiracy, and the whole thing was faked, then I could have been deceived too, so again my answer cannot be trusted.

I am as sure as I think one can reasonably be about anything that I was *not* deceived, but, of course, how could I possibly be *absolutely* certain, in principle?

—Paused.

Prosecution: Objection! Again the witness is failing to answer directly the question as it was put. I of course realize that the witness is a scientist, and maybe not so well practiced in the art of giving simple answers to simple questions, but may I have the permission of the learned judge to remind the witness that the question is this: Is it possible that the witness was deceived? Again a reply either in the affirmative or the negative is all that is required.

Judge: Objection sustained. The witness *will* answer the question with a simple yes or no. For the benefit of the jury, I will put the question to the witness again on behalf of learned counsel for the People. Is it possible that at the time, you too—and others like you—possessed as you were of all the undoubted skill and knowledge your chosen profession requires, might somehow have been deceived by NASA's alleged hoaxing of the Apollo Moon landing missions?

Defense: Should it please the learned judge, this witness, like all the others who have been or will be called to give evidence, has only provided written testimony to the court. I have read the wording of this piece of testimony thoroughly on behalf of the witness and have personally failed to see *any* ambiguity in it. May I, therefore, reply on behalf of the witness?

Judge: Indeed, you may. With a simple yes or no.

Defense: The answer, then, is yes. In other words, I understand the witness to be admitting that it *is* possible that the witness, and others of the same learned profession, could have been deceived by any alleged hoax. After all, a hoax is designed to deceive. I understand the witness to be admitting the susceptibility of *everyone* should any deception be done cleverly enough, and most certainly if done with the collusion of other scientists.

Judge: Very well. Your reply of "yes," on behalf of the witness, is duly noted. Your witness may proceed with giving testimony to this court.

Defense: My thanks to the learned judge.

Mad Scientist continued—

FAQ: Whom do I trust?

Answer: The question implies that I cannot know anything for absolute certain that I have not verified myself, and that all I can do is take the word of people I trust. So whom do I trust?

There is a lot of truth in this one, especially in principle. In practice, though, we can usually do quite a bit better, especially in the sciences; but the issue is not silly or unimportant, even so. The head of the government of South Africa, for example, is in serious doubt about whether the human immunodeficiency virus, HIV, causes AIDS, because he (probably sincerely, I guess) is in doubt about whom to trust; although there seems to be no serious scientific controversy about the issue. Millions of lives could be at stake as a result.

By way of example on this question of trust, let us take a more substantial look at the first question and my answer regarding the Van Allen belts. The idea is to outline the basic facts of the case, and then give you the materials you need to verify my statements, to whatever level of detail you wish. This is the traditional scientific way of answering a question. There are three basic issues here:

1. What is the actual amount and nature of radiation present in the Van Allen Belts?
2. How long would an astronaut be exposed to that radiation while passing through the belts on a lunar trajectory, and what dose of radiation would he/she receive?
3. What would be the likely health effects?

Regarding the Van Allen belts, and the nature of the radiation in them, they are doughnut-shaped regions where charged particles, both protons and electrons, are trapped in the Earth's magnetic field. The number of particles encountered (*flux* is the technical jargon, to impress your friends!) depends on the energy of the particles. In general, though, the flux of high-energy particles is less, and the flux of low-energy particles is more. Very low energy particles cannot penetrate the skin of a spacecraft, or even the skin of an astronaut. Roughly speaking, electrons below about 1 million electron volts (MeV) are unlikely to be dangerous, and protons below 10 MeV are also not sufficiently penetrating to be of concern. The actual fluxes encountered in the Van Allen belts is a matter of great commercial importance, as communications satellites operate in the outer region, and their electronics, and hence lifetimes, are strongly affected by the radiation environment. Thus billions of dollars are at stake, never mind the Moon! The standard database on the fluxes in the belt are the models for the trapped radiation environment, AP8 for protons, and AE8 for electrons, maintained by the National Space Sciences Data Center at NASA's Goddard Spaceflight Center.

[Janet] Barth (1999) gives a summary which indicates that electrons with energies over 1 MeV have a flux above a million per square centimeter per second from 1–6 earth radii (about 6,300–38,000 km), and protons over 10 MeV have a flux above one hundred thousand per square centimeter per second from about 1.5-2.5 Earth radii (9,500 km–16,000 km).

So, what would be the radiation dose, due to such fluxes, for the amount of time an astronaut crew might be exposed? In fact, this was a serious concern at the time the Apollo program was first proposed. Unfortunately, I have not located quantitative information in the time available, but my recollection is that the dose was roughly 2 rem (= 20 mSv, milli-Sievert).

The time the astronauts would have been exposed is fairly easy to calculate from basic orbital mechanics, although probably not something most students below college level could easily verify. You have perhaps heard that to escape from Earth requires a speed of about 7 miles per second, which is about 11.2 km per second. At that speed, it would require less than an hour to pass outside the main part of the belts at around 38,000 km altitude. However it is a little more complicated than that, because as soon as the rocket motor stops burning, the spacecraft immediately begins to slow down due to the attraction of [Earth's] gravity. At 38,000 km altitude, it would actually be moving only about 4.6 km per second, not 11.2. If we just take the geometric average of these two, 7.2 km per second, we will not be too far off, and get about 1.5 hours for the time to pass beyond 38,000 km.

Unfortunately, calculating the average radiation dose received by an astronaut in the belts is quite intricate in practice, although not too hard in principle. One must add up the effects of all kinds of particles, of all energies. For each kind of particle (electrons and protons in this situation) you have to take account of the shielding provided by the Apollo spacecraft and the astronauts' space suits.

Here are some approximate values for the ranges of protons and electrons in aluminum:

Range in Aluminum [cm]

Energy [MeV]	Electrons	Protons
1	0.15	~ nil
3	0.56	~ nil
10	1.85	0.06
30	no flux	0.37
100	no flux	3.7

For electrons, the AE8 electron data shows negligible flux (1 electron per square cm per sec) over E=7 MeV at any altitude. The AP8 proton compilations indicate peak fluxes outside the spacecraft up to about 20,000 protons per square cm per sec above 100 MeV in a region around 1.7 Earth radii, but because the region is narrow, passage takes only about 5 min. Nevertheless, these appear to be the principal hazard.

These numbers seem generally consistent with the ~2 rem doses I recall. If every gram of a person's body absorbed 600,000 protons with energy 100 MeV, completely stopping them, the dose would be about 50 mSv. Assuming a typical thickness of 10 cm for a human and no shielding by the spacecraft gives a dose of something like 50 mSv in 300 sec due to protons in the most intense part of the belt.

For comparison, the U.S. recommended limit of exposure for radiation workers is 50 mSv per year, based upon the danger of causing cancer. The corresponding recommended limits in Britain and Cern are 15 mSv. For acute doses, the whole-body exposure lethal within 30 days to 50% of untreated cases is about 2.5-3.0 Gy (Gray) or 250-300 rad; in such circumstances, 1 rad is equivalent to 1 rem.

—Paused.

Prosecution: Objection! This witness would appear to be attempting to blind the jury with science. Technical jargon—to use the witness's own words of earlier—such as "E=7 MeV," will have about as much meaning to 99 percent of the jury as Einstein's equation of "E=MC2" for his Theory of Relativity! While I accept that certain scientific principles have to be explained—indeed, I am in favor of *all* scientific principles pertinent to this case being explained— could I ask the witness, on behalf of the jury, to at least *attempt* to explain such principles in English, although not necessarily in words of one syllable?

Judge: Although I am mindful of the point the prosecution is making on behalf of the People, I am also mindful that *some* members of the jury *will* understand this so-called "technical jargon." They, at least, *will* be able to recognize the calculations and equations as being either genuine and fairly arrived at or so much gobbledy-gook deliberately and cynically designed to baffle the vast majority of us into submission and into giving the defendants the benefit of the doubt. Objection overruled, although I would strongly advise the witness to try to explain things a little more simply from this point onward.

Defense: My thanks to the learned judge and to my learned friend for his timely intervention. I must admit that *my* brain, too, was becoming more than a little addled. I shall do my best to get my witness to at least come to the point.

Judge: I feel that you will be in receipt of gratitude from every corner of this court if that were to be so.

Mad Scientist continued—

The point is this: the effect of such a dose, in the end, would *not* be enough to make the astronauts noticeably ill, but repeated low-level exposure could possibly cause cancer in the long term. I do not know exactly what the odds on that might be. I believe on the order of 1 in 1000 per astronaut exposed, probably some years after the trip. Of course, with nine trips, and a total of 3 X 9 = 27 astronauts (except for a few, like Jim Lovell, who went more than once) one would expect probably 5 or 10 cancers eventually in any case, even *without* any exposure, so it is not possible to know which, if any, might have been caused by the trips [into *outer* space, beyond the Van Allen belts].

Much of this material can be found in the 1999 "Review of Particle Properties," (see below) in the sections *Atomic and nuclear properties of materials, Radioactivity and radiation protection,* and *Passage of particles through matter.*

By this point I have no doubt told you much more than you really wanted to know about the Van Allen belts and the Apollo radiation problem!

—Paused.

Defense: Uh-hum…the learned judge, my learned friend, the members of the jury and indeed myself, would most certainly agree with you on that score, I think.

Prosecution: Hear! Hear!

Judge: Silence! I remind both learned counsel again that this is *not* the House of Commons or the Senate; it is a court of law. The witness will continue giving evidence *without* having to suffer sarcastic remarks from *any* quarter.

Mad Scientist continued—

Nevertheless, I have barely scratched the surface. True, I've waved my hands about a bit to make it seem less likely that I'm full of baloney. But in the end you always have to either do it all yourself, trust a stranger completely, or try to find some path in between, which maybe means trying to understanding a little science, so that you can judge for yourself if my arguments make any sense or not. Check for yourself a little. Think about it. Maybe even do a bit of research on your own from reference books if you are interested. The only other alternative is to trust no one and double-check everything personally, which, for a program like Apollo, is simply not possible. Sometimes, you have to accept the judgments of other people, who may be saints or sinners, honest or crooked, wise or insane. I hope you will try to find the possible, but not perfect, in-between path by learning a little science. It is hard, but it is fun and interesting, and it gives you your own power to think and evaluate for yourself, albeit in a limited and imperfect way.

—End.

§ § § § §

I thank the witness from *Mad Scientist* website for a much-valued scientific and technical contribution to these proceedings. I have reproduced for the benefit of the jury, a list of Internet sources the site strongly recommends for further reading/research on the subject of space travel and the effects of radiation.

For the jury's reference:

Health Physics Society. Professional society concerned with radiation effects and radiation protection.

University of Michigan *Radiation and Health Physics* page. Good general reference on radiation in the environment, including many links about radiation in space.

"Radiation Hazards to Crews of Interplanetary Missions: Biological Issues and Research Strategies," by the Task Group on the Biological Effects of Space Radiation, Space Studies Board, Commission on Physical Sciences, Mathematics, and Applications of the National Research Counsel; National Academy Press, 1997. About radiation hazards of possible long-term future missions in space. "Health Effects of Ionizing Radiation in Manned Space Activities." *http:// radefx.bcm.tmc.edu/ionizing/publications/space.htm* containing an extensive bibliography on the subject.

Standard reference for the Van Allen Belts is *AP8 & AE8 Models for the Trapped Radiation Environment*, NSSDC, GSFC.

The Radiation Environment, by Janet Barth of GSFC, 1999; available at *http://flick.gsfc.nasa.gov/radhome/papers/apl_922.pdf*.

An Annotated Bibliography of the Apollo Program, compiled by Roger D. Launius and J.D. Hunley. Published as *Monographs in Aerospace History*, Number 2, July 1994. *http://www.hq.nasa.gov/office/pao/History/Apollobib/contents.html*

Berry, C.A. "Summary of Medical Experience in the Apollo 7 Through 11 Manned Spaceflights." *Aerospace Medicine.* 41 (May 1970): 500-19. Described as, "This is a sophisticated scientific paper describing the results of biomedical experiments during the early history of Apollo. It is especially helpful in discussing the problem of radiation and other effect on the astronauts during the missions to the Moon of Apollo 8 and 11."

1999 Edition of the *Review of Particle Properties*, compiled by the Particle Data Group at Lawrence Berkeley Laboratory, and collaborators. *http://www-pdg.lbl.gov/1999 /contents_sports.html*

13

I, Clavius

Members of the jury, *Clavius* is an organization of knowledgeable professionals and dedicated amateurs devoted to the Apollo program and its claimed manned exploration of the Moon. A momentous endeavor, which *they* are in no doubt took place. The group's special mission is to comprehensively debunk the so-called conspiracy theories that claim such landings may never have occurred. Their website is named after the Clavius Moon Base in Arthur C. Clarke's novel *2001: A Space Odyssey*, which was made into a famous film bearing the same name by Stanley Kubrick.

I call *Clavius*.

THE TESTIMONY OF *CLAVIUS*

Source: *http:/www.clavius.org/envrad.html*. Article entitled: *Radiation and the Van Allen Belts*.

Begins—(some editing).

Challenge: There is too much radiation in outer space for manned space travel.

Response: This general charge is usually made by people who don't understand very much at all about radiation. After witnessing the horrors of Hiroshima and Nagasaki and the tragedy of Chernobyl, it is not surprising that the idea of radiation should elicit an intuitively fearful reaction. But when you understand the different types of radiation and what can be done about them, it becomes a manageable problem to avoid radiation exposure.

Challenge: It doesn't matter how difficult or expensive it might have been to falsify the lunar landings. Since it was absolutely impossible to solve the radiation problem, the landings *must* have been faked.

Response: This is a common method of argument that attempts to prove something that can't be proven, by *disproving* something else. In this case the reader is compelled to accept the conspiracy theory and all its attendant problems and improbabilities, simply on the basis that no matter how difficult, absurd, or far-fetched a particular proposition may be, if it's the only alternative to something clearly impossible, then it must—somehow—have come to pass. This false dilemma is aimed at pushing the reader past healthy skepticism and into a frame of mind where the absurd seems plausible. The false dilemma is only convincing if the supposedly impossible alternative is made to seem truly impossible. And so [pro-hoax] conspiracy theorists argue very strenuously that the radiation from various sources spelled absolute doom for the Apollo missions.

They quote frightening odds and statistics and cite various technical sources in an attempt to establish beyond doubt that radiation poses a deadly threat. But the truth is that most conspiracy theorists know only slightly more about radiation than the average reader. This means that a few people can dispute their allegations, so what the conspiracy theorists then do is simply dismiss them as being *part* of the conspiracy!

C: The Van Allen belts are full of deadly radiation, and anyone passing through them would be "microwaved."

R: This is a very simplistic statement. Yes, there *is* deadly radiation in the Van Allen belts. But the nature of that radiation was known to Apollo engineers and they were able to make suitable preparation to protect the astronauts against it. The principle danger posed by the Van Allen belts are high-energy protons, but these are not that difficult to shield against. The Apollo navigators plotted a course through the thinnest parts of the belts and arranged for the spacecraft to pass through them very quickly, thereby limiting the amount of exposure. The Van Allen belts span only about forty degrees of Earth's latitude—twenty degrees above and below the magnetic equator. The diagrams of Apollo's trans-lunar trajectory printed in various press releases are not entirely accurate. They tend to show only a two-dimensional version of the actual trajectory. The actual trajectory was three-dimensional. The highly technical reports pertaining to Apollo, accessible, even if not generally understood by the public, give the three-dimensional details of the trans-lunar trajectory.Each mission flew a slightly different trajectory in order to access its landing site, but the orbital inclination of the trans-lunar coast trajectory was always in the neighborhood of 30°.

Putting that another way, the geometric plane containing the trans-lunar trajectory was inclined towards the Earth's equator by about 30°. A spacecraft following this trajectory would bypass all but the edges of the Van Allen belts. This is not to dispute the fact that passage through any part of the Van Allen belts wouldn't be highly dangerous, but NASA conducted a series of experiments designed to investigate the nature of the Van Allen belts, which culminated in the repeated traversal of the Southern Atlantic Magnetic Anomaly (an intense, low-hanging patch of Van Allen belt) by the Gemini 10 astronauts.

C: NASA defenders make a big deal about the Southern Atlantic Magnetic Anomaly, but the Apollo spacecraft ventured into more intense parts of the belts.

R: True, but the point was to validate the scientific models using hard data, and to ascertain that a spacecraft hull would indeed attenuate the radiation as predicted.

C: We know the space shuttle passes through the Southern Atlantic Magnetic Anomaly (SAMA), but since the shuttle astronauts have time in each orbit to recover, the effects are not felt as strongly. The Apollo astronauts spent around four hours at a single stretch in the Van Allen belts. (Mary Bennett).

R: This is exactly the opposite of the recovery principle. If the shuttle astronauts spend 30 minutes of each 90-minute orbit passing through the SAMA, that sums to an exposure of 8 hours per day. The human body does not recover from radiation in a matter of minutes but rather hours and days. The damaged tissue must be regenerated. If radiation exposure is more or less continuous over several days, such as in the shuttle scenario, the tissue never has time to regenerate before being damaged by continuing radiation.

A short, intense exposure is safer than continuous or periodic exposure at lower intensity.

Even though the outlying parts of the Van Allen belts contain more intense radiation than the SAMA, a four-hour passage followed by days of relatively little exposure offers a better recovery scenario than days of accumulated low-level exposure. The four-hour figure is reasonable, but somewhat arbitrary. Since the Van Allen belts vary in flux and energy, it is not as if there is a clearly demarcated boundary. It's a bit like walking over a hill. If the slope gently increases from flat and level to 30° or so, where do you say the hill starts?

C: It would require six feet (two meters) of lead in order to shield astronauts from the Van Allen belts. The Apollo spacecraft had nowhere near this amount of shielding and so could not have provided the astronauts with anything like adequate protection.

R: The "six feet of lead" statistic appears in many conspiracy charges [*A.N.* Indeed, some say a *fourteen-foot* thickness of lead shielding would be required!] but no one has yet owned up to being the definitive source of that figure. In fact, six feet (2m) of lead would probably shield against a very large atomic explosion, far in excess of the normal radiation encountered in space or in the Van Allen belts. While such drastic measures are needed to shield against intense, high-frequency electromagnetic radiation, that is not the nature of the radiation in the Van Allen belts.

In fact, because the Van Allen belts are composed of high-energy protons and high-energy electrons, shielding involving any kind of metal is actually counterproductive because of a re-radiation phenomenon known as *Bremsstrahlung* that would likely be induced. (When beta particles impact larger, heavier atoms, this causes them to give off X-rays. Metal atoms are heavy, and so are especially susceptible to this kind of re-radiation problem). Metals *can* be used to shield against particle radiation, but they are still not the ideal substance. Polyethylene is the choice for particle shielding today, and various substances were available to the Apollo engineers in order to absorb Van Allen radiation. The fibrous insulation between the inner and outer hulls of the command module was likely the most effective form of radiation shielding.

Where metals *have* to be used, as in the construction of spacecraft for example (for structural strength) then a lighter metal, such as aluminum, is better than a heavier metal like steel or lead. The lower the atomic number, the less the amount of Bremsstrahlung. The notion that only vast amounts of a very heavy metal can shield against Van Allen belt radiation is a good indicator as to just how poorly thought out the whole pro-hoax radiation case is. What the conspiracy theorists say is the *only* way of shielding against the Van Allen belt radiation, actually turns out to be the very *worst* way of all to do it!

C: Official NASA documents describing the pre-Apollo studies of the Van Allen belts clearly state that shielding was recommended for the Apollo spacecraft, yet no shielding was provided. (David Percy).

R: To reiterate, "shielding" does not have to mean thick slabs of dense material like lead. Commensurate with the common perception of radiation as something both inescapable and deadly, is the mistaken notion that shielding against it must always be universally heavy and dense. Percy, and others it seems, are relying too much upon the notion that radiation shielding, if present, needs to be conspicuous by its bulk. As discussed a minute or two previously, shielding against *particles* is *not the same* as shielding against *rays*. To say that the Apollo spacecraft did not provide adequate shielding is to ignore both the construction of the Apollo command module and the principles of radiation shielding. And it must be kept in mind that shielding was only one element of a multi-pronged solution for safely traversing the Van Allen belts. It was never intended that the shielding in the command module would provide the *only* protection for the astronauts. The shielding was perfectly adequate for protecting the astronauts against the circumstances of the trajectory and exposure duration, as estimated and worked out by the mission planners.

C: NASA apologists [the pro Apollo lobby] come up with different numbers when estimating the amount of exposure the astronauts were subjected to while in the Van Allen belts. This suggests that they really don't know what the hell they're talking about. (David Percy).

R: All the estimates we've seen still lie within the same order of magnitude and generally outline a plausible method of computation. This is in contrast to the pro-hoax estimates, which generally have no quantitative support whatsoever. Computing the actual risks and precise exposure for the Apollo astronauts is very difficult. That's why the astronauts wore *dosimeters* to measure the actual exposure. The analytical factors involved in computing the amount of exposure expected include:

Exact trajectory: The Van Allen belts are not uniformly shaped. They have thick and thin spots. And the level of radiation is not constant at all points. Toward the center of the belt cross sections there is more radiation than at the edges. Most Apollo enthusiasts don't know the exact trajectory or how it relates to the location of the Van Allen belts. But because they don't know this, they do their computations assuming that the astronauts passed through the densest parts of the belts. Therefore, these computations err on the side of overestimating the exposure.

Exact velocity: Exposure time is very important when it comes to a correct computation of radiation dosage. Because the velocity of the spacecraft is constantly changing, the same ambiguity, which governs the geometry of the trajectory, likewise governs the rate at which the trajectory is followed. Most pro Apollo enthusiasts (but *all* conspiracy theorists!) lack the information and skill to precisely determine the velocity of the spacecraft during the Van Allen belt traversal, and therefore the exposure time.

Exact energy and flux: In any given cubic meter of the Van Allen belts there will be a "soup" of particles at various energy levels and fluencies. Energy describes the *velocity* of the particle, how far it will penetrate, and how much damage it will do if it hits something penetrable, like the hull of a spacecraft or the flesh and blood organisms inside it. Flux is the *density* of particles, or how many of them can pass through a given area in one second. Generally speaking, it is a case of the higher the energy, the lower the flux. Low-energy particles (*i.e.,* protons 30 MeV and below) can be ignored because they do not penetrate the spacecraft outer hull. But at each point along the trajectory through the Van Allen belts, there is a different continuum of flux and energy. It requires a lot of complicated mathematics to fully calculate it all. And since some of the variables are hard to determine, they can only be approximated to a figure estimated as fairly typical.

Probabilistic factors: Even should a high-energy particle penetrate the spacecraft hull to the interior, it will only cause problems to a human organism if it collides with tissue. It is quite possible for these particles to pass through a human body without actually colliding with anything, in which case they are harmless. The human body varies in density, so particles are more likely to collide with dense tissue, like bone for example. The amount of radiation absorbed by the body is a statistical probability based upon how much radiation is detected by dosimeters. To summarize then, a fully accurate analytical solution must first determine the exact trajectory of the spacecraft through the Van Allen belts. This will then give us a continuous function describing particle flux and energy at each point along the trajectory. At each point along the trajectory, we will have a function giving flux per given energy level. So a 100 MeV proton will have, say, a flux of 20,000 particles per square centimeter per second at that point in space. But, for other energy levels, the flux will be different at the same point. The total irradiation inside a spacecraft will be the sum of all the fluences at energy levels capable of penetrating the hull and shielding.And, at each point along the trajectory, the velocity of the spacecraft must be determined so it can be known how much time the spacecraft spends at that point. This is multiplied by the conglomeration of fluences to arrive at a dose. This dose is simply the amount of radiation present. It must be converted to a meaningful value that describes its likely effect upon human tissue. Again, energy and fluence come into play, because low-energy particles (although still high enough to penetrate the shield) are likely to accumulate in the outer layers of the skin, causing some damage, which is sloughed off harmlessly. High-energy particles, though, are absorbed in the bones and internal organs, causing much greater injury.The procedure for analytically and accurately computing a radiation dose is simple enough in principle, as outlined above, but of course is very difficult to actually carry out. This is why engineers generally don't bother to try computing the dosage to any great degree of accuracy ahead of time. They are usually quite happy simply to arrive at an order of magnitude, which provides adequate design criteria. The actual radiation exposure is always measured, not computed.

During the Apollo missions, each astronaut wore a personal dosimeter. The accumulated dose for each astronaut was regularly reported to Mission Control over the radio.

C: New evidence shows that the Van Allen belts are much stronger and far more dangerous than NASA claims. (Bart Sibrel).

R: Sibrel misinterprets the source article published by CNN. It was reported only that the Van Allen belts were slightly larger in places and slightly denser than previously understood. This is not a new reality, merely a refinement of existing figures. We are still studying the Van Allen belts and must occasionally revise our numerical models. The new findings have implications for the astronauts in the Alpha space station. Since these astronauts will be exposed to the fringes of the Van Allen belts for an extended period, it is prudent now to provide a bit of extra polyethylene shielding to the sleeping quarters. For transitory exposure such as in Apollo missions, the new findings add only a negligible hazard.Sibrel and others argue that NASA has under-reported the intensity of the Van Allen belts for many years as part of a cover-up. They argue that the real magnitude of the radiation is now being made known, and that it is strong enough to have precluded a successful Apollo mission. Unfortunately that's a very naive argument. The United States has never been the only space faring nation, or the only nation ever to study the Van Allen belts. Canada provided valuable data to the Apollo project, and the USSR not only duplicated all the U.S. research, but may even have conducted more. For thirty years, the same body of engineering data used to produce the Apollo spacecraft has been used by all nations when designing communication satellites, probes, and other devices intended to operate in and beyond the Van Allen belts. If this data had seriously under-reported the actual radiation present, the space vehicles engineered to those standards would all have failed prematurely due to radiation damage.This is a very important point, since it involves the financial interests (to the tune of billions of dollars!) of countries with no particular brief to protect the reputation of the United States. Had this data been *seriously* wrong, surely someone would have complained by now! Satellites are insured against premature loss, and insurers want to make doubly sure that spacecraft are engineered to the best possible standards. There is immense worldwide economic incentive in possessing the best available data on the Van Allen belts, so it is highly improbable that the U.S. has been intentionally providing erroneous data to the entire world for thirty years.

C: A secret study done by the Soviet Union and obtained by the CIA determined that at least a meter of lead would be required to shield against deep space radiation.

R: Many conspiracy theorists allude to this alleged report, but none can actually attest to having seen it. They argue that the alleged report is now securely held by the CIA and is therefore top secret. Thereby, the conspiracy theorists find themselves as safe from having their allegation refuted as they claim astronauts would be from radiation behind a several feet thick shield made of solid lead. How can anyone prove that a document doesn't exist, much less one that is allegedly classified by an intelligence agency? However, the up side is that conspiracy theorists cannot expect the world to accept an argument based purely upon evidence that they cannot produce. And here's another question: If it is *that* top secret, how do *they* know about it? Our guess is that no such document exists. If it does, then it is wrong. We already know that a great thickness of lead is *not* required to shield against particle radiation. We also know that Soviet science and engineering was excellent. We therefore conclude, with much amusement and no small amount of suspicion either, that the alleged report goes against commonly accepted principles of physics, and that it also bears a striking resemblance to the naïve assertions of certain conspiracy theorists who are not scientists, who claim that only thick sheets of lead are suitable for radiation shielding. A report containing such nonsense might be plausible to the lay reader, but would be laughable to a qualified physicist.

Also, Soviet lunar spacecraft designs clearly demonstrate that they also did *not* believe a meter or two of lead was required to shield against radiation. And again, history provides the final proof. Had the Soviets truly believed that a great thickness of base metal was required to shield against particle radiation, they would have questioned America's design of the Apollo spacecraft, but they did not. None of the Apollo spacecraft provided anything like a meter of lead shielding, yet NASA claims they successfully traversed the Van Allen belts with no noticeable ill effects being incurred by any of their occupants, either at the time or since. The Soviets acknowledged then, and continue to acknowledge, that the whole Apollo program was a resounding and unqualified success.

—Paused.

Prosecution: Objection! A resounding and unqualified *hoax*, more like! Of course the Soviets acclaimed the project. There is a strong belief by some that they were in on the scam from its very inception and that the whole idea actually came from them originally. When Nixon agreed to end the space race, but only if it could be made to look as if America had won it, the Supreme Soviet must have thought that all their Christmases at the Kremlin had come at once!

Judge: Overruled. Learned counsel for the prosecution has presented *his* case. It is now the turn of learned counsel for the defense, and he is entitled to bring forth witnesses—and very expert witnesses, it has to be said—who totally refute the People's allegations. The position of the defense is that no hoax was perpetrated in the first place, far be it that it should be the brainchild of the then government of the former Soviet Union.

Prosecution: As the learned judge pleases.

Judge: The witness may continue giving evidence.

Clavius continued—

In recent years, since the Berlin Wall came down, the Western world has been able to examine the Soviet spacecraft designs, which would have been used to carry cosmonauts to the Moon, had not the United States beaten them to it, of course. These Soviet designs clearly do not specify the utilization of a meter of lead for shielding their spacecraft either.

—Paused.

Prosecution: That's because any fool would easily have calculated that the things would never even have gotten off the fucking ground—let alone be able to fly to the Moon!

Judge: May I enquire of learned counsel for the People if he is raising yet another objection? Only I do not recall hearing the word mentioned. I did, though, hear another swearword that I trust will *not* be repeated.

Prosecution: No, learned judge. And apologies once more for the use of any strong language. But I refuse to allow the jury to be bamboozled in this way by this self-righteous…well, never mind. There never *was* any suggestion by anybody that a spacecraft with walls consisting of a meter or more thickness of lead might ever be considered a serious option. The point the People's case attempted to make in this area was that in order to convey living creatures made of flesh, blood and bone through the deadly radiation belts and then beyond these belts into even more deadly outer space *and* to then deposit them onto the surface of an another planet with no atmosphere, magnetosphere or any form of natural protection, thereby leaving them totally exposed to deadly radiation of all kinds, one would *need* to cocoon them in a craft bearing lead walls at least a meter thickness, not that it was *possible* to do so and still achieve lift off!

Judge: Is learned counsel now reasonably happy that the jury will almost certainly have taken his latest comments on board, even though some might not agree with them?

Prosecution: I can only inform the learned judge that I sincerely hope so.

Judge: I would also comment that an element of "self-righteousness" is not solely confined to one side of the argument in this case.

Prosecution: Accepted...with apologies.

Judge: Does Mr. Frank representing the People perhaps need to take another sip of water before I ask Mr. Frank for the defense to continue presenting his case?

Prosecution: No. But I thank the learned Judge Frank for his kind concern. I am fine.

Judge: Very well. The witness may proceed giving evidence for the defense. And I too *sincerely hope* that this will be *without* further interruption.

Clavius continued—

C: Former cosmonauts have been quoted as saying radiation was a very grave concern to the Soviet space program.

R: True. NASA officials have been quoted as saying exactly the same thing. Radiation *is* a very grave concern, but there is, essentially, a whole world of difference, literally, between a "concern" and an insurmountable obstacle. The pro-hoax argument relies entirely upon the radiation problem being completely insurmountable, but nothing said by either NASA or any of the old Soviet Union's former cosmonauts conveys the notion that these problems were ever considered to be insoluble.

Here Comes the Sun

C: Many published sources say quite unequivocally that solar particle radiation is a terrible hazard to astronauts.

R: And it is. But again, a hazard is not necessarily an insurmountable obstacle. Wet roads are a hazard to motorists, but people continue to drive on them anyway. There are many risks and hazards consistent with a voyage to the Moon, but the utmost care is taken to minimize these risks. But that doesn't alter the fact that it is still a very, very dangerous thing to do. There are people willing to brave the hazards of mountaineering, or of exploring deep caves or the bottoms of oceans. Similarly, there are those willing to brave the hazards of outer space. By repeating *ad nauseam* the statement that radiation is not merely hazardous, but deadly, the conspiracy theorists attempt to instill the notion that it is unavoidably fatal or always present in the same deadly amounts. It is not. Remember, this is the same area of space where dozens of countries operate sensitive communications satellites.

—Paused.

Judge: Excuse my ignorance regarding this particular area, but is the witness suggesting that these satellites have human pilots to guide them into their orbits and to maintain them in space? Astronauts or cosmonauts, for example?

Defense: Er, no, learned judge. I think the point being made by this witness is that the high degree of sensitivity of some of these appliances is not conducive to being constantly bombarded by high velocity particles from space, and that if they were so bombarded, they would soon go off line. Probably almost from the moment they reached orbit.

Judge: I see. So which side of the Van Allen belts then, which I now understand protect the Earth from most of this horrible, penetrating, highly damaging stuff, would the majority of these highly sensitive machines be operating on?

Defense: It is my understanding that the greater majority would be operating in orbits on the *earthward* side of the belts.

Judge: In other words "phony" space, not outer space?

Defense: That is indeed my understanding of the situation, yes.

Judge: (Sighing deeply). I must confess to seeing clearly now the point that was raised by the prosecution earlier. Will learned counsel for the defense care to remind his witness that I, also, will not tolerate attempts to "bamboozle" the jury with descriptions of what is possible with machines operating in "phony" space, and then trying to pass off such descriptions as clear evidence that human beings—who can likewise operate in the relatively radiation-free areas of so-called "space" situated on *this* side of the protective belts, hence *Mir*, *Skylab* and the space shuttles—have gone through the belts into outer space and landed upon the Moon. Any more twaddle of this nature, and the entire testimony of this witness will be stricken from the record.

Defense: Commensurate with the learned judge's wishes, I advise the witness accordingly.

Judge: Then literally by the skin of the teeth only, the witness may continue giving evidence to this court.

Clavius continued—

C: Experts say that during a solar maximum, about 15 flares per day emit detectable X-ray energies. (David Wozney).

R: The source cited by Wozney for this claim is no longer at the web URL he gives. But we must distinguish carefully between a *major* event and a *detectable* event. Just because instruments can detect the radiation of a solar event, that doesn't necessarily mean it presents a health hazard to a lunar-destined astronaut. We could draw the parallel between a detectable earthquake and a catastrophic earthquake. A seismometer can measure an earthquake so gentle that the people at its epicenter won't even notice it. But its occurrence would still be recorded as a detectable seismic event, and similar ones occur all the time.

C: According to records, more than 1,400 solar flares occurred during the Apollo missions.

R: This number represents the total number of *detectable* solar events, not major flares, which undoubtedly would have posed a danger to the astronauts. The records also show that no major solar flares occurred during the Apollo missions, but the conspiracy theorists don't bother to look *that* closely, do they? The impressively large number is all they are interested in. The closest call came when the Apollo 12 spacecraft's external radiation sensors detected a minor flare, but the interior sensors did not indicate any appreciable amount of radiation having penetrated the spacecraft's hull.

C: Major solar events last for hours, sometimes even days. (David Wozney).

R: Strangely enough, Mr. Wozney provides no reconciliation for these claims. On the one hand, we are told that these major events occur 15 times a day. Now we are told that a single one can *last* for days. Fortunately, we at Clavius *can* offer reconciliation. It is true that major solar events can indeed last for hours or even days. But the events that occur *15 times a day* during peak activity are the low-level events, which, as explained, pose no particular hazard to astronauts. They are strong enough to trigger off detection instruments, but nowhere near strong enough to warrant concern.

C: Solar flares produce huge amounts of radiation. One source says 3,000,000 rem for a one-year continuous exposure. Another source puts it at 100 rem per hour. NASA web sites say the radiation approaches 10 million electron volts! (David Wozney).

R: Is ten million electron volts a *high* energy level then? The reader isn't told, and is simply left to guess that it is. Sure, it *sounds* like a big number. Attach a nine-volt battery to your tongue [*A.N.* No, please don't! Clavius is merely attempting to clarify a point here] and you'll get an unpleasant but harmless jolt. You will see sparks from a 12-volt battery when you jump-start a car. Of course, we take great pains to shield ourselves from the 110-volt current in our homes because we know it can kill us. So *ten million* electron volts must be an enormous amount of unquestionably fatal energy, right? Well, no. Wrong, actually. The "electron volt" (eV) is *not* equivalent to the common "volt" that measures household electricity. Instead it is the amount of energy picked up by a single electron as it passes through an electrical potential of one volt. We realize that this is not a very helpful definition to the layman, but look at it this way: It takes the equivalent energy of about 620,000,000,000,000 million electron volts (MeV) per second...

To light up a 100-watt light bulb!

Bet you thought I was going to say a nuclear power station, didn't you?

No. The figure has obviously been cited because it looks like a big scary number. It is a bit like stating the weight of an automobile in milligrams—around 2.3 billion!—instead of more sensibly giving its weight in kilograms or pounds. A large number. True. But a large number of very small units. The very large figure given for the light bulb is explained by knowing that each individual electron that participates in the operation of a light bulb has a fairly small energy level, but there are billions and billions of electrons involved. In radiation terms this is called a high "flux." In space the individual electrons can have very high energy levels, but there aren't as many of them. The flux is much smaller.But major solar events do produce dangerous radiation. They have even been known to knock out communications satellites and even disrupt terrestrial communications. But in order to correlate the conspiracy theorists' numbers with the possible threat, we first have to know what *kind* of particle the number refers to. For example, a 10 MeV electron is relatively harmless, but a 10 MeV *proton* might be cause for concern. Again, though, the energy level is only half the story. You also have to know the particle flux (how many of them there are).The dosage figures, which take into account both energy and flux, are likely to be fairly accurate. But the conspiracy theorists make the fundamental error of multiplying worst-case exposure characteristics by the 15-per-day figure, or 1,400 total figure, which merely represents the number of *detectable* events, thereby arriving at what they claim to be the exposure level astronauts are subjected to on a typical mission to the Moon. If we stick with the earthquake analogy, it would be like counting the dozens of microquakes that occur on a daily basis and multiplying that number by the 7.0 or 8.0 Richter magnitudes for a single major earthquake, and then presuming that massive devastation must have taken place during those microquakes.

C: Various regulatory bodies have established that the maximum safe dosage for the general public is 1 millisievert (mSv) per year, and 5 mSv in special circumstances. The Apollo astronauts would have been exposed to several orders of magnitude more radiation than these figures allow. (David Wozney).

A: First it must be understood that this claim is based upon the improperly computed dosages described above. If you care to read the *radiation primer* [below] you will learn that it's very difficult in practice to compute dosages. Radiation dosages are measured rather than computed. The Apollo astronauts wore dosimeters to measure how much radiation they were exposed to. And sensors both inside and outside the spacecraft measured the amount of radiation present.

—Paused.

Judge: (Stifling a yawn). I must interrupt here. May I enquire of learned counsel for the defense if there is much more of this terribly boring, tedious scientific stuff to follow?

Defense: I have to inform the learned judge that there is, unfortunately, a *lot* more of this "terribly boring, tedious scientific stuff," as he puts it, to follow. Tedious for the majority of the members of the jury, for whom the learned judge is rightly concerned, it may well be, but absolutely essential if my clients are to refute at *every* level the quite ridiculous accusations being made against them. I remind the learned judge that he himself has correctly pointed out that there will be those among the jury who *will* understand exactly what is being said, and who will find it neither boring nor tedious, and who will be able to make a fair judgment regarding whether what is being said is true or not.

Judge: Very well. And this might be an appropriate time for me to remind the jury that this trial is *not* intended as entertainment. (There go any thoughts of a bestseller then!). Its sole purpose is to try to establish, one way or another, whether a hoax was perpetrated on an unsuspecting public with regard to the alleged Moon landings between the years 1969—1972, or whether these landings were entirely genuine. Such matters *cannot* be properly debated without discussing the science involved. So, for those of you who may be *au fait* (up to speed) with such matters, enjoy. For those of you who, like me, can barely understand a word of it, please do not grin, lest it be incumbent upon me to remind you all again that we are in a court of law trying a very serious case, but do try to *bear it* if you can. Better still, at least *try* to understand some of it too. The witness may proceed.

Clavius continued—

Radiation is a hard-working word in physics. It describes several diverse natural processes and their effects. As used in common speech, it means what physicists call "ionizing radiation," or that which can produce detrimental effects in materials and organic tissue. Ionization is the process of removing electrons from atoms, and when this occurs in biological tissues it disrupts the delicate chemical and physical processes that sustain life. This can happen through mutation, when the DNA of the organism is altered, or directly via the destruction of atomic bonds and the break-up of important molecules at the site of the ionization.

A basic division

We can agree two broad categories of ionizing radiation: that caused by electro-magnetic rays, and that caused by high-energy charged particles.

The electromagnetic spectrum (Figure A on Plate 24) is familiar to most people. What we call "light" is really a narrow band in a single phenomenon, which includes radio waves, microwaves, and X-rays. A wave's position in the spectrum depends upon its wavelength, or the distance between two adjacent "crests" of the wave. On the left of the diagram are "long" waves, such as radio, television, microwave, and infrared. On the right are "short" or "high frequency" waves, such as X-rays and gamma rays.

As you can see, not all electromagnetic (EM) waves are ionizing radiation. Generally, anything above the visible spectrum is considered to be ionizing radiation and thus harmful to some degree. Ultraviolet radiation from the sun is what sometimes causes skin cancer. X-rays and gamma rays are produced by nuclear reactions—atomic bombs, for example and, to a much lesser degree, nuclear reactors.

Non-ionizing EM radiation can still be dangerous, of course, if absorbed in sufficient quantities. Microwaves cook food by exciting the water molecules in the food until they vibrate and create heat. Quite obviously, they can also excite the water molecules in the human body and cause a similar effect.

The other category of ionizing radiation comprises high-energy charged particles. An alpha particle [Figure B on Plate 24] is the nucleus of a helium atom. It is composed of two protons and two neutrons. It has a charge of +2 and is very large and heavy.

A beta particle is an electron emitted from the nucleus of a radioactive substance. It has a charge of -1 and is much, much less massive than a proton or a neutron. A proton is, well, a proton. And neutrons are neutrons. Protons and neutrons have about the same mass, but the neutron doesn't have a charge, while the proton has a charge of +1.

Now having made a careful distinction between waves and particles, we note that many authors use the terms interchangeably (*e.g.* beta ray and beta particle). Since EM radiation is carried by the photon (a particle of light) and since equivalent energies can be computed for proper particles, there isn't any real need to maintain such a strict distinction. In fact, it is frequently useful to be able to share measurements between all the different kinds of radiation.

But when computing radiation dosage (the effect radiation has upon organisms) and when constructing shielding, the differences must be clearly understood. Pound for pound, particle radiation is much more dangerous than wave radiation. And he larger the particle, the more damage it is capable of doing.

Energy and flux

I mentioned above that physicists deal with radiation as a more abstract concept. In their terminology, it is one of the mechanisms by which energy is transferred from one place to another. Radiation is "energy in transit." When we speak of the "energy" of a wave we consider its intensity. The energy of a particle can be thought of as equivalent to its speed. High-energy particles travel very fast, while low-energy particles travel much more slowly.

Physicists use another measurement, "flux," to describe particle density. It is calculated by measuring how many particles pass by a certain point in a given length of time. If it is very many, we say the flux is high. If few particles pass, the flux is low. "Flux" is, of course, the Latin word for "flow."

If we take a cubic meter of space from anywhere in the universe, we will discover that it contains many particles of varying flux and energy. In general, flux and energy vary inversely. That means the higher the energy, the lower the flux. So if we look at the low-energy particles, we may find an enormous flux.

Where does radiation come from?

We can answer this question in two ways. We can say that charged particles come from the nucleii of various atoms that undergo nuclear decay. We can say that EM rays especially (and gamma rays) are emitted from those same nucleii, and we can note that any substance with sufficient energy, or heat, emits EM radiation as a method of releasing that energy. That describes the source of radiation at the microscopic level.

But the pressing issue is where in the universe we might expect to encounter these types of radiation, and in what quantities. The short answer is that radiation is all around us. EM radiation bombards us constantly, but thankfully not generally in the ionizing range of wavelengths. High-energy charged particles rain down on us from space, and are produced by the natural radioactive decay of many natural substances. The constant low level radiation which we encounter every day is [called] "background radiation."

—Paused.

Prosecution: With the learned judge's permission, the witness has just passed a remark that I would like clarified for the benefit of the jury.

Judge: Then feel free to cross-examine the witness regarding this point.

Prosecution: For the benefit of the jury, can I ask the witness whether if by the use of the words "produced by the natural radioactive decay of many natural substances," this therefore includes not only deadly particle radiation from explosions and disturbances on the surface of our own sun, but from billions of exploding stars, planets and God-only-knows-what, since the very Dawn of Creation, some of which is still traveling towards us?

Judge: Perhaps learned counsel for the defense might once again, on behalf of his witness, commit himself to a straightforward yes or no on this particular point?

Defense: That is my understanding of the statement by the witness. Yes, is the answer.

The Judge: Thank you. The witness may proceed.

Clavius continued—

Predictably, though, the chief source of all kinds of radiation in space is the sun. A full spectrum of EM waves radiates outward from it. Charged particles of all types emanate from it, especially during periods of extreme solar activity (*e.g.* flares).

Earth's atmosphere protects us from most ionizing electromagnetic radiation, both from the sun and other sources. Ultraviolet, X-ray, and gamma rays penetrate to some extent (enough to give us sunburns, for example), but in space there is a consistently higher level of all of these. But only during periods of extreme solar activity does this radiation exceed our ability to shield against it.

—Paused.

Prosecution: May it please the learned judge, but I must again ask the witness to clarify something before continuing. Will the witness kindly clarify for the benefit of the jury what is meant by the statement, "Earth's *atmosphere* protects us from most ionizing radiation"? Does the use of the more general term "atmosphere" more specifically mean the *magnetosphere,* of which the Van Allen belts are an important part, acting as they do like vast storage tanks for trapped, highly dangerous radiation that would otherwise have hit the Earth and snuffed out all existence? Because if it does, I feel certain the jury would like the witness to say so.

Judge: Again I agree. Learned counsel for the People is correct to ask for complete clarification of this point. "Atmosphere" is far too general a term. The witness must be more precise. Is the witness referring to the Van Allen belts in particular? Answer with a yes or no.

Defense: I can safely say that from my understanding of what the witness is saying, the answer to that question is yes. But if the learned judge will be gracious enough to allow the witness to continue, I feel that this particular point is about to be elaborated upon to the satisfaction of all.

Judge: Very well. The witness may continue.

Clavius continued—

Alpha and beta particles and protons carry electromagnetic charges, making them susceptible to magnetic fields. The Earth's magnetic field deflects the flow of these particles from the sun and other sources. But it also causes them to collect in two large regions of space surrounding the Earth—indeed, the Van Allen belts. We are reasonably safe inside the Van Allen belts. And as long as the sun remains reasonably quiet, we are even safe *outside* [beyond] them.

But when the sun acts up, the area outside [beyond] the Van Allen belts [outer space] becomes thick (*i.e.,* high flux) with dangerous, high-energy charged particles. A solar event was depicted in the motion picture *Red Planet,* forcing the crew of that mission to seek cover. But since the Van Allen belts themselves contain concentrations of charged particles, going through them presents its own hazard. We can think of it as crossing a barbed-wire fence: the fence offers protection, but can also snag us as we crawl through it.

We have left neutrons out of the picture up until now, simply because they just do not occur as high-energy particles *anywhere* in the universe in numbers great enough to worry about. Scientists even have trouble creating them in the laboratory.

How to shield against radiation

This is where the difference between various types of radiation becomes important. Wave radiation requires thick, heavy shielding to protect against it. However, it requires considerably *less* material to block particles.

In general, the shorter the EM wavelength, the thicker and denser the shield material must be. Ultraviolet (UV) can be blocked simply by placing a sufficiently opaque sheet of plastic between the source and that which you seek to protect. We are all familiar with tinted sunglasses, which block some 97% of solar UV rays. Not that much additional protection is required in space. X-rays and gamma rays, though, are admittedly another matter. Where intense X-rays and gamma rays occur, it might require several inches—not feet!—or centimeters of lead and/or concrete to provide adequate shielding.

Alpha particles are very large particles. As such they don't penetrate very deeply into many things. In fact, alpha particles will not even penetrate the epidermal (dead) layer of skin, and so present no particular hazard to humans. A sheet of reasonably thick paper will block all alpha particles.

Protons penetrate farther. But they can be shielded against by light metals or plastics in thicknesses of about a centimeter.

Beta particles are very small and can penetrate centimeters into the body. But, luckily, they are far too small to cause much damage if they hit anything. But there *is* a special problem here. When beta particles hit large atoms, the impact causes those atoms to give off X-rays. Metal atoms are usually quite heavy, and so are especially susceptible to this kind of re-radiation which is known by its German name "Bremsstrahlung." In fact, this is precisely how X-rays are produced intentionally for medical applications.

The best materials to shield against beta particles have lots of hydrogen atoms in them. Hydrogen atoms are extremely light, and so absorb the particles without giving off X-rays. Plain old water works very well too. In fact, 4 inches (10 centimeters) of water will block almost all background beta particles. But water is impractical for shielding in space, so high-density polyethylene (HPDE, chemical formula CH2CH2) is frequently used instead. This also effectively blocks protons.

An alternative to shielding

Radiation exposure is cumulative, meaning that the longer you are exposed to it, the worse the effect will be. It is very much like running through rain. We've discussed shielding, which is a bit like using an umbrella in the rain. But where an umbrella might be impractical because of high wind as well as rain, you can reduce your body's amount of exposure to the rain by running through it rather than walking.

Say you've forgotten your umbrella on a rainy day and have had to park some distance from your office. You can reduce your "exposure" to the rain by running from the car to the office entrance.But if you would rather be seen dead than have your staff see you running, and so walk smartly instead, you will spend more time in the rain and thus will get wetter. Bio-organisms can recover from exposure to radiation, just as you will eventually "dry out" after walking or running through the rain. But the wetter you happened to get, the longer it will take to dry out. Similarly, the more exposure to radiation, the harder it is to recover. The body *will* repair damage done to DNA or to other important molecules, but it is accepted that *it will be sick* in the meantime. It is actually better for the body to absorb a high dose of radiation *quickly* than a low dose over a more prolonged period. True, the higher dose sustained quickly may cause more problems in the short term, the low dose, received more slowly, will sometimes produce continuing damage and your body simply may not be able to keep up, even though the damage is only slight at any one moment.

When it comes to designing a spacecraft, the problem with additional shielding is that it means additional weight, which means your spaceship may have to travel more slowly. The answer to this trade-off then, by NASA in this instance, was to skimp on the shielding and go faster as a result. Of course, the radiation exposure would be much more intense, but the exposure *time* would not last as long. On balance, traveling faster through the Van Allen belts was the preferred option.

How to measure radiation

Most people are not too familiar with the various units and concepts used to measure radiation. It just isn't something the majority of us have to deal with. And so when conspiracy theorists [pro-hoax believers] describe radiation in terms of big, impressive-looking numbers they find in textbooks and elsewhere, the general public is not always equipped to understand what those numbers might truly mean.

The problem is further exacerbated by the fact that the Americans have their own system of units for measuring things to do with radiation, and the rest of the world has another system. So, not only do non-Americans see labels they might never have seen before, but despite them being in English, ones that might as well be written in a foreign language. Let us compare these terms:

Comparison of Radiation Terms

U.S. unit	International (SI) unit	Description
Curie (Ci)	Becquerel (Bq)	Amount of radioactivity produced by a given amount of a substance. 1 Ci = 37 billion Bq 1, Bq = 1 particle emission per second
rad	Gray (Gy)	Amount of radiation required to deposit a certain amount of energy in some substance. 1 Gy = 100 rad, 1 rad = 100 erg/g
rem	Sievert (Sv)	Amount of radiation required to deposit a certain amount of energy in human tissue. rem = rad × Q, 1 Sv = 100 rem
Roentgen (R)	N/A	Amount of radiation required to ionize a mass of air to a certain degree. 1 R = 0.93 rad

We measure two general phenomena when we measure radiation. We measure "activity" and "exposure." Activity is basically just how much radiation is coming out of something, be it particles or waves. Exposure is the most important factor. It measures the effect of radiation upon substances that absorb it.

Radiation activity is measured in an American unit called a "Curie" (Ci) or an international (SI) unit called a "Becquerel." The Curie is defined by how much radiation one gram of a radium isotope emits. The Becquerel just counts how many particles or photons (in the case of wave radiation) are emitted per second. The device used for measurement is often the familiar Geiger counter. If you put a Geiger counter over a gram of substance and count 3 clicks per second, the radioactivity of that substance would be 3 Bq.

Radiation exposure is measured in American units by the RAD, generally lower-cased to "rad," an acronym standing for "Radiation Absorbed Dose," and in the SI system by the Gray (Gy). The exposure is the amount of energy "deposited" in a substance by radiation. A rad is the amount of radiation required to deposit 100 ergs of energy in a gram of material. An erg is a very small amount of energy, but it takes only a very small amount of energy to ionize an atom. The number isn't important. The important *concept* is that exposure is measured by what radiation does to substances, not anything particular about the radiation itself. This allows us to unify the measurement of different types of radiation (*i.e.*, particle and wave) by measuring what they do to materials.

But what materials? Wood? Water? Human tissue? They all have different densities, so a gram in weight of one material may be bigger, thereby occupying more area, than a gram in weight of another material. Quite obviously, then, the bigger something is, the more surface area it presents for bombardment by rays or particles. What is required is some way of comparing exposure in various substances directly.

Enter REM. No, not the rock band [*A.N.* Oh, what a pity! *They* would have livened things up a bit!] but another acronym used for identifying the effects of radiation, standing for "Radiation Equivalent, Man." As with all measurements of exposure, the "rem" describes the effects of radiation on substances that absorb it, but in this case the substance is specifically human tissue. It's an American unit, of course, just like Michael Stipe's famous band. The corresponding SI unit is the Sievert (Sv).

Anyone "reading up" about radiation, is also likely to encounter the term "Roentgen." This is another unit of exposure peculiar only to the USA. It measures the amount of ionization that a certain amount of radiation will produce in air, and has been largely abandoned in favor of the rad. It can, therefore, be roughly equated to a rad for estimation purposes.

Above, we have discussed that different kinds of radiation are inherently more dangerous than others. By measuring exposure regarding how it affects surfaces, we can largely ignore other differences between various kinds of radiation. But in order to compute rems from rads we need to take into account that some kinds of radiation are inherently more dangerous to biological tissue than others, even if their "energy deposition" levels are similar. Types of radiation, therefore, carry a 'relative biological effectiveness' (RBE) factor, also called a 'quality factor' (Q).

For X-rays and gamma rays and electrons absorbed by human tissue, Q is 1. For alpha particles it is 20. For protons and neutrons, it is 10. To compute rems from rads, or Sieverts from Grays, simply multiply by Q. This is obviously a simplification. The RBE/Q factor approximates what otherwise would be very complicated computations. And so the values for Q change periodically as new research refines the approximations.

Exposure to radiation occurs over time, of course. The more rems absorbed in a unit of time, the more intense the exposure. And so we can express actual exposure as an amount over a specific time period, such as 100 rads per hour, or 5 millisieverts per year. This is called the "dosage rate," and is proportional to the flux of radiation in a particular situation.

How much is too much?

Conspiracy theorists exploit the natural "radiophobe" in all of us since the bombing of Japan with nuclear weapons, atomic testing by various nations, and the Chernobyl accident in the former Soviet Union. But now that we understand a little bit more about how radiation is measured, we can quantify the danger.

The U.S. government endorses the recommendations of various international regulatory bodies on the acceptable levels of radiation exposure in the workplace and among the general public.

If a worker must deal with radioactive materials in the course of his/her job, the legal limit is 5 rem (50 millisieverts or mSv) per year. If somebody works in the vicinity of radioactive materials but does not work closely with them, the limit is 0.1 rem (1 mSv). For persons younger than 18 and pregnant women, the occupational exposure permitted is 0.5 rem (5 mSv) per year. These are measurements above the natural background radiation limits, and are measured by dosimeters and other equipment in the area where the exposure takes place. (*Standards for Protection Against Radiation.* 10 CFR § 20.)

People usually get about 0.24 rem (2.4 mSv) in background radiation per year. (Jawororwski, Zbigniew. *Radiation Risks in the 20th Century: Reality, Illusions, and Risks,* Presented 17 Sept. 1998 at the International Curie Conference, Warsaw, Poland).

Dosage allowed by law

The maximum occupational dosage allowed by law is 1/700 the lethal dose for humans. The standard for a lethal dose is designated LD 50/30, defined as the short-term exposure (*i.e.*, over a period of a few hours or less) which would kill 50% of the human population within 30 days. The figure is around 350-400 rems (3.5-4.0 Sv). (Radiation Safety Office. *Radiation Safety Handbook.* Columbia University, s.d.).

The limits imposed by U.S. Federal Regulations are thus extremely conservative. The lethal dose is 700 times the amount of radiation acceptable per year for people who work around radioactivity. The regulations are very strict, because while it has been determined that even dosages up to 30 rems per year produce no visible effect, there is no such thing as radiation that possesses no harmful effects.

—Paused.

Prosecution: With the learned judge's permission, would it be possible for the last part of that sentence to be repeated for the benefit of the jury?

Judge: (Turning to defense counsel). Well?

Defense: No problem. My witness will repeat that last sentence.

Clavius: There is no such thing as radiation that possesses no harmful effects. It just so happens that for low doses the body is able to repair itself effectively.

Prosecution: Thank you.

Judge: The witness may continue.

Clavius continued—

Average Radiation Exposure For Apollo Flight Crews

Apollo Mission	Skin dosage (rads)
7	0.16
8	0.16
9	0.20
10	0.48
11	0.18
12	0.58
13	0.24
14	1.14
15	0.30
16	0.51
17	0.55

Source: Bailey, J. Vernon, "Radiation Protection and Instrumentation," in *Biomedical Results of Apollo*, Johnson Space Center.

—End.

§ § § § §

I thank *Clavius* for providing such a thorough and detailed testimony on behalf of my clients. Self-righteous? No, I don't think so. Self-assured, maybe. And with good reason.

No problemo!

Well, members of the jury *(stifled yawn)*, I don't know about you but *I* feel like going to sleep after that lot! However, it was all absolutely necessary, I assure you.

So it would appear that radiation in space—be it encountered on *this* side of the Van Allen belts or beyond them, (*i.e.* on the earthward side of them or the outer space side of them) *is* a problem, but not a *major* problem, according to the above statistics. So much for the prosecution's claim, then, that there is "safe" space and "unsafe" space. There are *no* safe areas in space, the same as there are no safe parts of Mount Everest, but brave men and women have climbed it in the past and will continue to do so in future.

If I understood any part of the above at all it was that the vast majority of the radiation that would have been encountered by the Apollo astronauts was *particle* radiation. Particle radiation, even though "pound for pound" the more dangerous of the Two Ugly Radiation Sisters—the other being *wave* radiation—is virtually harmless in the low dosages received provided one *speeds* through it.Better to have a smaller dose of the "wrong" kind of radiation than a much larger dose of the "right" kind, is what I think is being said.

I think I also managed to grasp some of the logic behind the fact that it is only the less harmful radiation—wave radiation—that requires very dense shielding against. But the important thing is that you, the jury, managed to grasp the general principle too. In case a few of you didn't, let me further illustrate the point with an over-simplification of my own. Say you lived on a tropical island and wished to turn a small lagoon into a natural swimming pool, safe from marauding sharks able to enter via the ocean at one end, and against piranhas, liable to attack in shoals thousands strong, entering via a river at the other end. Strong steel wire netting with a five or six inch mesh would be quite sufficient to keep out the occasional shark, which analogously becomes our "particle." But a net of far greater density—one equally as strong but with a much finer mesh—would be required to keep waves of hungry piranhas at bay, even though a shark is potentially a much more dangerous creature than a piranha.

So, it seems to me, then, although admittedly a total layman in these matters, that unless what we were told by that last witness was also an elaborate—not to mention incredibly boring—hoax as well, the main thrust—indeed the 'Saturn V' of the argument of the believers in a hoax—has well and truly turned out to be something of a damp squib. Radiation is indeed a hazard, but *not* an impregnable barrier totally preventing man from ever traveling in space. That means it is also *not* a barrier preventing man from going to the Moon. So, if man *can* go to the Moon anytime he wishes, why should *anyone* be disputing the fact that he has already done so?

And, maybe just to rub a little more salt into the wounds thus inflicted upon my learned friend and all those ready to mention my next witness's discovery in the same breath and sentence as the words "we didn't go," I shall let the world's foremost expert on the belts of radiation that surround the Earth—after all *he* discovered them!—deliver the final *coup de grace* regarding the supposed dangers posed by radiation in space.

I call Dr. James Van Allen.

Dr. James Van Allen's Testimony

Source: *http://www.clavius.org/envrad.html*. A blunt statement directly from the lips of the acclaimed scientist.

Begins—

"The recent Fox TV show, which I saw, is an ingenious and entertaining assemblage of nonsense. The claim that radiation exposure during the Apollo missions would have been fatal to the astronauts is only one example of such nonsense."—Dr. James Van Allen.

—End.

§ § § § §

On behalf of my clients, the defense thanks Dr. Van Allen for his small but extremely timely contribution—again via *Clavius*—to these proceedings. Members of the jury, we cannot get it any more straight from the horse's mouth than that, can we?

14

A Touch-Up With an Air Brush?

Members of the jury, my learned friend went to great lengths when representing the case for the prosecution, to provide you with photographic "evidence" of a fraud. Some of these "stills," as well as poorer quality film footage, are claimed to be obvious fakes. It has been further suggested that all were very likely taken in an aircraft hangar, duly converted into a film studio, at a top-secret desert location somewhere in the United States of America rather than on the Moon.

As proof, my learned friend introduced witnesses—some expert, I admit—who testified to the likely fraudulent nature of some of these photographs. The word "fake," would also seem to apply, at least in their eyes, to photographs that might merely have been doctored.

The airbrushing of photographs used in product advertising—photographs obtained at great cost to companies who often pay for the best photographers and the best models at the best and most expensive locations—is not only a very common practice but a lawful and accepted one. Large companies generally hire smaller advertising agencies to carry out this kind of work for them. In the U.K., there are strict codes of practice for advertisers, and if any product fails to perform as advertised the maker could face prosecution. Respectable companies have to operate within the law, and most would not knowingly hire advertisers who employed methods that could bring them into disrepute.

The practice of airbrushing a photograph can only be deemed fraudulent if the intention is to use the changed photograph to *deliberately deceive* for financial gain, or in order to obtain unfair advantage over a competitor. The problem for advertisers is that although their job is to present their client's product in the fairest light possible—a bit like a defense lawyer, as it happens!—even the most professional of models and the most exotic of locations appearing in photographs are seldom perfect on the day. An otherwise lovely background view could be spoiled by an ugly mobile phone mast, for example, which may not have been there the last time the location was used. Another example might be a very pretty female model developing a nasty red spot on the end of her nose on the very morning of the shoot! But the shoot still has to go ahead, because the expense of postponing it after bringing a top photographer, a top model, lighting people, sound engineers, make-up artists, wardrobe girls and a small army of other helpers expert in a variety of fields, hundreds of miles to the location, doesn't even bear thinking about! If the product the girl is advertising is a skin cream, then the company hiring her has a legitimate right to erase both the ugly phone mast and the annoying pimple from the picture. The photo would then show the girl with the fresh, clear complexion she *normally* enjoyed against a background unspoiled by the phone mast. In the U.K., providing the product being advertised can live up to the claim that it can give most other women a clear complexion too, if used regularly, then all is perfectly legitimate.

However, if the same altered photograph is later used in a tour brochure to advertise the *location* itself, the removing of the mobile phone mast could be construed as a deliberate attempt to defraud.

If NASA decided some of the photographs taken on the Moon needed a little re-touching here and there before being made public, then so what? The hoax believers themselves frequently remind us of the extremely difficult and virtually impossible conditions the cosmic cameramen were forced to work under! So, a piece of crosshair has vanished behind a rock, eh? That looks a bit dodgy. And there! Look at that! That shadow has obviously been painted in or enhanced to a darker shade. Now, that *is* suspicious!

Come off it.

Some shots may well have been re-taken in a studio environment simply because the originals from the Moon were so bad as to be completely unusable. Again, what does it matter? A shot might be "staged" and re-taken simply because the loss of the original otherwise leaves an unfortunate gap in the photographic record of a great historical event. What conspiracy theorists prefer to melodramatically refer to as fakes might merely be substitutions for ruined or missing originals. Believe me when I say that where the worlds of PR and photography overlap, that is not such a big deal. In any case, all is perfectly legitimate providing only the pictures are fake—*not* the event they are purporting to depict. Providing a retaken photograph fairly represents a scene from the Moon that would have been captured the first time around if only all things had been equal, then no offence has been committed. Certainly not according to U.K. law.

To accuse NASA of faking *six* whole missions to the Moon—wouldn't *one* have sufficed?—because of a few dodgy photographs, is patently absurd. One might as well accuse a politician of running a fraudulent election campaign simply because his face had been re-touched on a poster to make him look more pleasant than he actually was. And more electable? The truth is, though, that any criticism of the poster—or due praise!—would be better directed at the politician's PR people than at him personally.

Anyway, that's enough from me about the frequently combined arts of photography, PR and advertising—all subjects I truly know very little about, although that doesn't preclude me from understanding their function and purpose in the scheme of things, or from knowing enough to remind the jury that commercial photography is most definitely an art form, and no less so than a painting isn't quite finished until the artist lays down the brush for the very last time after signing it, neither is a commercial photograph necessarily considered ready for public consumption the moment it is developed. And, while we are on the subject, it is also worth noting that a newly discovered, but damaged, Rembrandt is no less an original masterpiece simply because the ravages of time may have necessitated some professional cleaning, retouching or even complete restoration by a skilled modern restorer. The name of the game in these circumstances is not fakery. It is repairing something almost sacred and utterly beyond price and attempting to bring out both its original beauty and intended message before putting it on display to the general public.

Before calling my next witness, let me first explain that although the following testimony is largely original in essence, and its some-times witty, occasionally aggressive—even confrontational!—"fla-vor" maintained as far as possible, it contains some essential re-writing by yours truly. I have, though, tried to keep such alterations and deviations from the original text down to a minimum. This should not be construed as an attempt to deceive or defraud anyone. On the contrary, my editing of it—no, I have *not* tampered with the evi-dence!—will hopefully have rendered this piece of testimony much clearer, less emotive than it originally was and, more importantly, far less ambiguous to the jury. In other words, I happen to now consider the following testimony—like some of those NASA photographs maybe—much more "presentable" to the public.

I call Red Zero.

THE TESTIMONY OF *RED ZERO*

Source: *http://www.redzero.demon.co.uk/moonhoax/photography.* © 2001 Red Zero.

Begins—[some editing necessary].

Red Zero is a pro-Apollo, anti-hoax website that has collected knowledge provided by literally hundreds of people over the years who have written to our site, explaining certain points, and clarifying and expanding on others. The material is Copyright 2001. But you are free to use anything you want. All we ask is a mention of the website when used.

Photography

What no stars?

FAQ/Accusation: None of the lunar photos show stars in the sky. The Moon has no atmosphere so why can't we see stars more clearly even than on Earth, let alone none at all? Is it because NASA faked the pictures and simply left the stars out as it was too difficult to attempt to replicate the ever-changing vista of the heavens as viewed from an exact spot on the surface of the Moon at different times, because the Moon—in reality—would still be orbiting around the Earth and the Earth around the sun? After all, the slightest error with a fake background that didn't accord with the provable mathematics of star positions as observed from a particular spot on the Moon at a particular time, would soon have been noticed by trained astronomers. One can almost envisage the headline that would have exposed any hoax: *"Hey! NASA's changed the universe!"*

Response/Answer: This theory isn't quite as dumb as some, but it is a close call. Anyone seriously proposing it, knows nothing about photography. Even so-called "photographic expert" David S. Percy is sometimes quoted as supporting this argument. How could anyone have the nerve to call himself an "expert," yet put forward this idiotic theory.

—Paused.

Prosecution: Objection! Perhaps this witness might be reminded that what is required by the People are answers to the questions put to the witness, not insults to either individuals or to the People as a whole. The People have a legitimate right to raise any doubts and questions they see fit to ask. That is the purpose of this trial.

Judge: Sustained. I shall indeed remind the witness to confine testimony to solely answering the questions put forth by the People, whom learned counsel is representing. Whether the questions are meritorious or not, in the opinion of the witness, is not for the witness to either decide or comment upon. I alone shall decide what is pertinent or otherwise to these proceedings. The witness will continue by directly answering the question.

Red Zero continued—

It is practically impossible to capture a brightly and harshly lit object in the foreground (like- say—an astronaut) and a dim object in the background (like a star) in the *same* exposure. Do a long exposure to capture the stars and you will get a fuzzy, over-exposed blob in the foreground. Do a short exposure to capture the astro-

Tut-tut! You can't trust those special effects guys. *That* one's definitely in the *wrong* place!

naut and the stars just won't appear. The Hasselblad cameras [Picture A on Plate 25] used by Apollo were small, and mounted on the spacesuits. To attempt stellar photography would have needed long exposures, with astronauts rooted to the spot for lengthy periods, and the end result would still only have produced very "wobbly" stars.

So why weren't such long-exposures taken of stars with the camera on a tripod, then? Answer—because NASA had *not* just spent 40 billion dollars for their astronauts to take pictures of stars. Telescope cameras on Earth—even thirty years ago—took far better pictures of stars, despite an atmosphere to cope with, than could possibly be taken by the tiny cameras used on the Apollo missions.

As for NASA being unable to fake a sky full of stars and their movements, if required, that will certainly come as a big surprise to planetariums the world over—because *they* have all been able to do it, often with NASA's help!

Small Earth?

FAQ/Accusation: In the pictures of the Earth taken from the Moon, the Earth should be bigger. Why does it look so small?

Response/Answer: This question shows a fundamental lack of understanding of camera focal lengths. Put simply, with a camera of usual focal length (*e.g.* 45mm) you get *tiny* images of distant objects. Try it yourself. Take a photo of the Moon with your camera. What you will get back from the processors is a photo of a tiny white dot in the middle of a black sky. The cameras used by the astronauts during the Apollo landings were not really that much different to the average family camera in this respect. Again they were not there—neither the astronauts nor the cameras—to take pictures of the Earth.

UFOs in the Moon's sky?

FAQ/Accusation: Some photographs [Picture B on Plate 25] have strange "blobs" of light apparently hanging in the Moon's sky like "UFOs." Are they UFOs? [*AN:* Not such an "idiotic" question, as it turns out, as I have already hinted here and there. But more later about that]. If not, *what* are they?

Response/Answer: Another point about some of the Apollo photos often held up by the pro-hoax theorists as "proof"…well, yes…of *what* exactly? Most pro-hoax theorists, it seems, cannot make up their minds between "UFOs" or "film studio lights" on a gantry. But if the latter is what they are, then aren't they so patently obvious that the Fakery Department of the National Aeronautics and Scam Administration would have airbrushed them out? And if the hoax believers are claiming that there may well be mysterious objects hovering about in the Moon's sky and these photos are proof of same, isn't that a bit like saying that the astronauts really must have been on the Moon in order to snap pictures of these strange objects? They can't have their cake and eat it! But it appears we shouldn't expect too much well thought out logic or consistency from the hoax theorists, should we?

Either way, the blobs of light are no real mystery at all and any photographer worth his salt should be able to identify them. Except those, of course, who may have made up their minds beyond all attempts at reasoning that hoaxing is the name of the game. These blobs are caused by lens flare. Effectively, very bright light reflecting off the interior of the camera lens. It is a quite common phenomenon when taking pictures of a brightly lit outdoor scene.

Home-made?

FAQ/accusation: Some photos show various paraphernalia and objects in shot that couldn't have been on the Moon. It proves these photos are fakes taken here on Earth.

Response/Answer: Not quite. It proves that they were taken on Earth. Has anyone, particularly NASA, ever claimed such photos were taken on the Moon? A lot of photos produced as so-called "evidence" of a fraud were definitely taken here on Earth, and nobody is denying that. In fact they were taken before any attempt to land on the Moon and simply show astronauts in training prior to a mission. The various pieces of apparatus "in shot"—lifting gear, girders, mock-up lunar surfaces complete with craters, and so on— were all items essential to that training if it was to be as thorough as possible. The Moon may well be barren and deserted, but many potentially dangerous scenarios caused by the rugged moonscape after landing and EVA, had to be practiced and rehearsed hundreds of times before a mission.

No one but an obsessed lunatic would ever claim that NASA had distributed such pictures as having been taken "on the Moon."

Above suspicion?

FAQ/Accusation: The famous "Man on the Moon" photo [Pictures B and C on Plate 3, both pictures on Plate 23 and Picture C on Plate 25] taken by Armstrong of Aldrin, is a view from above. This is ascertained by the fact that we can see Aldrin's head, yet the camera was mounted on Armstrong's chest, below eye level. Does this prove that Armstrong couldn't have taken this picture and that it is a fake?

Response/Answer: No, of course it doesn't. Of all questions and accusations made on behalf of the pro-hoax theorists they don't get much more stupid than this!

—Paused.

Prosecution: Objection! This witness continues to challenge the mental integrity of those I am representing and the validity of the questions themselves. The enhancement of the reflection in "Aldrin's" visor appears to show the shadow of a tall, slim, *third* person (See Plate 23A) seemingly taking the photograph in normal attitude, viewfinder to eye. The figure of Armstrong is also evident in the background, apparently walking away from Aldrin. Does the witness have an opinion on this?

Judge: I refer the question to learned counsel for the defense.

Defense: With due regard for the learned judge's views on the use of bad language in court, not one that is printable, I'm afraid. My clients, too, refute absolutely the claims of what some say can be discerned in the visor reflection. My clients maintain that these people are merely seeing what they *want* to see, nothing more.

Judge: Very well. However, the prosecution's objection is sustained. I, too, would like to hear the answers to *all* questions put to the witness and I am neither an "obsessed lunatic"—although I happen to admire the turn of phrase on this occasion—nor am I "stupid." I would remind the witness that this is *not* a cockfight or indeed a slanging match, it is a court of law, and any future attempt by this or any other witness to turn it into either of the former will be deemed to be in contempt of court. That will mean all testimony from this witness, however vital to the defense case, being immediately stricken from the record. The witness *will* answer the question originally put and will do so *without* hurling insults at those who ask it.

Defense: If the learned judge will permit me to say so, I feel I am partly to blame for these insults. I have attempted to get the witness to tone down the aggressive nature of this contribution to my clients' case, but it would appear that my attempts to do so have fallen rather short. However I am confident that from this point on, the witness will bear the learned judge's warning in mind.

Judge: I hope so. Very well, the witness may continue.

Red Zero continued—

Armstrong was obviously further up on an incline or hump on the uneven surface of the Moon, some way above Aldrin. Like the questioners, Armstrong realized he would need to do this to get a good full-length shot of his colleague when so close. Only if the shot had been taken on the flat, imitation dust-covered floor of a converted aircraft hangar "studio" would we likely be missing Aldrin's helmeted head, so this photo more proves it *was* taken on some kind of genuine irregular surface than the other way round! And there is no reason that makes a scrap of sense to suppose that this surface didn't belong to the Moon.

Aldrin is leaning slightly forward in the shot too. Due to the weight of the backpack, the moonwalkers had to do this to keep their balance on the Moon. Despite any other considerations, why would actors, or whatever is being suggested, carry heavy back packs around a studio for hours on end when ones full of light-weight popcorn would have looked just as realistic on film?

Where did those crosshairs go?

FAQ/Accusation: All right, explain this one then. In some photo-graphs the crosshairs on the camera lens actually disappear *behind* objects. This is not possible unless the pictures have been tampered with.

Response/Answer: Okay. It's a fair cop guv'nor. This proves fakery beyond a doubt. Sorry—*we* were duped as well. But say, hold on a moment…isn't it always the *bright* bits of the pictures that these Reseau marks tend to "hide" behind? [Picture A on Plate 25. See also Pictures B and C on Plate 1]. I think so, don't you? In which case there is a very simple explanation, especially for those who don't understand how photographic film works. All that has hap-pened is that the over-exposed bright white object has caught the light and "bled" into the thin black line of the crosshair. Not surprisingly, you won't find Reseau marks disappearing behind any-thing other than pale, brightly lit objects in any photographic image where the film has been over-exposed.

This is a well-known phenomenon, and you can easily experi-ment with it at home yourself. Take a length of black thread, pull it taut between your fingers and hold it up towards [not against!] a switched on light bulb, way above your head, and watch the thread "disappear" where it crosses the brightest part of the light.

The intriguing part of this particular allegation of fakery is the obvious suggestion that certain cut out images were stuck onto unoriginal backgrounds and then re-photographed to produce false overall images—because all this was long before the digital image manipulation performable on our PCs these days.

But why would anyone bother with all that, when it would have been a hundred times more simple to park the Moon Buggy, for example, on an artificial lunar landscape, dress up a couple of actors or male models in spacesuits and snap away to your heart's content? If a particular studio light reflecting off a bright section of the Rover was causing a nuisance and kept on blotting out the crosshairs—thereby causing the picture to look a little "dodgy"—then all that would be necessary would be to adjust either the light or the position of the Rover vehicle.

To demonstrate, finally, that the missing crosshairs phenomenon as it applies to Apollo was not down to "cutting and pasting," either in the time-honored fashion or by computer in more modern times, take a look at photograph AS12-48-7071 [Picture B on Plate 26]. This was taken on the Apollo 12 mission and shows astronaut Charles "Pete" Conrad on EVA.

—Paused.

Prosecution: If it should please the learned judge, I once more feel an urgent need to interrupt this witness at this point.

Judge: Very well. Are you raising another objection, or do you wish to cross-examine the witness?

Prosecution: With the learned judge's permission, I would simply like to address the jury regarding a matter of interest that the witness has just this moment alluded to.

Judge: Then by all means do so.

Prosecution: Members of the jury, the witness *Red Zero* has drawn your attention to a photograph on Plate 26 of your case papers, which purports to be a photograph of Charles Conrad of Apollo 12 on EVA on the Moon. On reflection—literally—maybe it is; but then again, maybe it isn't. It is rather difficult to tell, isn't it? It is also difficult to tell whether or not we can see two opposing horizons—one in the background and one reflected in the helmet visor—of the Moon itself. They could just as easily be opposite ends of the floor of a huge aircraft hangar, couldn't they?

Judge: Looking at the picture, I must say I agree with learned counsel for the prosecution. Perhaps the witness can point out grounds—not an intentional pun, I assure you—why this photograph should be looked upon as having been genuinely taken on the Moon?

Red Zero continued—

Well, at least it is *supposed* to be Conrad, although I agree there is no way of telling that it is really him or that the very even-looking horizons—yes, we can see both via the visor reflection—aren't just the flat floors of a giant hangar. But we are discussing disappearing crosshairs. Look at the reflection in Conrad's visor and you will see that the white object placed just in front of the astronaut taking the picture—which *should* be Conrad's colleague Al Bean!—has bled almost directly into the middle of a Reseau mark.

So someone at NASA "pasted" a reflection of a completely unnecessary white object right across the middle of a crosshair, did they? To precisely what end, exactly? This particular photo would have looked much better *without* that piece of equipment in it. Yes, airbrushed *out,* in fact, to simply leave the one astronaut and the reflection of the other in his visor. In other words, why would anyone—certainly anyone in his or her right mind—decide to superimpose something so unnecessary onto an otherwise perfectly good and poignant image! Don't you agree?

Too good to be true?

FAQ/Accusation: Aren't the Apollo photographs just a little too good to be true? Think about it. The supposed astronauts couldn't use a viewfinder and were wearing bulky gloves, yet they still managed to get the all the photos in focus and framed perfectly! This suggests they brought in professional photographers as part of the conspiracy.

Response/Answer: No they are not, no they didn't, and no it doesn't, are the answers to first the question, then the assumption, and finally the accusation. On the Apollo 11 mission alone, Armstrong and Aldrin took around 1,400 photographs! Like the majority of holiday/vacation snapshots taken by amateurs, most of these exposures were complete rubbish. Boring rock formations, cut-off heads, out-of-focus images, blurred images, images suffering from camera-shake, and a variety of accidental as well as under or over-exposed shots. But naughty NASA, for some strange, warped reason known only to the darkly-motivated inner sanctum of that organization, decided that only the very *best* of an otherwise "bad bunch" should be released and published.

Oh, you spoilsports, NASA! How dare you deprive us of that out-of-focus snap of Neil's sexy thumb! And what about that fuzzy shot of Buzz Aldrin's nice, space-suited bottom coming down the LM's ladder! How *could* you?

Whatever demon possessed NASA in order to decide against publishing such rubbish, though, would appear to be beyond the comprehension of the average believer that there was a hoax. Obviously the sort of people who, although patently amateur—a fact that certain inane questions leave us in not too much doubt about anyway—take 144 pictures while on vacation in Florida and every last one is "so perfect" that it goes straight into the family snapshot album!

As for those very few pictures that *were* good enough to publish [and four of these are Pictures A and B on Plate 27 and Pictures A and B on Plate 28, which include the Lunar Rover on various missions] the wide-angle lenses of the Hasselblads undoubtedly played their part here. All an astronaut had to do was aim the camera lens roughly in the direction of the scene he wished to capture. This often resulted in there being lots of empty space around the intended object, but trimming a photo, enlarging it, and even "tampering" with it a little after being developed, merely to bring out the best and most interesting features caught on camera, does *not* constitute fakery in anyone's language. That's simply making the best of an unavoidably bad job.

Half baked?

FAQ/Accusation: All right, see if you can squirm your way out of *this* one! The temperatures on the Moon would have baked the film in the cameras until it frizzled up. You can't take photos as good as that with frizzled up film. Go on—get outa that one! You can't, can you? *GOTCHA!!*

Response/Answer: Oh, dear. I'm sorry, but it really is all rather pathetic. If only people would do some half decent—instead of half-baked—research before sneering down from their high horses, seemingly *enjoying* their attempts to debunk the greatest achievement in all the history of mankind.

—Paused.

Judge: Although I am still concerned about the less than respectful tone of this witness towards Apollo critics, I must say that I agree with that last remark. It is a pity, though, that much more isn't made of it. Such marvelous achievements as indeed all the Moon landings were, if genuine, surely should be celebrated a little more often. As an Englishman, and with the best will in the world, I am bound to make the casual observation, not a criticism, that it is not at all like our American friends to be backward in coming forward whenever the opportunity arises to pat themselves on the back, blow their own trumpets and wave the flag. We British like to do that, too, on occasions, but we tend to be just a shade more reserved when we do it, generally speaking. Still, all credit to the Americans, I suppose, for being so self-effacing—dare I say bashful even?—regarding this particular slice of their history. I do, though, sincerely hope that Apollo hasn't set a trend for the future. We British prefer our Americans to *be* Americans. We *expect* Americans to be loud, brash and boastful and to make lots of noise whenever there are good reasons to be happy and to celebrate. When they do *not* do so after such a tremendous and quite unmatchable achievement, and instead appear for some strange reason to be noticeably keeping their heads down about it all, we find ourselves not only curious but genuinely *worried* about our dear and much admired friends, and are left wondering what could possibly be wrong...

The witness may continue. I remind the jury that the question was asked about possible damage to the film in the cameras due to the surface temperatures on the Moon.

Red Zero continued—

It is true that the surface of the Moon can reach as much as 280 degrees Fahrenheit at mid-lunar day! That is *very hot* indeed. It is fairly certain that just about everything—including the astronauts themselves!—would "frizzle up" in such a temperature.

But what the hoax subscribers either omit to explain to those they wish to convert to their faithless "religion of non-belief," or simply *do not know* through lack of research, is that a *lunar day* does not last 24 hours like an Earth day. It last *two weeks*! Not only that, but that two-week long "day" is actually a year—yes, a *year!*—in Moon terms. The Moon rotates on its axis once every 29 and a half Earth days. It also orbits around the Earth in the same amount of time. Because of this, one Moon day is exactly as long as one Moon year. This is why the same side of the Moon faces Earth all the time.

The Apollo missions were carefully timed to land and return either during a lunar "morning" or "evening," or you could say in "spring" or "winter." One mission would land, spend a few days, and then blast off again on the *same morning*. Another mission would land, spend a few days, and then take-off on the *same evening*. So, rest assured that the Apollo astronauts were *never ever* on EVA in the scorching temperatures of high noon, snapping away with their Hasselblads like "mad dogs and Englishmen out in the midday sun," as Noel Coward once sang. And that had nothing to do with preserving film in cameras. That was the very *least* of everyone's worries. It was, of course, to do with the preservation of the *astronauts* themselves!

It also has to be kept in mind that because there is no air on the Moon, there is no ambient temperature or convected heat either. If you are out of direct sunlight, and therefore radiated heat, you will be quite chilly without a spacesuit. The film at all times would either be in the LM, or within the camera. Both had surfaces designed to reflect as much heat as possible rather than absorb it. So the film wouldn't even pick up much conducted heat from its container. So that's no convected heat, little radiated heat, and little conducted heat. There are no other methods of receiving heat. There was actually much more chance of the film in the cameras *freezing* than frizzling!

Who filmed Armstrong?

FAQ/Accusation: Who filmed Armstrong coming out of the Lunar Module and down the ladder to the Moon's surface? Was someone already—ha, ha!—on the "Moon" to do this?

Response/Answer: Anyone around at the time will clearly recall that a worldwide audience of many millions was told to expect to see the astronauts emerging from the vehicle and coming down the ladder by way of a video camera mounted and extended from the side of the LM. It had been positioned there especially for the purpose of capturing what would arguably be the most historic of all historic moments.

But, at first, we could see nothing recognizable until someone at Mission Control asked Armstrong, by now already on the LM's ladder, to deploy and adjust the camera—which was still in its storage position—to give us all a better view of what was happening. Note by looking at a still from the video footage [Picture C on Plate 26] how the left-hand side of the image is obscured by both the side of the LM and the arm of the storage compartment that lowered it. (The horizontal black bar across the middle is interference flickering across the picture). The way the camera was mounted meant that these first pictures were upside-down and had to be "flipped over" once received on Earth.

The same video camera was later removed from the LM and mounted on a stand nearby in order to take footage of the astronauts' activities on the lunar surface.

Hoax story? Maybe just a ghost story!

FAQ/Accusation: On some video footage you can see right through the astronauts to the background behind them, as if they were ghosts! This must prove it was all a fake!

Response/Answer: Is anybody there? Sorry…I mean, is anybody suggesting NASA enlisted *supernatural* assistance to fake the film?

There was no fakery, but there were indeed plenty of spooks—or "gremlins"—in the primitive state of video technology at the time. This "ghostly" footage was not taken by a film camera, but by a state-of-the-art video camera. The very best available at the time, though absolute rubbish by modern standards. When it was pointed at a bright stationary object, the image could 'burn' into the electronic receptors. It's a temporary effect, just the same way your eyes will still see the image—even if closed—of a light bulb for some time after you first stared at it and then turned away.

So the "old" image of the LM, for example, could be "burned" onto the camera and remain in full view even as the "new" image of a seemingly "phantom" astronaut was "materializing" in front of it.

No one has to take NASA's word on this. Just ask anyone you know who may have purchased one of the very first video cameras on the market in the early seventies.

Take off or take on?

FAQ/Accusation: Who filmed the Lunar Lander ascent stage taking off from the Moon's surface and panned the camera upwards to follow its rise?

Response/Answer: A reasonably sensible question, this one, despite the implications of the use of the word "who" rather than "what." This often shown footage was taken by a remotely controlled video camera mounted on the Lunar Rover, which was left behind. Everyone knew precisely how fast the LM's ascent stage would ascend, so they also knew how fast the camera had to pan upwards. It could all be set up in advance of the take off and triggered remotely.

—End.

§ § § § §

I thank the witness from *Red Zero* for this valuable contribution to the case for the defense, although I can understand why certain objections were raised here and there by my learned friend and fully concur with the words of warning for the witness from the learned judge. However, I feel the court will be disposed to forgive any noticeable irritation caused by one or two of the questions posed.

Judge: Indeed, a splendid and extremely informative contribution once the witness calmed down and decided to answer the questions directly and in detail instead of beginning each answer with an insult. The witness most certainly has *my* forgiveness and I think I can safely speak also for learned counsel for the People and the jury.

Defense: I thank the learned judge on behalf of my last witness. That is most gracious.

Refuting the claims

So, members of the jury, there you have it. I will recap briefly. Thus far, we have managed to clearly refute, scientifically, any claims that the Van Allen belts themselves, and radiation beyond them in space, were a major problem to the Apollo program. Thus, the main argument being brought to bear by the pro-hoax lobby is destroyed.

Now, also, I think it fair to claim, one of the main supporting acts of the pro-hoax lobby's overall argument, *i.e.* "fake" photographs and film footage, has indeed been riddled with holes if not well and truly shot down in flames! Perhaps I should say "lens flare" rather than flames on our airless Moon?

There can be no smoke without fire. That is a universal law. And most of us now understand the reason why there can be no fire on the Moon. But, members of the jury, is it beginning to look to you, as it is to me, that there is not even a puff of smoke to be had in this case either, other than the foggy clouds of technological ignorance?

15

Outfoxing the Fox

I shall call my next witness for the defense. He is a planetary scientist from the University of Arizona and you will be familiar with his name already.

I call James Scotti.

JIM SCOTTI'S TESTIMONY

Source: *Comments on the FOX Moon Landing Hoax special* by Jim Scotti © 2001. Created by Jim Scotti, 2001 February 15. Last updated by Jim Scotti, 2001 August 17.

Begins—

FOX Special—Conspiracy Theory: DID WE LAND ON THE MOON?

Fox aired a special on the [alleged] Moon landings hoax on Thursday night, February 15, 2001 at 9:00PM ET (8:00PM PT). I had hoped the special might treat the hoax claim with skepticism, but all hopes were dashed as I watched this program unfold. They presented the arguments of the "true believers" [in the hoax] without any significant rebuttal. Below are some of my comments, many made on the fly as I watched the program. The program claims to "let the viewers decide for themselves" about whether there was a hoax or not, but failed to present a balanced program of pro and anti-hoax views, giving the viewers a highly biased pro-hoax set of evidence on which to base their conclusions. [*A.N.* That is an imbalance I am attempting to redress, hopefully with an entirely level playing field for both sides].

THE PRO-HOAX AND ANTI-HOAX LINEUPS

Pro Hoax

Bill Kaysing, Moon Hoax Investigator.

Brian O'Leary, NASA astronaut (…candidate, but I nitpick).

Paul Lazarus, Producer, Capricorn One.

Ralph Rene, Author/Scientist.

David Percy, Royal Photographic Society—photographic "expert."

Bart Sibrel, Investigative Journalist.

Jan Lundberg, Project Engineer, Hasselblad.

Scott Grissom, son of Gus and his widow, Betty Grissom who thinks accident was murder.

Boris Valentinovich Volinov—Russian cosmonaut.

Anti Hoax

Brian Welch, NASA spokesperson.

Howard McCurdy, Space Historian, American University.

Julian Scheer, Former NASA spokesperson.

Paul Fjeld, NASA LM Specialist.

THE FOX CLAIMS RE-STATED AND SCOTTI's COUNTER CLAIMS:

Stars

Fox Claim: There were no stars in the images. With nil atmosphere on the Moon there should have been lots of stars visible.

Scotti's Counter Claim: We see no stars in the images because the images are exposed for the bright sunlit scenes. The stars are too faint to show up on the images due to their short exposure.

Odds against success

FC: The likelihood of success was too small. Kaysing says odds of one successful Moon landing were 0.017 per cent, let alone six.

SCC: Kaysing's odds of a successful Moon landing being 0.017 per cent must have been from some early report, based on who-knows-what assumptions. Does anyone know where he may have gotten that estimate or did he make the number up?

Capricorn One

FC: The film *Capricorn One* scenes were copied.

SCC: No, it was actually the other way round. The Apollo landings occurred between 5 and 8 years *before* the film and were used as a model *for* the film in order to make it look more convincing to the movie-going audience. Making claims that the technology was in place, the FOX producers seem to think that film-making technology of the 1960s was up to creating such a convincing hoax despite vast amounts of evidence to the contrary, as a simple examination of even the best modern sci-fi movies will demonstrate. Sit next to any "science geek" (like me!) in your favorite sci-fi movie and you will hear innumerable comments about how this scene or that is [scientifically and technically] incorrect.

Area 51 off limits

FC: The pro-hoaxers claimed the Apollo film sets are still there at Area 51, and this is real reason why the "black project" [ultra top secret projects] infamous experimental area is off limits.

SCC: Oh, I wondered why that might be, considering the top-secret research that is apparently being done in the area. [*A.N.* Don't worry, Jim. At least it's a new one compared to the usual back engineering of UFOs claims!].

Astronaut deaths

FC: The show claimed that 10 astronauts died "in mysterious circumstances" during Apollo.

SCC: "Mysterious" apparently includes accidents in high performance jet aircraft and accidents in new untested spacecraft. Astronaut deaths: Ed Givens (car accident), Ted Freeman (T-38 crash), C. C. Williams (T-38 accident), Elliot See and Charlie Bassett (T-38 accident), Gus Grissom, Ed White, and Roger Chaffee (Apollo 1 fire). So who are the other two? According to the show, two other pilots were shown, but they weren't astronauts, at least by NASA standards. One was X-15 pilot Mike Adams who was the only X-15 pilot killed during the X-15 flight test program. Mike Adams, though not a NASA astronaut, had flown his X-15 above 50 miles, which is considered space, and technically, he could be considered an astronaut along with a number of the other X-15 pilots. The other was Robert Lawrence, a would-be Air Force Manned Orbiting Laboratory pilot who died in a jet crash shortly after reporting for duty to that program.

Later in the show, they claimed that the Apollo 1 fire might have been a murder conspiracy to silence Gus Grissom's outspoken criticism of the Space Program. The Fox producers obviously have not researched Gus Grissom at all. He was enthusiastic about the program and very aware of the dangers of spaceflight while trying to make his spacecraft as safe as possible. His hanging a lemon on the Apollo simulators has been widely misinterpreted as dissatisfaction with the entire Apollo program.

They also claimed that the death of NASA worker Thomas Baron was murder and an attempt to cover-up a 500-page report on the Apollo 1 accident. A more accurate version of the story is of a deeply depressed man who committed murder/suicide in distress over the loss of the crew.

It is true that Thomas Baron released a report critical of various issues dealing with the development of Apollo hardware, and that he was somewhat outspoken after the tragic fire which killed the Apollo 1 crew. But claims that his report has vanished are simply false, as parts of that report (if not the entire report) are available on the Internet. Baron and his family died when their car was struck by a train at a [level] crossing. Was it an accident or was it suicide? The stress on all who worked at NASA and were involved in any way in the spacecraft was tremendous and that many more NASA employees did not do the same as Tom Baron and commit suicide in response to the accident is nothing if not surprising.

These unfounded claims of murder and conspiracy are worse than libelous, particularly as they lack any significant evidence to support them.

—Paused.

Judge: Is there *any* evidence that the witness is aware of, to support these claims of murder—be it even of an insignificant nature?

Defense: Like others, the witness has supplied only written testimony to the court, but my research into this matter probably places me in an even better position to answer this question than the witness himself.

Judge: Then learned counsel may indeed answer on behalf of the witness.

Defense: I thank the learned judge. The answer is no.

Judge: Very well. Then I would like the record to reflect that fact as clearly as possible. That both you and your witness are actually saying that despite certain claims, as far as you know there is *no evidence* whatsoever linking the defendants to the very serious allegations of murder or manslaughter involving anyone. Is that correct?

Defense: Indeed it is, yes.

Judge: Just another small point regarding the mention of unfounded claims of a conspiracy. We do not know for certain that these claims *are* unfounded, do we? This court has been convened expressly for the purpose of testing such allegations before a jury and it remains to be seen whether or not the evidence relating to same is significant, most importantly in the opinion of the jury. Bearing that in mind, the witness may continue giving his evidence to the court.

Scotti continued—

No Blast crater under the Lunar Module

Fox Claim: The hoax believers claimed during the show that the LM descent stage used its full thrust of 10,000 pounds at lunar landing and that it should have excavated a large blast crater under the LM.

Scotti's Counter Claim: At landing in the low lunar gravity (which is $1/6^{th}$ of Earth's gravity), the LM only needed to throttle down to about 3,000 pounds of thrust. The blast of rocket exhaust is not nearly as large as the 10,000 pounds claimed and results in a scouring of the topmost layer of lunar soil along the ground path and under the LM. The LM had 6 foot long landing probes under 3 of the 4 footpads and when any of the probes contacted the surface, the crew shut down the engine so that the LM would fall the last few feet to the surface, so the engine was more than 6 feet above the surface at its closest. You can even see effects of the blast in some of the lunar images including any taken under the LM, and one set taken on Apollo 12 which shows a disturbance along the ground path of the LM before landing. The dust is clearly visible flying out at high speed away from the LM prior to touchdown in all of the lunar landing films taken from the LM cabin windows during approach and landing. Given that the descent stage engine bell is about 5 feet across at the bottom, and that thrust of the engine at touchdown was about 3,000 pounds, that blast pressure of the rocket exhaust was only about 1 pound per square inch.

Why would we expect to find a blast crater under the LM? Does a garden hose sprayed at high pressure into the dirt create a blast crater? It certainly blows away some of the surface dirt in a radial direction and will create a small depression or hole, but not a crater in the form that the hoax proponents suggest. There is even an earthly example of a rocket landing on dirt. The DC-X was a test flight program of a vertical takeoff and landing rocket. On one of its last flights, it made an emergency landing outside of the pad area. Despite the hydrogen/oxygen engine producing a thrust of some 60,000 pounds, the engine produced a mark on the desert floor that was barely recognizable.

No dust on the LM footpads

FC: During the show, Bill Kaysing cited the lack of dust on the LM footpads as evidence of fakery.

SCC: He should consider the high velocity of the dust blown away by the descent engine. That dust flew far away from the Lander and very little of it settled near the LM itself. Consider the flight of a dust particle blown off at an initial velocity of 100 meters/second (a little over 220 miles per hour) and at an angle above the horizon of 10 degrees. Its horizontal initial velocity is 92.5 meters/second while its upward initial velocity is 17.4 meters/sec. In the atmo-sphere-less 1/6 lunar gravity, it would fly upward for 10.6 seconds before reaching its maximum height of 92 meters above the lunar surface. About 10.6 seconds later, it impacts on the lunar surface almost 2.1 kilometers away from the Lunar Module!

Lack of sound from the LM descent engine

FC: Bill Kaysing further claimed that you should hear the sound of the descent engine in the audio from the landings.

SCC: There are several obvious problems with this hypothesis. First, the engine is many feet away in a vacuum so that the sounds would have to be transmitted through the spacecraft structure itself. Second, the microphones used are insulated inside of the spacesuits worn by the astronauts. Third, the microphones are worn next to the astronaut's mouth and are designed only to pick up sound from its immediate vicinity.

Footprints

FC: Footprints appear around the LM despite the rocket engine scouring the lunar surface during landing.

SCC: This question is related closely to the question regarding the thrust level of the LM descent engine. The hoax proponents exaggerate the thrust level of the engine during landing, claiming that it fires with a thrust of 10,000 pounds when in fact, the engine was throttled and only had to fire at a thrust level that nearly balanced the 1/6 gravity weight of the LM at the moment of touchdown. Dust was blown away, but the regolith on the lunar surface was found to be many meters thick while the engine would have blown perhaps a few centimeters of dust away from the area immediately under the engine at the moment of landing.

Pictures of space-suited crewmembers inside a building

FC: There are pictures of crewmen with background walls, overhead lights, hoses, tiled floors, etc. This is evidence for the hoax.

SCC: These photographs are common and were obtained during crew training for the actual flights. No attempt is made by NASA to claim that these images were taken on the Moon. And the LM, Rovers, experiments, etc., are all replicas used for training or flight spares, rarely actual flight hardware. Some flight hardware also appears, often while it is being stowed for flight or when it is being fitted to crewmen. That the hoax proponents claim that these training photos are evidence of a hoax shows just how little research the hoax believers actually have done on how NASA actually carried out the Apollo program.

The "Protoype LM" accident

FC: It was claimed in the show that a "Prototype LM" was tested on Earth by Neil Armstrong and that during one pre-mission test flight, Armstrong was unable to control the vehicle and had to eject!

SCC: The vehicle in question was strictly a training vehicle with a jet engine to simulate the 1/6 lunar gravity and to simulate the thrust of the LM descent engine. The Apollo astronauts used the LLTV (Lunar Landing Training Vehicle) to learn how to maneuver the actual LM. They also found flying helicopters to be a useful analog to flying the LM.

Again, a small amount of research would have shown the producers of the Fox program that the LLTV was not a prototype of the LM, but instead a training device.

Fox continues by asking how the "untested" LM could land flawlessly 6 times when the "prototype" had so much trouble on Earth. The LLTV was very different from the LM, not a prototype and the "untested" LM was far from untested. Every component of the LM was tested over and over again during the development of the LM. The descent and ascent engines were perfected through a test-firing program carried out at a NASA White Sands test facility. Other components, like the landing gear, were tested under simulated load conditions. The LM flew flawlessly to the Moon because of the hard work of thousands of workers over many years during the design, development and construction of the spacecraft.

[*A.N.*It also has to be remembered that whether the "LM" used in training was exactly like the ones used for the missions or not— and Scotti says not—it would still be *six times heavier* on Earth than on the Moon. It would also have to be flown in prevailing weather conditions, like in a very stiff breeze for example! There are not too many of *those* on the Moon! I mention this all too obvious but sometimes overlooked fact because it is important to get this right once and for all. The pro-hoax lobby have put forward one or two very good arguments—which the jury will either find have been refuted to their satisfaction (or at least put into proper perspective), or not—but *this* simply is not one of them. The Lunar Module was designed to fly and land *on the Moon*. Its capabilities and maneuverability in a weightless vacuum and within the Moon's gravitational pull were thoroughly tested by the Apollo 9 and Apollo 10 missions and found to be satisfactory. There was actually only *one way* to properly test the LM's ability to land on the Moon, and that was to actually land it on the Moon.

This, we understand, is precisely what highly experienced test pilot Neil Armstrong did for the very first time on July 20, 1969. And no matter what mishaps, misadventures and "near misses" may have occurred during training here on Earth—which is, after all, what training is all about—we are led to believe that this time, thank God, he managed to do it *without* crashing it!].

No rocket plume in the video of the ascent stage lift-off

FC: Kaysing claimed we should see a rocket plume from the engine of the LM ascent stage during lift-off video footage.

—Paused.

I need to interrupt my witness for a moment. If you recall, members of the jury, one prosecution witness actually claimed he thought the Apollo missions were fraudulent *because* he has a piece of footage showing a "rocket plume" emitting from one of the ascent stages blasting off from the Lunar Lander! So how inconsistent is that of those who believe there was a hoax? Let us get something else straight. There can be no such thing as fire, flames, or indeed "plumes of rocket fire" on our airless Moon. There can be no fire without oxygen and there is no oxygen on the Moon. But there can be heat, of course, and heat can emit *light!* But I will let the planetary scientist continue to explain it all properly.

Scotti continued—

SCC: The LM ascent stage engine was a hypergolic rocket engine, which burned fuels that burn on contact with each other, making them very reliable since they don't need an igniter. These fuels burn with only faintly visible exhaust plumes. If one looks up the engine bell, you would see probably a bright blue *light* in the combustion chamber, but the "plume" itself is nearly transparent. The Titan rockets, which launched the Gemini spacecraft, also used the same type of fuel. With very much larger thrust levels, this rocket produced a plume that was nowhere near as spectacular as the plumes we saw on the Saturn V rocket or on the spectacular Space Shuttle (which is dominated by the solid rocket boosters at lift-off). The faint plume that would probably be visible to the human eye if someone were there to see it is not obvious enough for the lower quality of the TV camera used to capture the images of the lunar lift-offs on Apollo 15, 16 and 17.

Flags waving in the breeze

FC: In the show, the pro-hoaxers ask: "How can a flag wave in the vacuum of the Moon?"

SCC: The flags only "wave in the breeze" as an astronaut touches and manipulates the flag and flagpole. Notice in each example of the flag waving [if you have access to original footage] the astronaut is still moving the flag or pole or has just finished adjusting the flag or pole. The flag wobbles for a moment as the force applied to the flag and pole damps out and then it comes to rest. There is a film from one of the lift-offs from the LM cabin which shows the flag waving in the breeze of the rocket exhaust as well (and perhaps you can see the flags move in the rocket exhaust from the Rover TV cameras, but those are far away and the cameras are trying to follow the ascent stage).

The flags also look as if they are waving in the breeze when *not* being adjusted or blown by the ascent engine thanks to a metal rod that runs along the top of the flag that holds it out as if being blown in the breeze. This is a well-documented piece of equipment.

Poor quality of video

FC: Claims were made that NASA purposely provided very poor video footage of the first moonwalks to disguise any flaws or "bloopers."

SCC: NASA didn't pay too much attention to using video cameras early in Apollo—the first such camera was carried aboard Apollo 7 almost as an afterthought, but the public ate it up, so they added it to later flights. The camera used on Apollo 11 was a black and white camera. Later missions used better cameras, but the portable video cameras of the day tended to be bulky and power hogs. Weight and power were at a premium on the lunar surface.

At the double!

FC: Double the speed of lunar video and it looks as if it was filmed on Earth.

SCC: Well, sure, that looks quite good, but does it really work? It turns out that you can't accurately simulate the lunar flight of objects in a vacuum on Earth without modern computer graphics techniques. If you shoot film on Earth and slow it down by a factor of two, the $1/6^{th}$ lunar gravity is not simulated properly. Imagine for example a particle of dust thrown at a 45 degree angle off the Rover tires at the speed the Rover was traveling, say about 10 kilometers per hour (I haven't actually measured the ejection direction of dust off the tires, but this is a good first approximation to estimate the height of the Rover's rooster tails of dust). 10 kph is 6.2 mph or 9.1 ft/sec. Thrown at a 45 degree angle, the upward velocity is then 6.4 ft/sec (as is its horizontal velocity). In the 1/6 lunar gravity, it should then fly upward for 1.2 seconds to a height of 3.8

feet. It would fly outward for twice this time before landing back on the surface about 15.5 feet from its launch site. In the case of a simulated film as the hoax proponents suggest running at half speed, the same film would have the Rover traveling twice as fast on Earth with full Earth gravity in effect. So the initial launch velocity of the dust would be 12.8 ft/sec. It would fly upward for 0.4 seconds in Earth gravity (or 0.8 seconds in the slowed video) and reach a height of 2.6 feet, landing some 10.2 feet away. In other words, you can tell the difference if you actually measure the speed of the dust or thrown object.

How can an astronaut [or indeed an actor!] on Earth wearing a bulky fake moonsuit, run at the high speeds you would need them to run in order to simulate the lunar imagery? The highest documented running speeds on the Moon were about 5.4 km/hour. That's a rate of about a 9-minute mile in a bulky suit. I have never seen any astronauts trying to run in their suits in straight 1-g [Earth] conditions.

—Paused.

Prosecution: Objection! That last statement by the witness is misleading. The jury should be reminded that "a bulky fake moonsuit"—to use the exact words of the witness himself—would essentially be just that: Fake. It could have been designed to look rigid, bulky and heavy when in fact of lightweight, pliable construction enabling much greater freedom of movement.

Judge: A worthwhile observation, I think. Objection sustained. The jury will disregard the remark made by the current witness concerning any ability to run on Earth in what might correctly be termed "a bulky spacesuit." The witness may continue.

Practice makes perfect?

FC: There is great difficulty using cameras in spacesuits. How come the photographs were so "absolutely perfect"?

SCC: 3 words: Practice, practice, practice. The Apollo astronauts trained over and over again on Earth before their flights. By the time they flew, they had shot hundreds of practice pictures, learning how to line up photos without a viewfinder. They also had pre-set exposures and focus positions easily settable with their bulky gloves for different types of shots. Also, they used fairly high f-stops on their camera lens to maximize the depth of field, so focus wasn't as critical when setting up a shot. Given all that, the photographs are far from perfect. Exposures are uneven, mostly due to lighting issues and there are plenty of badly composed and out of focus images amongst the thousands obtained. The pictures we see are often the best of the lot—that's why they were picked for public release. An examination of the images in the *Apollo Lunar Surface Journal* will show you what most tended to look like.

Wrong way

FC: Shadows go in wrong directions, not parallel.

SCC: The hoax proponents apparently don't understand simple convergence—the disappearing point which elementary school art students learn about in order to draw roads or railroad tracks disappearing into the distance. The shadows, though parallel from overhead, look to be going in different directions from the perspective of a person on the ground. You can see the same effects here on Earth. The most used image by the pro-hoaxers is from Apollo 14 where the Lunar Module appears in the distance to be casting a horizontal shadow while the shadows of the rocks in the foreground are angled towards the camera [Picture D on Plate 2]. However, if you look closely at the LM shadow and the LM itself, you'll see the LM partly lit—similarly to the rocks in the foreground along the same direction and you can see that the shadow is not horizontal, but is greatly foreshortened. As usual, just a casual examination of the evidence contradicts the pro-hoaxers' argument.

The lunar surface is also very undulating with hills and craters in great abundance. Shadows appear longer if they go down a slope on the sunward side of a crater or hill or appear shorter on a slope that faces into the sun. Hills and craters can also change the apparent direction of a shadow to make it look non-parallel with adjacent shadows.

A matter of detail

FC: You can see details in the shadows

SCC: Although the sun is the primary light source on the Moon (the Earth also lights the Moon somewhat since it is about 13 times the surface area as viewed from the Moon compared to the Moon from Earth, but also is 5 times brighter per unit area), the shadows are lit up by scattered sunlight off of the surrounding lunar surface and equipment. You don't need an atmosphere to fill in the shadows on the Moon. Any photographer knows how to use objects to reflect light into the shadows to make the shadowed areas more visible. Photographers use efficient items like reflecting umbrellas or white boards. On the Moon, the surrounding lunar surface, mountains, astronauts, Rover, and even the LM itself scatter light into the shadowed areas of the Moon and equipment. How come neither of the two photographic "experts" used by the Fox show to examine apparent photographic anomalies were able to think of this obvious solution to this shadow "problem"? [*A.N.* Because, in all probability, they had already closed their minds to any possibility that the Apollo landings might be genuine, I'm afraid, Jim].

Identical

FC: There are identical backgrounds in photographs from different places

SCC: The hoax proponents cite cases of finding the same exact background mountains in images taken from vastly different places around a landing site. In the case of the Fox special, they showed a picture of the LM with mountains in the background and a second image without the LM with the "exact" same background mountains. The mountains in question are several miles away from the LM. Two pictures taken a few hundred feet apart can have a vastly different foreground (like LM and no LM) while having what appears to be exactly the same background. There's no mystery to that. Tucson has mountains surrounding the city and amazingly, from the University of Arizona Campus area, one can travel from one end of campus to the other and find not only very different buildings in front of you but what appears to be the exact same mountain backdrop in the background. Apparently, the mountains around Tucson are a giant background painting if one follows the logic used by the hoax proponents.

Location, Location

FC: Location on video identical on two days at two different sites

SCC: If one watches the actual video from the Lunar Rover's TV camera, you'll find that the two identical images were obtained within minutes of each other from the exact same site. A small amount of investigation easily solves the mystery. The source of this error is a NASA documentary film in which the film editor mistakenly claims the two film clips were taken on different days at different sites when in fact they were not.

Reseau marks

FC: Cross-hairs [Reseau marks] disappear behind equipment.

SCC: Film, like most all imaging media, is not a perfect recorder. When a very bright part of the subject appears next to a dark area, there is often a saturation effect, which appears as bleeding into the darker areas, such as the Reseau marks. It is a well-known photographic effect, which the two Fox photographic "experts" both conveniently "forget" to suggest.

Van Allen Belts

FC: Radiation was too high—especially regarding Van Allen belts and solar storms.

SCC: The hoax proponents consistently exaggerate the effects of radiation in space. Radiation was a definite concern for NASA before the first spaceflights and they invested a great deal of research into it before flying the first astronauts into space. The most dangerous part of the journey to the Moon for radiation exposure was during the passage of the spacecraft through the Van Allen belts. This is a zone from about 1000 kilometers up to about 20,000 kilometers. The Apollo missions flew through this zone at very high speed—outbound starting around 40,000 kph and inbound at about the same speed. They only spent a few hours within the Van Allen belts and estimates of the total exposure during their passage is about 2 rems, which is the equivalent of about 100 chest X-rays or about 40% of the maximum permissible dose of radiation according to OSHA standards.

—Paused.

Prosecution: Objection! Is the witness not being far too glib and dismissive of the dangers posed by submitting oneself to a 100 chest X-rays? I put it to the jury now. Would you, those of you of sound mind that is, agree to receiving a 100—I repeat again, *100*—chest X-rays all in one go, one after the other, even if necessary to detect a suspected severe illness, let alone volunteer yourself to undergo such a process in the cause of scientific endeavor?

Judge: Objection overruled. I understand the witness to be including both outward and return journeys through the high-intensity

Van Allen belts when arriving at the figure of one hundred. Therefore the figure, when applied to radiation dosages, was *not* "all in one go," to use prosecuting counsel's own rather glib turn of phrase. The witness confirmed, in the same sentence, that this amount of exposure still falls well below the maximum permissible industrial dosage allowed under the laws of the United States. If the witness *is* being somewhat...well, yes..."glib" regarding any possible damage to a human body sustainable by absorbing such an amount, then I can only presume he personally feels that the amount of such damage, if any, would be negligible.

Prosecution: If it is not being too impertinent, may I ask the learned judge a personal question?

Judge: Providing the question is itself not impertinent—and *too* personal—by all means.

Prosecution: In view of the interpretation placed by the learned judge upon what the witness is saying, would the learned judge himself be prepared to subject his body to a 100 chest X-rays in the proportions described? That is, fifty on one day and a further fifty, say, one week later?

Judge: Er...my word! Oh dear, now *that* has placed me on the spot! Let me think for a moment...No.

Prosecution: Supposing, after this trial, NASA offered you *one million dollars* to do so. Would you then?

Judge: No. Absolutely not. One's health, especially at my time of life, is far more important than large sums of money. Even if a doctor I trusted implicitly were to explain that he really needed to do that amount of X-rays in order to quickly discover the cause of a current health problem, I would still decline and opt for the chance that the problem might still be diagnosed by using an alternative, if slower, way.

Prosecution: So, in the learned judge's considered opinion then, why did *these* men volunteer to subject *their* bodies to a process that *you* would avoid even if it was to be used diagnostically for your possible benefit, as well as you then receiving $1,000,000 into the bargain?

Judge: I have no idea. I suggest you put that question directly to learned counsel for the defense.

Prosecution: Very well. I ask of my learned friend if *he* has any idea what would possess young men in the very prime of life, perhaps with young families and everything to live for, to subject themselves to one hundred chest X-rays—in two lots of fifty—as patients in hospital, let alone *volunteer* for the same dosage to inevitably ravage their bodies on an adventurous but essentially unnecessary journey to the Moon and back? A journey so perilous and fraught with danger already, that if any single one of many other thousands of things that could have gone wrong had done so, then overdosing on radiation might have been the least of their worries.

Defense: To answer my learned friend I immediately throw back at him that same observation. Because overdosing on radiation *was* the least of their worries, that's why. At the same time, one might as well ask why thousands of young men who once upon a time also volunteered, crouched for months on end in stinking trenches full of mud, blood and excrement, with bullets whistling overhead and shells bursting all around them? Most of *them* had young families back home too. Just like those aforementioned selfless young men, the astronauts were also trained to do a job; a job they were prepared to do for the honor of their country and for the entire free world. My learned friend makes the all too familiar mistake of presuming that these men were just like him, or me, or—with respect—the learned judge. They were not. These were incredibly brave—indeed heroic—men of precisely the same ilk as those in those trenches as they prepared to "go over the top" to face the raking machine gun fire and mustard gas—not radiation—yet again. Such men do not think of self. They think only of the common good of all mankind and are prepared to sacrifice themselves totally to that end. I tend to think that is just about as far from being just like my learned friend as it is possible to get!

Prosecution: Indeed, it is. And my learned friend should know!

Judge: Now, now. Stop this bickering in court. You have both made your respective points very well. Learned counsel for the People raised a very interesting question. With regard to it he first gave *his* opinion, then asked me for mine, and then defense counsel for his. But the fact is that the witness is also entitled to *his* opinion, and I ask him to continue giving it, glibly or otherwise.

Prosecution: As the learned judge pleases.

Scotti continued—

Doses of 100-200 rems cause a person to experience nausea several hours after exposure and fatal doses occur above about 300 rems. Solar flares were a concern as well, but typical doses due to flares that the astronauts were exposed to were only a few rems. The crews wore dosimeters which were read back roughly daily during the flights.

Too hot?

FC: Temperatures on the lunar surface were too hot.

SCC: Daytime [remember, a lunar "day" is two Earth weeks] temperatures reach about 250 degrees F. Night time temperatures sink to a chilly—270F. The landings occurred within a day or two of local sunrise so that the sun angles were low and the surface had not heated up to its full daytime levels. With no atmosphere, convection does not transport heat from object to object. Conduction of heat occurs only when a hot surface is contacted and thermal radiation is the only other source of heat. Film in a camera is protected from direct sunlight except during exposures and a light colored or silver camera [because it reflects light] does not absorb heat efficiently. The lunar EVA suits were designed to withstand temperatures of +250F.

And then it all came to an end

FC: No plans to return to the Moon and the Russians have never sent anyone.

SCC: Despite the apparent ease with which NASA landed 12 men on the lunar surface between 1969 and 1972, traveling to the Moon was difficult, dangerous and enormously expensive. The advanced planning and preparation of the spacecraft and crews resulted in spectacularly successful missions, which succeeded despite the dangers and the inherent malfunctions of manmade equipment. The United States landed men on the Moon while the Soviet Union failed in its attempts to build a lunar program despite its hard work. Once the U.S. succeeded, the Soviets primary reason for going to the Moon was eliminated and residual work dwindled. Despite the official word of the Soviet Union claiming that they were never in a race to the Moon, the post-Soviet Union evidence demonstrates otherwise with lunar landing hardware and the huge N1 booster program as well as training programs for its cosmonauts. To fly to the Moon today would be nearly as difficult and probably more expensive (even accounting for inflation) than it was in the 1960s. Until there is enough motivation to do so, we are unlikely to mount any new missions to the Moon in the near future.

Where did all the equipment go?

FC: Earth based telescopes should be able to see the Apollo equipment. They can't.

SCC: A telescope's diffraction limited resolving power depends linearly on the aperture of the telescope. Ground based telescopes also have to look through the murky and turbulent atmosphere, so without corrective techniques that are just now becoming common in large telescopes (called adaptive optics), a telescope's resolution is limited by the atmosphere to about 0.5-1.0 arcseconds (3600 arcseconds are in one degree and 360 degrees around the whole sky). That limits ground-based telescopes to a resolution of about 2 kilometers on the Moon. From space, a telescope is limited by its diffraction limited resolution. For the Hubble Space Telescope, that is a little less than 0.05 arcseconds or about 90 meters at the distance of the Moon. To resolve the LM descent stage which is about 10 meters across, one would need to have a resolution better than 10 meters, perhaps 2-3 meters which means we need a telescope some 30 times larger than the HST in orbit around the Earth to resolve the largest equipment left on the Moon.

—End.

§ § § § §

My thanks to Jim Scotti.

So there we are. Despite my learned friend's objection to the glibness of certain remarks made by the planetary scientist during his testimony, I again prefer to use the adjective *self- assured* to describe his delivery throughout. Jim Scotti appeared extremely confident that whatever he was telling us was not only of a highly informed nature, but also correct.

And, speaking of equipment...

One of the facets of the Moon landings program conveniently overlooked by the pro-hoax lobby is the fact that every Apollo landing mission placed equipment on the lunar surface—presently visible or not—which still sends valuable data back from the Moon. One such example are the ALSEP (Apollo Lunar Surface Experiment Package) radar reflectors which NASA maintains had to be placed in pretty precise positions or the experiments wouldn't have worked. The McDonald Observatory Laser Ranging Station near Fort Davis in Texas, USA, regularly sends a laser beam through an optical telescope to try to hit one of the reflectors. The object of this exercise is to always be able to tell the mean speed, and distance from the Earth, of the Moon at any given time.

A recent TV show about the Apollo landings featured a spot debunking the hoax theorists. In it, a spokesperson for the McDonald Observatory said, "We do this sort of thing all the time using the reflectors left by the Apollo crews. These ignorant people keep harping on about man not going to the Moon but nobody thought to ask us."

Prosecution: Objection! My research before I undertook to prosecute this case on behalf of the People has left me far from ignorant on this subject, and has given me to understand that all such pieces of equipment could have been placed there robotically.

Judge: That is my understanding too. Objection sustained.

ALSEP radar reflectors still send information about the Moon back to Earth.

Defense: Fair enough. But if that last little lot failed to take the wind from my learned friend's sails, then he should try this:

Tracing a Hoax?

On this same show a spokesperson for the world famous Jodrell Bank Radio Observatory in Cheshire, U.K., which uses the world's biggest radio telescope to "listen" to the stars, held up to the camera a graph printed out by the observatory's radar tracking system in July 1969 of the Eagle's final descent to the surface of the Moon. "We tracked its computerized course all the way down after it separated from the Columbia," the spokesperson said. "You can clearly see here..." he indicated a series of bumps in the otherwise straight line of the Eagle's electronically traced path, "...where Neil Armstrong took over manual control of the LM to steer it further along from the landing site originally selected by the computer to an area in the Sea of Tranquility where the ground was considerably less rocky."

Yes, indeed, I still recall—although not word for word—Armstrong's frantic cry, which went something like:

"Down...down...No, no! Boulders! Big boulders! Up, up, up!"

Which means, members of the jury, that despite the fact that the prosecution continues to protest that only machinery ever went to the Moon, had not the LM been under manual control at the moment of its landing—that means under the physical control of a human being who was actually present in the spacecraft and able to assess the viability of the landing site for himself with his very own eyes—it would almost certainly have crashed.

Judge: Well, on the strength of that, I must confess to having had much neutral wind well and truly blown from my sails. I would enquire of learned counsel for the prosecution whether or not his are still billowing?

Prosecution: This was Apollo 11. I inform the learned judge that indeed my sails are still billowing by way of an appropriate eleven words. I will write them down for the benefit of the jury. Placed in a sentence together these eleven words look like this: *Remote control by a human being at Mission Control, Houston, Texas.*

16

2001: A Space Mythology

Oh dear. Some people just cannot take it on the chin, can they, members of the jury? I shall shortly be calling various witnesses who will give brief testimony each. The views expressed may give us an insight as to how these cynical little schools of thought come into being and eventually flourish and grow into fully-fledged "universities of doubt" challenging authorized or official versions of major historical events. The eventual stranglehold gained by the choking weeds of cynicism, though, can often be the result of what might originally have been a few comparatively innocuous seeds of suspicion having somehow fallen upon fertile ground, but which were never intended to suggest that the event in its entirety was fictitious, only that it maybe didn't quite happen as the version in the history books would have it. Which is fair enough.

After all, the Holy Bible is a history book, after a fashion. And although Darwin's theory may have occasioned many to doubt much of The Old Testament, the Gospels remain sacrosanct. The "gospel" according to NASA regarding whether or not men have landed on the Moon claims that they, NASA, did, not once, not twice, but six times in all, deposit human beings upon that planet's surface to the tune of twelve total, and then returned them alive. This great achievement is not ancient history, some of which might be questionable in modern times; it is recent history. The problem, though, is this: After the elapsing of such a long period, during which technology has advanced dramatically, the death toll in relation to space oriented activities has increased equally as dramatically, but without mankind gaining a single further foothold in space. Not one that would seem to be of any great consequence after that first momentous "giant leap" anyway. Even one further visible "small step," perhaps, would have been acceptable.

But that further small step actually has been taken. In fact, many, many more small steps have been taken, even if not overwhelmingly spectacular in the visual or mass participatory sense, as undoubtedly was that first sensational landing and stepping upon the surface of the Moon. But in terms of studying man's ability to survive and endure in a weightless environment for long periods of time, gigantic leaps rather than small steps have been made. And, of course, we must not forget, either, the probes being sent to other worlds to photograph and sniff out the chemistry of their atmospheres and environments. All this groundwork is absolutely essential before manned expeditions to such places can be even remotely considered.

But despite the absolute fortune being spent on all of the above, it hasn't prevented the hoax theorists from querying why NASA hasn't been back to the Moon in the meantime. Why can't they accept that my clients cannot afford—indeed, humanity cannot afford—to keep going to the Moon and back for no real and proper advantage? Did my learned friend not say it himself? "Adventurous but essentially unnecessary journeys" was how he described those manned expeditions to the Moon, the integrity of which he disputed. For God's sake, my clients have already been there and back six times! How many more billions of dollars need to be expended before the penny eventually drops? The pro-hoaxers themselves have admitted it.

There is nothing there.

No oxygen to breathe. No running water to drink. No life. There has probably never been any life on the Moon. That means no great carboniferous forests eventually fell and decayed to be crushed into coal under the tremendous weight of layer upon layer of rock; no bodies of tiny sea creatures became buried in their countless billions of trillions under sediment, which the passing eons petrified into heavy layers of rock to crush those innumerable tiny corpses into vast underground oceans of oil.

And, believe it or not, the very last things we need to discover are huge veins of gold in the Hadley Mountains or for the Sea of Tranquility to turn out to be a great field full of diamonds. Such discoveries would destroy the economy upon which our whole financial and business world is structured. The more gold and diamonds there are, the less they are both worth as commodities.

All the current signs are that any such vast underground resources of coal and oil that would see us well into the future, if they be accessible anywhere else in the solar system at all, might well be found on the planet Mars. But the Moon is a stone's throw away compared to Mars. Mars is almost as far away from Earth as our planet is from the sun. We are talking something within the region of around ninety million miles, although Mars' orbit around the sun does bring it closer every so often. To journey to Mars and back, employing technology currently available, would take around three years. The Moon, on the other hand, is only a quarter of a million miles away and entails a return journey of about a week, depending upon how long one stays. So, with no pressing or viable economic reason to return to the Moon, is it any great wonder that mankind—in the form of NASA—should decide to spend just a little more time thinking about, and planning, that next giant leap? And, as much of such thinking and planning consists mostly of the kind of technicalities that seem to have already bored the pants off my learned friend—and no doubt the jury, too—is it any great wonder, either, that much of this takes place behind the scenes, as it were? When I say "a little amount of time," I am of course speaking in relative terms. I don't just mean the past thirty-five years. A whole century, at least, would make much more sense.

Precisely what do they not believe?

The Moon hoax proponents need to be challenged regarding precisely *what* it is they do not believe about the Moon landings? A nonbeliever in Christ can be tackled directly by asking him or her for specifics. What is it about Christ that you do not believe? Do you not believe that He existed at all? One might get the response "Oh, He existed, all right—but I don't believe He was the Son of God, *or* that He managed to feed all 5,000 people at a rally with a few loaves and a few fish!" Well, okay. Perhaps it was a few *hundred* loaves and few *hundred* fish and something was lost in the translation after two millennia. So maybe it wasn't exactly a miracle; but it was still a tremendous feat of organization nonetheless. But the mechanics of the situation are not the point. The point is that something of great consequence *must* have happened. Why? Because it seems that five thousand people *witnessed* it, four of Christ's contemporaries *wrote* about it, and *two thousand years later,* millions of us are still *talking* about it!

What exactly is the pro-hoax lobby claiming? That only Apollo 11 was a sham? And if so, designed to fulfill murdered President John F. Kennedy's vision of landing a human being on the Moon by the end of the sixties decade? Or are they suggesting that *all six* of the successful Moon landing missions were faked? Because if they are, they may as well suggest that all the alleged manned missions right from Apollo 7 were largely acted out and filmed in a studio, too, because *all* missions, with the exceptions of Apollos 7 and 9, would have involved sending three living, breathing human beings *through*, *beyond* and then *back through* the Earth's protective, but highly dangerous, Van Allen shields. So, in that case, why not fake Apollos 7 and 9 too, even if they were only to take place on the user-friendly side of the radiation belts? To have not done so would have meant those two "genuine" missions costing much more than all the other nine "faked" missions put together!

The main point I now make on behalf of my clients, members of the jury, is this: Why fake six missions, thereby making the risk of being found out six times greater, when surely just the one faked mission, possibly two at most, would have sufficed? Why two? Well, one for a victory and two for total triumph.Just to rub it in.

If NASA had decided that enough was enough after Apollo 12, there would have been no need to fake the near-tragedy of Apollo 13 or follow it up with another four faked "successful" missions. Not only that, but billions upon billions of dollars would have been saved. Making movies is an extremely expensive business—although admittedly not as expensive as genuinely putting men on the Moon—and it has to be said that by Apollo 15, or *Man on the Moon Part IV,* NASA could be said to have already made one box-office disaster too many anyway.

Against this background of pure logic, the argument of the pro-hoaxers and the ridiculous charges made against my clients, somehow do not hold water, do they? Just like the Moon itself, according to these same theorists. NASA's considered belief, though, is that the Moon *may* contain some water—in the form of ice, of course—and therefore the potential for possessing both oxygen and moisture. The pro-hoax lobby insists that there is absolutely no H2O on the Moon whatsoever. Not in the air, they say, because there is no atmosphere, not even one so thin as to maybe register to the most infinitesimal degree, or in the soil in the form of tiny particles of ice. They claim that this is why it is impossible to leave a footprint on the Moon.

If the learned judge pleases, I invite them to double-check their all too sweeping conclusions regarding the physics and chemistry of the Moon and to ultimately present their findings to this court after nothing less than an extensive period of research. Hopefully, like NASA, this research will be carried out with the very best equipment available at a cost of trillions of dollars, and I shall expect it to be backed up by the opinions and views of hundreds of qualified scientists and astronomers. And while they are at it, one can only trust that these same pro-hoax theorists will be good enough to provide the court with details of their own individual scientific credentials and qualifications. And I do not mean the kind that one awards oneself once one has availed oneself of a $200 pair of binoculars, a set of Wolfe jars and a Bunsen burner and sets these items up in one's outhouse or backyard shed. I mean ones that would give them the right not only to forward their theories, but also expect them to be taken seriously. Not just by me or by this court, but by anybody.

Meaning absolutely no disrespect to the profession itself, people normally employed as door-to-door insurance salesmen and whose knowledge of space revolves around an ability to build model rockets in their spare time, just will not do. With that, I shall call my next witness for the defense.

I call James Oberg.

JAMES OBERG'S TESTIMONY

Source: *Wired* Magazine Issue 2.09—Sep 1994. Extract from article by Rogier van Bakel entitled ***The Wrong Stuff***. © 1994-2002 Wired Digital, Inc. All rights reserved.

Begins—

Poll or no poll, even James Oberg [anti-hoax], a nemesis of Kaysing [pro-hoax], conservatively estimates that the disbelievers may number between 10 and 25 million Americans.

Oberg works for NASA contractor Rockwell International as a space-flight operations engineer with the space shuttle program. He writes as a second profession, covering all aspects of space activity, with a special interest in space folklore.

"Myths have a way of blossoming in the fertile soil of scientific discovery," Oberg notes. "Every age of exploration is the same in that respect—from the time of the Phoenicians to Marco Polo, and including mermaids and unipeds and all these mythological creatures that lurk at the edge of our exploration.

"To me, it's extremely humanizing to have this typically human reaction—this denial, this myth making—to our lunar adventure. I'm not at all surprised that these stories or interpretations exist. Actually, I'm surprised they aren't more widespread."

Nonetheless, hoax believers can be found in many parts of society, here and abroad. According to Oberg, Cuban children are officially taught that "Yankee" space technology failed miserably and that NASA was reduced to pitifully faking every single lunar landing. Some New Agers also contest the possibility of the Moon landings, as do the Hare Krishnas. Non-mainstream Christians at the Flat Earth Society—a Lancaster, California-based anti-science group of about 3,500 members—contest the entire field of astronomy (not to mention the Moon landings). They liken the towering launch pads [at Cape Canaveral] to the Tower of Babel [in the Old Testament].

The eccentricity of such convictions certainly intrigues Oberg. "I respect these people's dedication to their view of the world. One reason they fascinate me is that they're a constant reminder to me that we can't rest on "common knowledge"; we can't be complacent with our traditional interpretations of things—even though these interpretations are almost always right. But I also find their pathology of reasoning, or *non-reasoning*, compelling. We define health by the boundaries of pathology, and I try and define rational thought by looking at cases that go over the edge."

That's damning praise indeed. So it's no surprise that Bill Kaysing doesn't much care for James Oberg, whom he dismisses as "a NASA agent."

—End.

§ § § § §

My thanks to Mr. Oberg. I call Linda Degh.

LINDA DEGH'S TESTIMONY

Extract from ditto source. Degh is a retired lecturer on U.S. folklore who taught at Indiana University in Bloomington, and has recently published a book entitled *American Folklore and the Mass Media*.

Begins—

Linda Degh is reminded of the film *Capricorn One*. Released in 1978, Capricorn One tells the story of a staged flight to Mars. The astronauts grapple with the moral implications of the giant charade and fear they might be killed to keep them from blowing the whistle. Sure enough, they find themselves hunted down by bloodthirsty government thugs; only one of the astronauts makes it to freedom and reporters' microphones. Degh recalls that it was "quite a slanderous movie, pretending that the government had been killing people," and she believes that it must have given a powerful boost to the Moon landing hoax theory. "The mass media catapult these half-truths into a kind of twilight zone where people can make their guesses sound as truths. Mass media have a terrible impact on people who lack guidance."

—End.

§ § § § §

My thanks to Ms. Degh. I call film director Peter Hyams.

PETER HYAMS' TESTIMONY

Extract from ditto source. Hyams was the director of the film *Capricorn One*.

Begins—

Peter Hyams, *Capricorn One*'s director, agrees that mass media can be very powerful—dangerously so, in fact. "My parents believed that if it was in *The New York Times*, it was true. I was part of the generation that grew up believing that if we saw it on television, it was true. And I learned how inaccurate newspapers were, and I realized that TV is just as inaccurate, or it can be. So I said, wouldn't it be interesting if you took a major event where the only source that people have is a television screen, and you showed how easy it would be to manipulate everybody." [*A.N.* Methinks George Orwell already succeeded in doing this with *1984*]. Hyams insists that he made *Capricorn One* "for entertainment, for fun," not because he was making not-so-veiled references to the alleged Apollo hoax. "I was aware that there were people who believed that we never walked on the Moon, but I never read their books or consulted with them. And frankly, I think they are being totally ludicrous." (Nevertheless, an invitation to a sneak preview screening at the time of *Capricorn One*'s release said: "Would you be shocked to find out that the greatest moment of our recent history may not have happened at all?").

—End.

§ § § § §

I thank the renowned film director. I call special effects expert Denis Muren.

DENIS MUREN'S TESTIMONY

Extract from ditto source.

Begins—

"The thing is, though, it [the Moon] wouldn't have looked the way it did. I've always been acutely aware of what's fake and what's real, and the Moon landings were definitely real," Muren stipulates. "Look at *2001* or *Destination Moon* or *Capricorn One* or any other space movie: everybody was wrong. That wasn't the way the Moon looked at all. There was an unusual sheen to the images from the Moon, in the way that the light reflected in the camera, that was literally 'out of this world.' Nobody could have faked that."

—End.

§ § § § §

I thank Denis Muren.

Members of the jury, duly note the final comment of the *visual effects expert* at Industrial Light & Magic, a division of Lucas Digital. Quite unequivocally, he said:

Nobody could have faked that.

The Defense

17

The Reluctant Hero

Much has been made, members of the jury, of astronaut Neil Armstrong's apparent reluctance to speak of his tremendous and heroic achievement in having been the very first human being to set foot on alien soil. Indeed, the reclusive stance he appears to have taken since Apollo has unsettled some of his staunchest admirers and the firmest of believers in Apollo, and has also given the pro-hoax lobby even more ammunition with which to label virtually the whole manned space program an elaborate confidence trick.

But Armstrong had a partner on the Moon who joined him on the surface of that other world within minutes. Admittedly, Buzz Aldrin didn't have the honor of being first, but even to be second to step into the limelight—or even the earthlight!—of man's greatest all-time achievement, would certainly do *most* of us, I think.

Aldrin, even on an off day—and we are all entitled to one of those now and again!—unlike Armstrong even on a good day apparently, is the most amiable and approachable of men. He has spent a huge amount of his time over the years since Apollo—in between acting as president of three Los Angeles-based companies and as chairman of the National Space Society—attending and opening functions, guest appearing on TV and radio shows, making speeches and presenting awards. He is the sort of affable, unassuming guy who seems just as at home opening a school fete as he is standing beside the President of the United States shaking hands with important people from all over the world—or indeed cavorting about on the Moon!

True, there *is* a story about him retreating from a banquet, at which he was guest of honor, with tears in his eyes after being asked to describe how it felt to be the second man to set foot on the Moon. But every time one reads an account of this event one gets a slightly different version. But no matter what version one believes, it makes not a scrap of difference either way. There is absolutely nothing to suggest that Aldrin was overcome by anything other than good, old-fashioned emotion and was merely embarrassed that the other guests at the banquet might think him not quite so tough were they to see him sobbing like a baby. He should have realized that for most of us "mere mortals," to witness such a display of normal human frailty from a revered, almost *super*human figure in our estimation, could only serve to enhance the image of this extraordinarily brave man that most of us already have. Second to step onto the Moon maybe. As a hero second to none.

But returning to Armstrong. The prosecution's star "character assassination" witness was apparently a NASA worker himself, and in no way a subscriber to the "It was all a hoax" philosophy. Armstrong has always been an intensely private man, as attested to by his own astronaut colleagues; he is also a very modest, unassuming and self-effacing man, too.

In the short and admittedly extremely rare interviews he has given to various sections of the media, Neil Armstrong has always played down his own role in that first momentous Moon landing. Deferentially, he explains that all he—and Buzz of course—had to do, was to carry out certain routines at certain times that they had already rehearsed hundreds of times before. He always claims that all the real credit be given to the scientists and designers of the technology and the guys and girls who made it all work, as well as plotting all the trajectories and courses with brilliant mathematical precision. He meant, of course, all the people, both on camera and behind the scenes, at Mission Control in Houston, Texas.

And he is correct, of course. Up to a point. As an analogy we could use the Channel Tunnel Rail Link that now joins Britain and France. It was looked upon at the time of its official opening in 1994 as a tremendous feat of engineering; the bringing to fruition of an original idea dreamed of centuries ago. Indeed, there had been several previous sets of plans for a tunnel under the English Channel drawn up, and even one or two eventually aborted attempts to dig such a tunnel, including one during Napoleonic times. Now, though, at long last, Britain once again had some kind of land link to the continent of Europe for the very first time since the last Ice Age 8,000 years ago!

In 1990, the all-important breakthrough of actually joining two tunnels—one being dug from France and the other from England—being excavated under the English Channel, was accomplished. The

NASA's "backroom" people at Mission Control were the real heroes, Armstrong insists.

circumferences of the two tunnels met to within centimeters. It was an engineering triumph. A modern miracle achieved by the combination of the unerring mathematical precision of computers, gigantic excavating machines and a lot of sheer hard graft. But despite the many thousands of English and French workers involved—and those of many other nationalities, too—the honor of drilling a hole through the remaining thin wall of rock that still technically—and physically— separated Britain from France as it had done for the past 8,000 years, fell to one man. Members of the jury, can any of you remember his name?

No?

Now, why doesn't that surprise me as much as it should? Anyway, I will tell you who he was, even though *I* only know because I have had both the time and the need to research the fact. He was a Frenchman, and his name was Philippe Cozette. Like the names of Armstrong and Aldrin, the name of M. Philippe Cozette will forever live on in the history books. Unlike those of the two first moonwalkers, though, his never has been, nor ever will be, a household name. Not even in France!

One cannot but get the impression that Neil Armstrong not only strongly identifies with M. Cozette, but often profoundly wishes that if only he, too, like that Frenchman, could have "done his little bit" for history, got his name in the record books, so to speak, and then been allowed to return with impunity to the obscurity he obviously prefers. His public statements leave few in doubt that he considers himself to be nothing more than a nerdy engineer. Someone who, at the time, had merely been required to press the occasional button or wield the occasional tool, or to pull a lever here and there, although in his case at the direction of Mission Control, not Tunnel Control.

No doubt Cozette would agree with Armstrong. He would certainly agree that all due accolades for the overwhelming technological triumph of the Tunnel belonged rightly to the utterly brilliant men and women who had worked late into the night for many years on both sides of the English Channel. Those who had not only designed every inch of that tunnel but also the machinery to excavate it as well. Just how many times had they been required to frantically *redesign* either or both whenever an unforeseen problem had been encountered? Like Armstrong, Cozette would consider himself to be no more than the man who had been selected to take that final cosmetic step purely for the sake of the cameras, on what had been a long and difficult journey. A final gesture for the film, photographic and historical record to show that it had all worked, that it had all been worthwhile. Just like Armstrong's "one small step" had done for Apollo.

Apollo 11, like all other great engineering projects down the ages, had to be meticulously planned down to the most minute of details. Even the design of a nut or bolt and the type of metal it was to be made from would have been of the very utmost importance.

Members of the jury, simply because Neil Armstrong doesn't particularly enjoy signing autographs does not have to mean he feels his celebrity status is undeserved. It does not have to mean he was part of a great conspiracy. It could just as easily mean that he genuinely feels that he was a relatively insignificant little cog—albeit the most visible one—in an enormous triumph. A momentous endeavor, the real praise for which belonged to a group of the most brilliant and gifted scientists, technicians and engineers that humanity has ever assembled together in one place for one purpose. To Armstrong's way of thinking, any one of these people was far more deserving of honor, accolade and praise than himself, Aldrin, Collins and all the other astronauts that followed them to the Moon put together. As I said earlier, he is right.But again as I said earlier, only up to a point.

Beyond that point, though, he could not be more wrong.

Without heroes like him and Aldrin, who down the ages have been prepared to risk life and limb attempting to fly "on a wing and a prayer" all manner of ungainly-looking contraptions—the products of the "awesome minds" of their time—all the candles burned at both ends and all the barrels of midnight oil would have been wasted. And those heroes of days gone by at least *knew* what they were landing— or crashing—upon once the flights of the aerial devices they were testing were over. But Armstrong and Aldrin were for the first time in history required to land upon God-only-knew-what until they actually landed on it and got out. Without those two incredibly brave men being prepared to do that, all the best endeavors of those thousands of scientists, technicians and engineers would have been absolutely for nothing.

It is one thing to spend long, sleepless, coffee-driven nights rack-ing one's brains over the niggling problem of some figures that don't quite add up on a blueprint. But it is something in a different league entirely to actually sit inside the machine your blueprint eventually designs, fly it a quarter of a million miles, land it safely, and then step down for the first time in history onto the surface of another planet.

Despite the modest assertions not only of Armstrong but of many of his fellow astronauts that it was "all in the training," how can any amount of training prepare anyone to set foot upon another world? We humans are intelligent creatures with a built-in desire to survive; not only as a species but as individuals, at least until we reach our allotted "three score and ten" years. Fear is an instinct we strongly retain from our animal heritage that helps us to survive. Fear causes a build up of adrenalin in the bloodstream that, at the first sign of danger, gives us added strength to either deal with the threat or flee quickly from it.

No amount of training could have prevented that sudden catching of the breath, the quickening of the heartbeat and pulse, the rush of adrenalin in both men as Armstrong now used the LM's manual con-trols to settle it down onto the Moon that very first time, because it would have been instinctive. Something quite beyond their control as human beings. This was *not* training. This was *not* simulation. This was *real*.

"Picking up some dust...."

What if...? What if...? What if...?

And then that same stomach-churning ordeal of having to make the same decision to stay and fight when all you really want to do is "cut and run" would have to be endured a second time by Armstrong. My learned friend earlier asked you all to try to imagine being one of the first astronauts standing upon the surface of the Moon. Well, I ask you to go back a little further and to try to imagine being Armstrong after you have just landed the *Eagle* safely. You climb awkwardly out of the spacecraft. Doing so without forcing the edges of your backpack and other pieces of equipment attached to your bulky, uncomfortable spacesuit through the restrictive hatch area is a feat in itself, but it keeps your mind otherwise occupied for a few more precious seconds. Free of the hatchway, you now step gingerly onto the ladder and commence making your way down it. It feels so strange because you are no longer weightless but you are only a sixth as heavy as you would be on Earth. It is an uncanny feeling. You reach the last rung and prepare to jump from the ladder onto the lunar surface, which is made of…well, *what* exactly?

You know the sampler drones of earlier missions have picked up some fairly harmless stuff in the past…but not from *this* precise bit of the Moon. Again those negative thoughts keep on coming…but you try your best to block them out. Again the adrenalin rises…again your pulse races like it had during the landing…which had been like at no other time you can remember…again your heart is thumping under your spacesuit…

Boom! Boom! Boom!

What if…? What if…? What if…?

But you can't back out now. There is no cut and run option left. Not at this stage. You've reached the absolute point of no return. Family, friends and colleagues back on Earth are watching your every move on TV. The whole human race is watching. Half of them are women. You are a man. You mustn't let the women see that you are afraid. Worse. That you are shaking like a jelly inside your spacesuit. What if the ground beneath your feet gives way and you sink up to your knees in sulphuric acid? What if…? What if…? What if…?

Fuck it…you're going for it anyway! *Ooops!* My apologies.

Judge: Not to worry. I think we can all appreciate the sentiment on this occasion.

Defense: Indeed.

The soles of both your boots touch the Moon's surface at the same moment. Perhaps you would have been wiser to lower just the one leg first, because that way you might only lose *one* foot! Despite your misgivings, though, the ground remains firm underfoot. It seems to be supporting your weight. Both your feet are still there…

Thank God!

You struggle to regain your composure…and your breath, because you have a few important words to say; just a couple of lines that you have rehearsed over and over again hundreds and hundreds of times. But somehow you still manage to get them ever so slightly wrong…

"That's one small step for man," you say. "One giant leap for mankind."

So, in effect, you have used a pluralistic alternative for humanity twice. What you should have said—the version you have repeated to yourself during calmer moments over and over again so that it was virtually impossible to get it wrong on the day—was:

"That's one small step for *a* man. One giant leap for mankind."

But there is no second chance. No one shouts, "Cut! You fluffed it, Armstrong. Okay, everybody, stand by for Take Two." There can be no second take, even if you wish there could be, because this is *not* a film set. This is the Moon. You've said your piece and it can only be said the once and, after all, it was *almost* perfect, wasn't it? And back down on Earth they would all know what you *meant*, anyway.

And back down on Earth we all knew something else too. That we had been privileged to witness, as it happened, the single most heroic act ever performed by a human being since the very inception of mankind, and I do not say that lightly.

That Neil Armstrong himself should dismiss this as "nonsense" because if we only knew what a "coward" he really was, is typical not only of all reluctant heroes but of the man himself. Only a complete, gibbering idiot is unafraid when confronted by extreme danger. To be unafraid in such situations is not heroic, it is unutterably foolish because one is not arming oneself with the right amount of adrenalin needed to combat or deal with any lurking danger. A hero is a person who despite being absolutely terrified, a coward even, somehow manages to overcome the fear and continues on to complete the task that needs to be done, even be the most horrible death the likely outcome.

But, if the Apollo 11 commander's own naturally modest, self-effacing, personal point of view is that he is *not* a hero then he is entitled his opinion. Armstrong becomes an even greater man because of this obvious desire to be treated just like an "ordinary bloke" again. My clients insist that they are telling us the truth. If this is the case, then even to refer to Neil Armstrong as a "great man" must mean that the English language is short of suitable epithets. Whether or not he personally enjoys or appreciates his fame for the remainder of his days is quite irrelevant. History will ensure that his name, like Aldrin's, will go on forever.

Providing I, too, do not later find myself persuaded by any of the People's arguments as presented by my learned friend, or by any

Aw! Come on Armstrong. "That's one small step for a man..."

assessments made by the learned judge during his summing up, which follows, if I were ever privileged enough to meet Neil Armstrong myself, *I* would request an autograph too...

But only after getting down upon my knees, all the while sobbing like Buzz—quite unashamedly in my case!—and licking the great man's boots. And if Neil were to consider my attentions too much of an annoyance, I would, of course, consider it an even greater honor to be kicked away by one of the first two feet ever to have been placed upon another world!

Members of the jury, learned judge, my learned friend on the opposing bench, with those recent observations of mine hopefully to remain firmly etched in the minds of you all, I conclude the case for the defense.

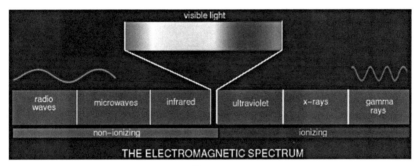

Figure A

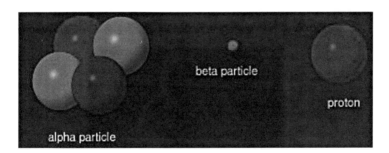

Figure B

Plate 24

Picture A

Picture B—*UFOs hovering in the Moon's sky? No, OLFs—Only Lens Flare.*

Picture C—*Let's face it—you always were way above me, Neil.*

Plate 25

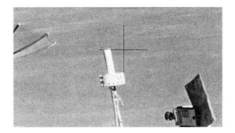

Picture A—*Perfectly legitimate. There was no double cross.*

Picture B—*Hey! Get that cross outa my face, will ya? Who d'you think I am— Count Dracula?*

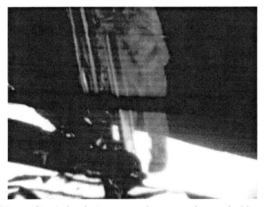

Picture C—*A ghostly Armstrong about to set foot on the Moon.*

Plate 26

Picture A—*Apollo 16 Lunar Rover. The LMs that carried the Rovers were modified accordingly.*

Picture B—*Apollo 17 Rover beside the modified LM that had carried it.*

Plate 27

Picture A—*Apollo 17 landing site. Does this look like a $30 billion film set?*

Picture B—*Or this? That's the Apollo 17 Rover at Plum Crater. Doesn't look too much like a studio shot either, does it? The light in the background? Lens flare.*

Plate 28

Picture A—*Another fake photo? That's the Eagle (minus lander) about to dock with Columbia and that's Earth in the background.*

Picture B—*This beautiful shot of Earth was taken from the Eagle just prior to landing on the Moon.*

Plate 29

Picture A—*No pictures of Neil? Aldrin took this after their first moonwalk.*

Picture B—*And this is Aldrin now turning his back on the camera. But the two men were not on the catwalk. They were on the Moon...and there was work to be done!*

Plate 30

Picture A—*It would have needed a pretty big JCB to bring that huge rock into the studio, especially as a much smaller one would have sufficed!*

Picture B—*Shot of Apollo 12's LM taken from CM prior to landing on Moon. Some "faked" shot this one, huh?*

Plate 31

The Case of the People versus NASA and the United States
Government
Concludes with a Summing Up by the Judge

Before the Grand Jury of the People:
The Case of the People versus NASA and the U.S. Government
The Summing Up

18

Claim versus Counter-Claim

Well, there you have it, members of the jury. Who is telling the truth? Difficult one to judge, eh? But I shall now attempt to do just that as once more I don *his* wig and silken robes and prepare to shoulder the awesome responsibility of acting as learned judge for the very last time in this case. After my summing up, I shall not be asking any of you to reach anything but your own individual, private verdict. This you will either arrive at, or you will be undecided.

If you find you have actually *become* undecided as a direct result of these proceedings, or continue to remain undecided after having had doubts anyway, then, by default, your verdict will be Not Guilty. Under English law, there is no such verdict as Maybe. But if you find you have completely changed from being a believer in the Apollo Moon landings to a disbeliever, then your verdict will be Guilty. If you were already a confirmed disbeliever and will have only had your opinion further strengthened by these proceedings, then again, of course, your verdict will be Guilty.

The sole purpose of this trial has been—and continues to be—to air both sides of a now fascinating argument as fairly as possible. There is little or no evidence to back up either argument that stands on its own as indisputable proof. Not even the quantities of "moon rock" allegedly picked up from the Moon's surface, because it is claimed that they are fakes also and that they were either manufactured in secret laboratories or were maybe even pieces of genuine meteorites that crashed to Earth. All we have, fundamentally, is a claim we cannot prove and a counter-claim, which equally we cannot prove.

Any opinions given and comments made by me when acting either as prosecuting or defending counsel or as judge, have merely been—and again will continue to be—attempts to ask any missing questions, discuss the issues, or do my level best to interpret the sometimes all too technical answers provided by those who know far more about these things than most of us do. Such experts, so remarkably clever as they are in their chosen fields, sometimes lack a certain skill in putting what they know into more simplistic or analogous terms that the population at large might more easily understand. For the sake of fairness, I have bent over backwards, in places, to put into lay terms my own tentative grasp of the thoughts that maybe drive both the pro-hoax and the anti-hoax lobbies.

But, as a rule, everyone is either "pro" this or "anti" something. There are very few issues that people are truly able to remain neutral about if brutally honest with themselves. When I commenced presiding over this case, I was firmly in the *anti*-hoax camp. But after hearing the People's case, I actually found myself in the *pro*-hoax camp! I suspect that by far the greater majority of you also, members of the jury, likewise had both feet firmly planted in the anti-hoax camp at the commencement of this trial. But what about after hearing the evidence presented to you by learned counsel for the prosecution? Had you by then already "crossed the floor of the House," as they say in parliamentary circles? Or did you keep a tentative foot still placed in both camps while thinking about defecting altogether?

Of course, even after having heard the prosecution case, there would still have been those among your number who steadfastly refused to entertain even for a second that so many of us could have been so easily fooled—albeit $40 billion worth of fooling!—into believing that man had landed on the Moon when in fact he had not done so. But their problem is being prepared to accept only one side of an argument. Problem? No, it is their prerogative. But for the majority of you who were prepared to allow it all to be challenged, you will no doubt have thus far experienced a roller coaster of emotions. There is nothing wrong with that. It is called being human. Upon receipt of certain information, one's opinion regarding something might change. But, upon receipt of a differing set of facts and figures, one's opinion might then revert to a position previously held, if now rather tenuously after having suffered such a nasty knock!

With all due respect to those with fixed views regarding this case, for they are entitled to those views, but isn't it all a classic illustration of why one should *never* simply listen to just one side of an argument? Only now, after having heard the case for the defense, do I find myself not so much back in the *anti*-hoax camp as sitting uneasily astride the fence between the two arguments. However, I desperately *want* to dangle both feet over the anti-hoax side rather than treat that fence like a saddle—Oooh! So uncomfortable!—but something inside keeps telling me to dangle both over the other instead. I don't really know or understand why. I just have this awful gut feeling that there is something not quite right about this whole affair and that some very serious questions desperately need very serious answers. But that I had a once-upon-a-time "confirmed" opinion, then changed it, then almost changed it back again but with reservations, at least gives me encouragement that this trial to date, may have succeeded in representing *both* sides reasonably fairly. Unfortunately, the result is that I am now as confused as all of you out there must be! But an English court of law, even an imaginary one like this, is no place for confusion.

Members of the jury, you have a distinct advantage over a jury sitting in any normal court because at the end of *this* trial you will be allowed to consider your verdict for as long as you wish without being dismissed for deliberating too long. You might even feel the need to read your case papers again if you are "almost" convinced of your verdict, but not quite. The most important thing to remember, though, is that you cannot abstain or find the charges against NASA and the United States Government Not Proven. There is *no middle ground*. Your verdict has to be Guilty or Not Guilty. For as long as you remain undecided, your verdict, according to English law, will be Not Guilty.

But it is hardly make your mind up time yet, members of the jury. Before we reach that critical stage, we need to go over it all—or very nearly all, because some of it, frankly, isn't worth repeating—just one more time…

POINTS RAISED BY THE FOX TV SPECIAL

There can be little doubt that this program was heavily biased in favor of the pro-hoax lobby. Having said that, the points the show raised were perfectly valid, although you may already feel that they mostly all received adequate explanations from planetary scientist

Jim Scotti of the University of Arizona. When I say "adequate," I mean that his explanations sufficed. Certainly in general, broad terms, which are the only ones that we, as mainly lay people, can be expected to understand.

Let us look again at some of these points (I repeat, not all because, frankly, some of them are not worth wasting any more time on), and examine the answers and explanations provided not only by Scotti but others as well. Let us see if we cannot, between us, get a clearer picture of this whole affair. We will do this by sometimes reading between the lines, by sometimes applying a little common sense here and there and, occasionally, even by drawing upon any elementary scientific knowledge we happen to possess through being reasonably well-educated people.

No stars

There is little point in anyone bombarding the vast majority of us with statistical evidence or too much technical detail because most of us simply do not understand it, but Scotti gave us an explanation to the challenge that no stars showed up in either footage or photographs taken on the Moon that was easy to comprehend. He told us that it was impossible to take a photograph of a nearby object that you wished to be in focus and at the same time obtain perfect focus on more distant objects—and they don't come any more distant than the stars, do they? And what he said was true. The astronauts were not on the Moon to take photographs of stars, were they?

Elsewhere in his testimony, the planetary scientist tells us that all the Apollo missions took place either in the morning or the evening of a lunar day, which is a fortnight long in Earth terms, and also happens to be synonymous with a year in Moon terms. The sky is only *black* during this lunar day, or lunar year, because the Moon has no atmosphere, unlike Earth. Otherwise the sky would be blue, or some other pigmentation—as is the case with the red skies of Mars. This pigmentation is caused by the reaction of certain elements or chemicals in a planet's atmosphere to the process of absorbing, reflecting or diffusing the sun's rays.

On the Moon, the sunlight and temperature is all the more intense because of the total lack of atmosphere, but less so during the morning and evening, just as it would be on Earth. And because of this "light pollution"—not just from the sun but from a four times larger Earth disc (*i.e.* four times larger than the Moon) also reflecting sunlight back down onto the lunar surface, it is easy to understand why the stars become invisible or certainly very faint. They do exactly the same on Earth during the day. All those great constellations we see on a clear night, like the Great Bear, Cassiopeia, Hercules and Orion are still there, just as they are at night. We simply cannot see them because their brightness is diminished by the sunlight. And light pollution is not something that occurs only during the day either. Try looking at the stars on a clear night from a city center. Then get into your car and drive ten miles out of the city to a dark, quiet lay-by in the countryside. (No jokes please, boys! This is all deadly serious, scientific stuff!). Now get out and take a look...Ahh! That's much better!

Some of you will be unhappy with the explanations given for the absence of stars in the footage or photographs allegedly taken on the Moon. And I confess to still being a little baffled myself that it did not occur to anyone to take a long exposure from a camera affixed to the LM on at least one of the six "successful" missions. Such a picture would have provided indisputable proof that the LM was "parked" at a specific location on the Moon. Just one irrefutably genuine photograph from any successful Moon landing mission, which showed the stars in positions an astronomer would expect to see them at the precise time the photograph was taken from that pinpointed location on the Moon's surface, would have been enough. Not only that, but the stars in their mathematically correct positions above an *identifiable* landscape on the Moon would have served as additional proof, should any more have been required.

The problem, though, is that there would still be those who would claim, even after the photograph had been proven *not* to be a fake, that the picture was simply taken by the camera of an earlier unmanned probe, like Surveyor III, for example, with merely the time and date of the exposure faked. Some would further claim that Apollo 12's alleged landing site, the Ocean of Storms, was deliberately chosen (again purely by way of example) in order to back up a photograph already in NASA's possession, but which had—thankfully for NASA—never before been released. So it would still be a "heads, you win, tails, I lose" situation for NASA.

But remember, members of the jury, that it is the People who have brought this case on behalf of the pro-hoax lobby. The burden of proof lies not with NASA but with the People. The charges against NASA and ultimately its masters, the United States Government, have to be proven beyond a reasonable doubt. The defendants are obliged to prove nothing. As regards this particular accusation, as opposed to the overall charge of a hoax having been perpetrated, may

I suggest that the question of "no stars" has not been proven, and that the element of doubt must therefore favor the defendants.

No blast crater, no dust on LM footpads, no engine noise

I think the greater majority of us would be inclined to accept Jim Scotti's given explanations in all three cases.

The "Prototype LM" crash

The difficulty of testing a mere prototype of a machine *six times heavier* on Earth than it would be in the environment in which it was designed to operate, surely needs little or no explanation. It will be clear to most of us that only certain rudimentary tests—like "can we get it off the ground at all?"—of the prototype of the Lunar Module could be carried out here on Earth. There could never be any adequate testing ground to test its flight and landing capabilities other than the Moon itself. Therefore the explanation given for brilliant test pilot Armstrong's accident, when trying to control the ungainly vehicle during training here on Earth, has to be acceptable to the majority.

No rocket plume

There can be no fire without oxygen. Notwithstanding the supplies aboard the LM, there is no naturally occurring oxygen in a gaseous (breathable) form on the Moon. That is not to say there is none at all in any way, shape or form, like in molecules of ice (H_2O), for example. The answer to that is that we simply do not know whether there is or not. Combustion for the LM ascent (take-off from Moon) stage was occasioned by two fuels that ignited upon contact with each other.In a vacuum, this combustive process would leave no fiery tail as the ascent stage blasted off, although a barely visible "exhaust plume" would possibly be discernable, but not necessarily. Any combustive process, though, will almost always cause light to be emitted.

In short, those expecting to see fire in a vacuum, will not; they might, though, see a hazy tail of exhaust. Those who cleverly say, "*Ah!* But on a TV clip of one of the missions supposedly taking off from the Moon, I saw *fire*, so it *must* have been filmed on Earth," have to ask themselves how long the "fire" lasted, or was it simply a *flash of light* from the initial combustive ignition process?

Waving flags

The "atmosphere" of the Moon is a vacuum. Study any piece of Apollo footage of the American flag "blowing in the wind" and it reveals the flag only moves when its pole is moved or knocked against, or when the flag itself is manhandled for any reason. It is easy to move even very heavy objects in a vacuum because there is no air resistance. The phenomenon of complete weightlessness, however, does not totally apply on the Moon itself as it does in space, because the planet's gravity of equivalent to one-sixth Earth gravity plays its part, whereas there is zero gravity in space. Give a fellow astronaut who had once been an eighteen-stone wrestler a gentle dig in the ribs while on EVA in space without a safety line (not that this would ever happen), and he will continue floating away from you until he reaches, or makes a grab for, some fixture that he can hold onto. If such a fixture or obstacle happens not to be there, or is not to be found anywhere along the path your little push has pre-ordained for him, then he will continue traveling away from you at exactly the same rate for…well, yes, precisely…

For all eternity!

Poor picture quality

Video technology was in its infancy at the time of Apollo. The very best cameras were very weighty and needed to use a lot of power in order to function properly. Weight, preferably a *lack* of it, and the use of electrical power were subject to strict rationing on all Apollo missions. Video cameras were, in any case, never considered an essential part of the equipment to be placed on board Apollo. According to Scotti, it was more an afterthought that such a camera was eventually carried by Apollo 7.People suggest that a budget of $40 billion, combined with the undisputed expertise of the world's leading cinematographic nation, should perhaps have led to slightly clearer pictures emerging from the Moon's surface. Although I can of course see where they are coming from, isn't it being ever so slightly sanguine and pedantic for anyone to expect the video picture quality of thirty years ago to be anything like as good as it is today? As far as I am concerned, video picture quality is not all that brilliant even now, although I have been advised to purchase one of the brand new, state of the art Digital Video Disc players (DVD) to enjoy a better quality picture.

But these were too perfect

I refer, of course, to the claims that the still photographs taken by the Hasselblad cameras affixed to the astronauts' spacesuits at chest height, were far too good for the circumstances they were allegedly taken under. But Scotti says thousands of shots were taken by the cameras (which had largely automatic settings) over the course of six missions. Of these, Scotti says, only a tiny few were considered good enough to print and publish.

The majority of us can easily accept this explanation by virtue of our own experience when taking photographs on a holiday or vacation. We invariably allow in too much light (over-exposure) or too little light (under-exposure). I can even recall a time when I snapped away like David Bailey (a famous U.K. fashion photographer of the sixties and seventies) for a full thirty-six exposures, feeling ever so pleased with myself. But upon opening the camera, I discovered I hadn't even wound the film onto the spool! But even I cannot compete with my wife. Over the years, she has managed to chop off more heads with a shutter release than the Queen of Hearts and Madame La Guillotine put together!

But after each holiday, a few, just a few, of our combined amateur efforts, as more than likely NASA also found with the odd one or two of the Moon photographs, were always good enough to warrant a place in the album.

Shadows wrong

Not a particularly sound argument when one examines Scotti's explanation, but one worthy of being raised and discussed all the same. The pro-hoax lobby was correct to call "multiple shadows" into question. The simple question for you to consider, members of the jury, is whether or not you find Scotti's explanation for this phenomenon satisfactory. I must confess that the last time I saw as many shadows flying off in different directions from a single figure was at an evening kick-off football game, with powerful floodlights blazing down from *all four corners* of the stadium!

Identical backgrounds

Personally, I found myself not too impressed by Jim Scotti's answer to this one, particularly as far as the two photographs C and D on Plate 6 of your case papers are concerned. Admittedly, distant mountain ranges, for example, would remain in the background if one wandered about snapping pictures from one end of town to the other; mountains or hills are pretty big things and would tend to dominate any landscape. But here, we are not merely discussing an identical *distant* landscape; we are discussing an identical *foreground* as well as an identical background. One that apparently contains the LM and one that doesn't! Picture C shows nothing but an identical foreground and background to that in the presumably later Picture D, which now—miraculously it would seem—appears to contain the LM, as you can see for yourselves.

Are the pro-hoax lobby correct to claim that the only way these photographs could genuinely have been taken on the Moon would have been if someone or something with a camera had *already been there* in that precise spot from which both pictures were taken, *before* the LM had actually landed?

Impossible? Well…yes, precisely. (On second thoughts maybe not, as you will see later. But it's unlikely we'd ever get to see the nega-tives!!) But it is conceivable that one or the other of these two photos is a more recent fake, especially with the technology we have avail-able today. Is it by sheer accident that Picture D, which contains the LM, is rather more fuzzy and over-exposed than Picture C? It would be quite simple to add a fuzzy LM to an over-exposed bare landscape shot and pass it off as a *real* photograph. Far simpler, I would have thought, than erasing the LM from a *bone fide* picture of it sitting on the surface of the Moon. But who would want to fake such a picture in the first place? Whose argument would benefit from such a picture? Certainly not NASA's, that's for sure! And I have yet to come across anyone who believes the Apollo landings were faked who didn't *truly* wish that he or she were mistaken, let alone start forging pictures to prove their argument!

The picture portfolio submitted by the defense contains various photographs that would appear to be undeniably genuine and to have been taken on the Moon. They purport to show a very wide expanse of "moonscape" in the foreground, with hills and mountains in the background, as well as evidence of man's activities also in the fore-ground. "How could these have been taken in a studio?" the defen-dants ask.

But I think you will find that the prosecution would expect you, members of the jury, not to take the word "studio" too literally on this occasion and to allow it to become but a euphemism for "on location" in the case of these particular photographs. Indeed, prosecuting coun-sel has submitted as evidence, from the People's own picture portfo-lio, a daylight photograph, Picture A on Plate 20, of what could easily have been the true location—here on Earth, of course—of the picture that appears in the defense portfolio, *i.e.* Picture B on Plate 28. True, the defendants make no claim that the "light in the sky" in the back-ground of Picture B on Plate 28 is the Earth. In fact it again looks very much like the phenomenon known as "lens flare." But how do we know that for certain? The People would be correct to protest that it could just as easily be a large, bright light shining down from a gantry.

Similarly, the two photographs [Pictures A and B on Plate 29] the defense submits to disprove pro-hoax claims that not a single photograph from *six* purportedly successful Moon landing missions clearly show an astronaut upon the Moon's surface and something that is clearly the Earth in the sky above, are, at first glance, impressive. But we must remember that the People aren't claiming that the *hardware—i.e.* space vehicles and peripheral equipment—never went to the Moon. They are claiming that human beings, very much the *software* in this case, did not *accompany* said machinery. And as all such machinery can be controlled remotely, especially cameras, it makes us all the more suspicious why the defense, with the photographic results of six missions costing untold billions of dollars at their disposal, are apparently unable to come up with even one single picture, which shows a *man,* not a machine, indisputably standing upon the surface of the Moon with the disc or half-disc of a recognizable Earth in the sky behind him?

Like the following one, for example, which *is* a fake. Nothing to do with NASA, though. Someone posted it on the Internet, no doubt also questioning why NASA has thus far failed to produce such a photograph for real. It clearly shows the beautiful, full disc of Earth in the sky behind Aldrin. Unfortunately, though, on this occasion, Earth has only been cleverly superimposed onto that famous photograph.

I will tell you why NASA hasn't yet released such a photograph for real.

Because there isn't one.

That's right. Not one. Not from six whole missions!

Mathematics

I am a great believer in the science of mathematics even if I personally have never been particularly good at it. But the simple question: "Does it all add up?" can often prove whether or not a story is true. Unfortunately, there *are* no provable numbers in this story either way. Only two sets of quite contradictory figures. Particularly as far as the radiation dosage question is concerned. But we will look at that question again shortly. I cannot stress enough, though, what a terrible oversight—or deliberate mistake?—it was for NASA not to have ensured that at least one picture from the Moon should have shown the LM, an astronaut, and either the Earth or a mathematically provable pattern of stars—preferably both—in the background.

But enough of photographs and film footage. I am actually—rather late in the day, I know—on the verge of ruling both inadmissible as evidence. We all accept that the camera never lies in itself. But the images it produces can be *made* to lie with an ever- lessening chance of detection, it seems.

In a serious attempt to establish the truth about the Moon landings as this is, that is not good. Not good at all.

Reseau marks/cross-hairs

Again this is to do with photography and the idiosyncrasies of cameras and reflected light; I refuse to deliberate upon the matter any further. I refer you to Scotti's explanation of earlier and trust you will simply decide if it "adds up" or not.

Move a little to your right, Buzz...(click!)...that's it! Nice one!

Effects of radiation on living tissue

This is the only argument that warrants any *real* attention at all. Testimony from creditable sources would suggest that the danger presented by passing through the intensely radioactive Van Allen belts, as well as to go beyond them into outer space, is acute. However, other testimony from equally creditable sources suggests the danger of passing through the belts is comparatively minimal, even though most of the radiation they contain is the "wrong" kind of radiation—namely, particles instead of rays—because they are passed through the "right" way, which is very quickly.

But that begs a question. Can there ever be a "right" or "wrong" kind of radiation or a right or wrong way of dealing with it, especially if one has an earnest desire to stay fit, healthy and free of cancerous tumors? The answer is obviously no. But the anti-hoax experts claim there are such things as "tolerable dosages" and that the Apollo astronauts were subjected only to such tolerable dosages. Without going over all those unutterably boring reams of statistics again, members of the jury, it is simply a question of deciding which group of experts and which set of statistics one is inclined to believe.

Where do I personally stand after all my extensive research into this case? Good question. My answer, and not purely because I am currently acting as judge and attempting to be fair to both sides, is that I honestly do not know. Because of the undoubted expertise of these people, whichever side of the argument they speak for, I am inclined to believe *all* of them—but only up to a point. Is it too dangerous for man to travel through and beyond the Van Allen belts into outer space and stay for long periods? The answer must be yes. But is it even too dangerous to attempt for relatively short periods lasting about two weeks or less? Was the former Pentagon based army officer, Philip Corso, implying precisely this when he said, "Mankind *cannot* travel in space"? Maybe—maybe not. I find myself unable and unqualified to give a definitive personal answer to the second question.

But what I will say is this, and it is a personal observation rather than an answer. Of the twelve men who allegedly walked upon the surface of the Moon thirty or more years ago—and this is to the best of my knowledge at time of writing—nine are still alive and enjoying relatively good health. And this despite the fact they are all now aged around seventy. As already mentioned, Irwin, Conrad, and Shepard, sadly, are dead. The first named died from heart failure, the second from injuries sustained in a motorcycle accident, and the third, Alan Shepard, America's first astronaut and the commander of Apollo 14, is the only one to have died in anything like "suspicious circumstances" possibly relating to his alleged journeying beyond the Van Allen belts. However, there is no evidence, given the compensation culture of the United States, of Shepard's offspring holding NASA responsible in the slightest for the death of their father.

So, in any normal way, without including the six command module pilots who allegedly went to the Moon, but didn't actually land upon it and walk upon its surface, and the crews of Apollo 8, 10 and 13, the majority of whom also still survive after having apparently ventured through and beyond the Van Allen belts to orbit the Moon, these surviving nine men alone should be enough living proof that going into outer space and receiving a tolerable dosage of radiation, might actually be…

Beneficial to the human body rather than bad for it!

In fact, I am very surprised that no medical person of any note has yet remarked upon this phenomenon. Again, certainly not to the best of my knowledge, and God only knows how many times I have trawled the Net for any linkage between the words "Apollo" and "cancer." Even allowing for a certain margin of error in my research and the fact that many people aged seventy and over—including former astronauts—could be lying on their deathbeds even as I write, we are still talking twenty-plus men of around seventy years of age here— out of a total of twenty-seven—who are still alive and as well as can normally be expected at that age. Is there a potential cure for cancer here? Does exposure to a tolerable dosage of radiation then, say between the ages of twenty-five and forty-five—the approximate age range of the Apollo crews—allowing the body to then naturally repair or replace the cells thus damaged, largely *prevent* most if not all forms of cancer from developing later on in life?

After all, it wouldn't be the first time that a small dose of the disease itself was found to be the actual vaccine that prevented it from ever re-occurring, would it? If Apollo was indeed the genuine article, shouldn't teams of medical scientists now be investigating this possibility? Unless there is some vital point I am missing, surely they should actually have *begun* taking a serious interest once all twenty-seven extra-Van Allen belt astronauts reached their mid-fifties?

Whatever the apparent advantages, though, I do not imagine there will be that many people aged between twenty-five and forty-five volunteering in future to act as guinea pigs to receive 100 chest X-rays in the course of one week, do you?

Telescopes should see some evidence of Apollo

Well, Jim Scotti gave a fairly detailed scientific answer in order to explain why they cannot. I don't think we need to go over that again either, even if we actually understood it the first time around. Scotti is a highly credentialed planetary scientist. If he says we would need a telescope *thirty* times larger than the Hubble Telescope, in Earth orbit, in order to see the largest pieces of equipment left on the Moon by the Apollo crews, then that is good enough for me. But I remain mindful that NASA have not, in the past, found it too difficult to place human technology into Moon orbit. It is only a question of whether or not a human being has, at times, been on board such technology that is in dispute. A further unmanned device thus deployed would surely not need to carry equipment anywhere near as sophisticated or as expensive as the Hubble. Nor would placing such technology into Moon orbit be anywhere near as expensive as sending another manned mission either, and it would resolve the argument one way or the other once and for all.

I wholeheartedly agree that NASA could hardly be expected to ask the American government to fund a project primarily designed to prove a point to a bunch of "looney-toons," but I am equally mindful that NASA and the American government are hardly likely to go to such an expense in order to prove…

That maybe they *lied* to the world a third of a century ago?

Why has no one gone back in more than thirty years?

I was tempted to say, "because it's too bloody dangerous!" but in the light of what occurred to me earlier regarding the possible beneficial effects of radiation, I am now not so sure.

Again, who can say with any degree of certainty why no one has been back, or why the iron has now not merely cooled but is icy-cold to the touch? I think learned council for the defense got it just about right when he informed us that there is no oil or coal on the Moon or anything else worth having. Journeying to the Moon and back is an expensive business. Russia and Europe simply cannot afford it in any meaningful way even if they pooled resources and technical know-how and were helped by the Americans. The Americans *can* still afford it, of course, and should still possess the original capability plus thirty or so years' worth of additional improvements in technology. But they seem to have decided that it's simply not worth it, certainly not economically, and have concentrated any continued investment in space to operations undertaken strictly on *this* side of the Van Allen belts.

However, judging by the loss of the space shuttle Challenger and its crew of seven in 1986, and the equally tragic loss of the shuttle *Columbia* and its crew, again of seven, in January 2003, fully paid up members of the pro-hoax lobby could be forgiven for sniping that journeys into outer space to the Moon and back are obviously a piece of cake compared to the obvious dangers of messing around in *near* space!

Have they a point? Let us compare the apparent risks:

Shuttle astronaut fatalities*:* Fourteen—All on active service. All in neo-Earth environment.

Apollo astronaut fatalities*:* Three—All three on active service in Apollo 1. Fatalities occurred while still in Earth environment.

Combining the two totals, I make the score zero to seventeen in favor of "dangerous outer space," don't you? So-called "safe space," "near space," or "phony space" (anyway, that bit situated somewhere

Shouldn't telescopes see some evidence of Apollo?

between where our planet's atmosphere and gravity end and the Van Allen belts begin) looks like a complete no-no to me!

I am sorry if my use of the term messing around, which has a somewhat different meaning in America than here in Great Britain, where we use it to suggest a person or persons not pulling their weight or not getting on with the real task in hand, offends anyone, particularly the families and friends of the bereaved shuttle astronauts. But the real task, according to many, should be to get back out there and start mixing it in space. After thirty-five years since Apollo 11, we all expected to have bases not only on the Moon but on Mars as well by now!

So what on Earth happened? Yes, that *is* a deliberate pun. *Why* is NASA still playing about in near space at the continued expense of the lives of courageous young men and women? Lack of funding by governments we can all understand, but not to the extent where it involves sending talented young people up in "past their sell-by date" if not altogether ancient, two-tone (white paint and rust?) re-cycled jalopies that, again according to some, were thrown together on a shoestring budget *even when they were brand new* and were obviously putting those valuable young lives at risk from the very outset!

I have visited the Kennedy Space Center in Florida. True enough, the tourist areas are as pristine as one would expect, but from the bus that took us out to the observation station from where can be seen the famous Launch pad 39A at Cape Canaveral, we passed areas containing what appeared to be scrap-yards full of discarded or otherwise redundant equipment, although the guide on the bus informed us that some of it was still in use. Some of this gear looked very run-down. There was lots of missing paint. And where the paint was missing it had inevitably been replaced by lots of ferrous oxide.

I surveyed Launch pad 39A through my binoculars from the closest point unauthorized persons can get to it. On the pad awaiting imminent take off, within days apparently, was the ill-fated *Columbia* herself, but my party weren't staying in America long enough to witness the actual launch. I now couldn't be more thankful that we didn't. I ran the binoculars over the whole assemblage of gantries and giant rockets, the white shuttle clutched to those rockets like a huge, mating albino bullfrog being carried around on the back of its even more enormous mate, which was ruddy-brown in color. Although the shuttlecraft itself looked to be fairly immaculate from this distance, I could still see a considerable amount of missing paint, staining and rust on everything else.

Fourteen astronauts have died while flying the space shuttle.

"So, what's a bit of missing paint and a spot of rust here and there?" I hear you ask. I don't know, except to say that you wouldn't find any missing paint, sea stains or rust on a British warship about to leave port and go into action after a refit—even if the rust was still there somewhere—and as the U.S. Navy is based on the same strict seafaring traditions, I suspect you wouldn't find any visible rust on a newly fitted out American warship either. "Everything shipshape and Bristol fashion," is the naval term.

But these observations, although made at a time when I had already started this book (sufficient enough reason to visit the Space Center!), did not actually develop into anything concrete at that early stage. Back then I had no doubt that men had landed on the Moon, although I was familiar with the rumors of course. Now, though, I have an open mind about it all even if it is no longer as open as I'd like it to be! But any "negatives" contained in the overall scene I found myself surveying that day were not conspicuously apparent at the time. To be perfectly truthful, I am only critical of certain things now with the benefit of hindsight. It was only after completing the case for the prosecution for the book, especially after the loss of the *Columbia* and those seven so innocent and unsuspecting lives, that little things I hadn't taken too much notice of at the time began to resound in my head like those awful air-raid sirens I heard as a tiny child, rising and falling and echoing and re-echoing their ghastly warning over the rooftops of a London about to be bombed for the umpteenth time.

No, at the time, just being there and seeing it all for myself first hand was something else. Whatever criticisms may have entered my head at a later stage, to me, then, this was a magical, wonderful place. The headquarters of NASA and the American space program. I was proud of America. I was proud to be *in* America. Proud of our bold, brash, sometimes overly flamboyant but wonderful cousins, who had taken *our* beloved English language to the Moon. And they had done so from here. From *this* place. And what I was looking at through those binoculars was no Tower of Babel. Oh no. This was a sacred place. A place of worship for all mankind. My friends some distance away, I allowed the tears to flow unchecked for a brief moment or two…uh-hum…oh dear, I must apologize. I'm filling up again…

Prosecution: Would the learned judge care to adjourn for a few minutes and retire to his chambers in order to compose himself?

Defense: Yes, that might be a good idea. The learned judge looks to be most upset.

Judge: No, no. I thank learned counsel for their kind concern. But these proceedings are far too important to be delayed by a few tears from a silly old man. I will be fine in a moment or two.

Prosecution: "Silly" is not a word I would ever use to describe the learned judge.

Defense: Echo that.

Judge: You are both most gracious. I shall continue:

It was just all those wonderful memories flooding back, that's all. Learned counsel and many members of the jury will recall that all the tears we shed in those heady days were ones of sheer joy and pride. They still were as I viewed Launch pad 39A that day some years ago now. But what they represent this time I cannot be sure. Nostalgia for a time gone by when the future looked so bright maybe, but then again, maybe a deep and profound regret for a lost cause doomed from the very beginning…

…fourteen…thirteen…twelve…eleven…ignition sequence start… nine…eight…seven…six…main engine start…four…three…two…

one…lift off. We have lift off at Zero minus five seconds…and counting.

It was the dream we all had as little boys. Lift off, to the greatest adventure of them all. And, via the miracle of TV, many of us watched it all happening. That colossal machine, mighty beyond description, literally groaning as it slowly, almost agonizingly, heaved its enormous bulk up off the ground and into the clear blue yonder beyond the fire, smoke, and a temperature at blast-off to rival that of the surface of the sun, its proud nose aimed at an unseen object somewhere up in that same sky. The *Moon!* Or so we all thought…

I was suddenly jolted back to the present day by the voice of one of my friends. It was time to return to the Visitor Center. Again we passed the scrap-yards and various huge appliances in dire need of a lick of paint and looking as if they had not been moved or used in ages. Unfortunately, the overall impression—although I repeat, this is with hindsight—was of something with the very guts ripped out of it. Something on its last legs that after only having just learned to walk and with seemingly everything to live for, no longer had either the will or the energy to even think about running. Something desperately short of cash, merely kept ticking over for no other real purpose than to maintain the image of America as the world's greatest superpower and undisputed leader in space. And although both are very much true, of course, the latter mentioned part of this image can appear a little tarnished if one manages to get up close.

I do recall, though, that even then—blown away as I nonetheless was—and chiefly because nothing truly exciting had happened in space for many years, a rather nasty word somehow lodged itself into my perception and, although unwelcome, it kept on creeping back to the forefront of my mind as I later browsed around the NASA shop—yes, shop!—at the main Visitor Center…

Redundant

Redundant. Yes, that's the word. And a very nasty word it is too. A once proud organization that I was convinced had landed the first men on the Moon, seemed now to be reduced to selling cheap trinkets and souvenirs of its great accomplishments to visitors. And, as I recall, a very sparse amount of visitors, too. Those of our bus party not already making their way back to their respective hotels virtually had the shop all to ourselves!

My personal opinion is that the American government should either once again put everything it's got into the space program or, sadly, abandon it altogether. Space—even near space—is far too dangerous a place unless it does. Like the sea, one treats space with contempt at one's dire peril. Despite it having been alluded to it as such to separate it from outer space at various times during this trial, there is no such thing as "safe space," anymore than there is a safe part of the sea. NASA says useful and valuable experiments are being conducted aboard the shuttles in space; the kind of experiments one presumes will be conducted full-time in the laboratories of the new International Space Station, which the shuttles also supply.

It isn't just the apparent total lack of interest in space by successive post-Apollo U.S. administrations that fuels the fires of the conspiracy theorists. The pro-hoax skeptics simply cannot fathom out how an organization which—despite budget restrictions—still has such a wealth of modern technology at its disposal, could lose seven young people, literally in one fell swoop, whilst merely bringing them back into Earth atmosphere from near space orbit? Fairly they ask: how could NASA place twelve astronauts upon the quarter of a million miles distant Moon and bring them all back alive using relatively primitive technology, yet somehow manage, since then and with technology much more advanced, to lose fourteen astronauts who either ventured only a few hundred miles up or were in the process of doing so?

And even if we were to largely dismiss much of the pro-hoaxers so-called evidence of a fraud, we would indeed be stupid, members of the jury, not to concede that they have a point. After those six amazingly successful landings, most of us would have been right to speculate, at the time, that we would not only have bases on the Moon by now, but would be going on holiday there, too!

But thirty plus years after the last Apollo flight there are still *no* bases on the Moon. Not even one. Not only that, but we cannot see any *traces* of man ever having been on the Moon, not even with the most powerful telescopes. Only you can decide whether or not Jim Scotti adequately explained why this is.

After Apollo, perhaps even while it was all still going on, something happened that for some reason appeared—and here I will borrow an appropriate expression from learned counsel for the defense—to take all the wind out of NASA's hitherto billowing sails. It is this aspect more than any other that concerns people; not least the American people themselves.

There we were, humanity, well and truly "on our way" in space. All set to go for bigger and better things. Science fiction was rapidly

Seven...six...main engine start...four...three...two...one...lift off! We have lift off at zero minus five seconds...and counting.

becoming science *fact* with every passing day. Via the Americans, it seemed, we had already reached the staging post for continuing the Great Adventure…

And then, quite suddenly, it all stopped.

No plans to return to the Moon in the near future. No plans to return even now, over a third of a century later! There has been media speculation about making Mars the next target, but nothing concrete has ever materialized as far as any kind of definitive manned program is concerned.

The anti-hoax people claim—possibly correctly, at least in part—that lack of funding for space is still the root of the problem. Governments are sadly not run by planetary scientists, rocket scientists, geologists and plain old fashioned, swashbuckling explorers, they claim.Governments will not keep pouring money into expensive projects that bring little or no short-term return, with no guaranteed long-term return either. Banks offer a very low rate of interest for moon rock apparently!

The anti-hoaxers claim that if the Moon or Mars showed any signs of containing large deposits of oil, coal, gas or any other type of fossil fuel, it might all have been a different story. But in order for fossil fuels to exist, there has to be fossils. And for fossils there has to be life.

As there is so far no evidence that life in any format ever existed on either the Moon or Mars, this is a very sound argument, and perfectly believable *prima facie* (at first glance). Very few expeditions down the ages have been financed solely to advance the knowledge of mankind. The primary objective has always been the eventual furthering of the bank balance of an individual, a group or a country.

In short, greed.

The simple truth is that no one expected to find life or even any evidence that it may once have existed, on our Moon. It is, after all, simply an asteroid. An oversized lump of rock trapped by Earth's gravitational pull billions of years ago. Mars, though, is a different kettle of fish entirely. It is nowhere near as large as the Earth (about one tenth of Earth's mass) but it does possess an atmosphere and a Martian "day" is roughly the equivalent of our own. There is also evidence that it contains water—although this may only be in frozen form—and that this water once flowed freely across its surface. If the miracle of life is not an accident that only happens in a fifty thousand billions to one chance but simply occurs naturally in a given set of circumstances, then most scientists believe that Mars has all the necessary ingredients to support—or to have once supported—some kind of life as we know and understand it. But until we get up there and have a really good old rummage around for ourselves instead of sending little robot probes merely to scratch the surface here and there, we will never know, will we?

Taking chances and confronting great dangers, but with the possibility of encountering riches beyond one's wildest dreams, used to be what exploration was all about. True, it would appear that there is nothing worth having on the Moon. It may turn out that there is nothing of value on Mars either (if the phrase "nothing of value" can be applied to a one-tenth Earth-sized plot of spare real estate!). But the problem is that we need to go back to the Moon first and then get ourselves to Mars before we can actually go anywhere *else* where there *might* be something worth having. And what is worth having most of all is somewhere else for us to go if anything happens to this good old Earth of ours.

As an intelligent species our prime concern, as learned counsel for the prosecution reminded us earlier, is survival. Whether or not the human race will still be around in ten million years from now, or even ten thousand years from now, is debatable, but we like to think that we might be. Some ambitious minds actually think that in some much more evolved form we might still be around when the Earth one day comes to an end. And in some way, shape or form it *will* come to an end. But Mars is even older than the Earth. It is sometimes referred to as a "dead planet" rather than the "Red Planet." There is little point in speculating that we might one day be able to evacuate the Earth and go and live there instead if and when the ultimate disaster strikes. By that time, Mars may no longer be there in any habitable form either.

So, the all too important question is this: When the Earth meets its eventual fate, do we, like the wives and servants of ancient pharaohs, come to an end with it simply because there is nowhere else for us to go? Unfortunately, if the world were to end quite soon, that would still be the unenviable position we would find ourselves in.

If we ever *do* get down to exploring space in earnest—a process which most of us thought had started between thirty and forty years ago!—it may take centuries, thousands of years even, to find proper alternative accommodation as opposed to glass or Perspex "bubbles" in constant need of re-supply by shuttles from Earth. For all we know, the Earth may last for another billion years. Or suddenly come to an end as early as next week! My point is that we need to get a *move on* in space. Time *could* be of the essence. We just don't know. It seems to me that we have already largely wasted more than thirty years. Just imagine that at some future time, scientists are able to reliably predict that the Earth will come to an end within a year. Would there be enough time to evacuate everybody? Hardly, but even if there *was* enough time, evacuate to *where?*

Further imagine that in this future time we have become skilled space explorers, having solved all sorts of problems that beset us now, but have yet to find an alternative planet with anything like the necessary natural resources to sustain us as does the Earth. Visualize then being told by these same scientists that all the computer forecasts and calculations confidently predict the discovery of such a planet in around 30 years. So, there we are, a million years from now, say, with around ten billion people all suddenly looking for a new home. The Moon, Mars and other planets have been colonized by way of plastic bubble settlements with a capacity to maybe accommodate half a million maximum. There are twelve months left in which to try to build many more domed cities, which, although containing a measure of self-sufficiency, will find it virtually impossible to survive long-term without continued supplies from Earth, which, after one year would become a dead planet itself, upon which we could only exist in…well, yes, precisely. Plastic bubbles!

But the plastic bubble refugees would be the lucky ones. They would number around a million at most. With them into these veritable Noah's Arks would go a wide variety of plant seeds and the capability for cloning many species of animals, birds, and fish. Sheer logistics and the amount of time available, would mean the excluded 9,999,000,000 human beings and all existing wildlife being left to whatever fate in God-only-knows-what horrible fashion. Will it be quick? Again, who knows? And all because long ago that ancient organization called NASA, in that ancient place called America, *wasted more than thirty valuable years* at the very commencement of man's voyage of discovery into space. A period of time which, had they not so criminally (in that future assessment) allowed to slip by with nothing of any consequence happening, could have led to mankind finding that alternative Earth more or less at just the right moment. An alternative "home from home" that maybe would have provided *all* the population, not just a chosen few, with continued refuge and succor for at least the *next* million years!

It is a cruelly aimed prophesy, I know, and I do not expect NASA or its masters to take it *too* much to heart. But quirks of fate do have a nasty habit of being cruel by themselves without any assistance from yours truly! I do not believe for one moment that mankind—even in the representative forms of NASA and the United States Government—has lost either the will to survive or the desire to explore. Neither do I believe that the U.S. Government would have let a little thing like money put them off. Not thirty plus years ago and not now. "Lack" and "money" are two words that quite simply do not appear together in the vocabulary of the United States.

Pull the Plug?

But if Apollo was indeed genuine and successful, and money truly no object, why did America suddenly "pull the plug," so to speak? Surely it wouldn't have been anything to do with waning public interest? Public interest always wanes when something becomes the established norm. It is a mark of a project's success, not its failure. In any case, Apollo was supposed to be a serious scientific quest, not a circus! And talking of circuses—shouldn't that be *circi?*—most managers of big projects are quite happy once the media circus moves on to other things because it allows them to concentrate quietly and efficiently on the proper job in hand without worrying too much about whether it all looks good on camera.

So, maybe it something else that put them off the whole idea. If so, what? Radiation? Not a serious problem according to Scotti and the others. The possibility of bringing some deadly bug back from the Moon, then? *Nah!* They should be so lucky! Bugs mean life. And life means the possibility of fossil fuels. Once we had buried the dead and found an antibiotic to kill the bug, the "gold rush" to the Moon would have reached stampede proportions. So, what then?

There *is* another possibility. It's a long shot. So much so, that even as a firm believer in the likelihood of other intelligent life forms existing elsewhere in the universe, maybe even in our own Milky Way galaxy—maybe closer than that even—it stretches even my normally very wide parameters of tolerance to the absolute limit. Whether or not we went to the Moon and back six times is, of course, debatable. If we did, then the reason *why* we suddenly stopped doing so when everything seemed to be going so well is not simply debatable, it is a mystery. If one is trying to solve a mystery, I am a great believer in allowing all shades of opinion and all theories to have sufficient "air time," as it were, be those opinions and theories plausible, implausible, or even…

Patently Absurd

This theory isn't merely absurd; it's w-a-a-y out! So way out as to be completely out of sight, in fact. And one I would have much preferred to avoid discussing in what you have thus far, hopefully, considered to be a very serious case. But deliberately avoiding testimony that should otherwise be aired simply because one's first natural reaction is to brand it "crazy," will not help us to establish the truth, even if we later have to throw it out with the garbage as at best unhelpful, at worst complete rubbish.

Although we shall examine this theory shortly, suffice it to say for the moment that most of the evidence to back up all this "silly stuff" has, on at least two occasions, apparently come from the mouth of he whom one would least expect to hear such ridiculous nonsense...

Neil Armstrong.

The Summation

19

Bogeymen?

Was something observed on "our" Moon that so profoundly shocked Armstrong, Aldrin, and Mission Control after the Eagle had landed that it actually led to the premature end of the Apollo program and, in the final analysis, man's long term ambitions in space? That is the question. Was something discovered about the Moon that no one could either have bargained for or in any way have been prepared for in anyone's wildest imaginings? Unless, of course, Isaac Asimov had co-written the script for NASA with Arthur C. Clarke acting as an advisor! Did something, after making its presence known, literally "hang around" and watch as our two first moonwalkers impersonated a couple of spastic Michelin Men in their bulky white spacesuits, but which all the while made no attempt at contact or tried to interfere in any way?

Please do not desert your benches and walk out of court in disgust at this point, members of the jury (in practicality, that means closing this book) although I would not blame you if you did. Furthermore, as this is not a real court, I have no powers to prevent you from doing so if you wish. The majority of you though, will hopefully choose not to do so, especially as we are about to get to the really good stuff! In fact, this could actually be the most important part on this whole trial for deciding, one way or the other, The Truth about Apollo, and you will see why shortly.

During the presentation of the People's case, you will recall the testimony given to this court by witnesses David Childress, Bill Brian and indeed Phil Corso. If not, then I suggest you have a browse back through your case papers. But the following preposterous suggestion that extra-terrestrials might have played a part in the eventual abandonment of the Apollo project does not this time come from any of the aforementioned creditable, if not necessarily credible, sources. Neither does it originate from that other fanatical brigade (as judge I will resist the temptation of using the term "lunatic fringe") who seem to spot squadrons of UFOs every time most of us—even those of us with open minds on the subject—see flocks of geese flying with the setting sun reflecting off their wings. No...

This time it comes from the astronauts themselves.

And, because the source is the astronauts themselves, it raises two serious questions, which first have to be asked and then discussed in detail before we attempt to answer them. If we can manage to answer them correctly, then we will also have our answer regarding whether or not the Apollo Moon landings were genuine. Members of the jury, believe me when I say that this part of our trial is as serious as that!!

OCCUPIED MOON?

And the first question is this: Have we not gone back to the Moon because something else, some other intelligence, according to statements by the astronauts, is already occupying it?

On the face of it, it seems a ludicrous question, doesn't it? If it were true, why are there all these discussions about whether or not there is life on other planets? Why are we sending robots to Mars to look for signs of life? Why do we have organizations like SETI (the Search for Extraterrestrial Intelligence) scouring the heavens with their radio telescopes waiting for that first identifiable signal from E.T.? For those of you currently thinking this is all becoming much too far-fetched, I agree with you. As I have said, I believe in the possibility of life in outer space, not that there necessarily is any. To be perfectly honest, I am not particularly bothered about whether there is or not. I am neither a "UFO nut" nor a complete skeptic in that regard. However, when I came across the source article below—from which I shall reproduce relevant extracts only—I was immediately reminded of something I heard with my own ears at the time of the alleged live broadcasts from Apollo 11, as I may have hinted while acting as prosecuting counsel earlier. I was only in my mid-twenties at the time, so my ears were comparatively young and, unlike now unfortunately, were in perfect working order.

I clearly heard a conversation between Neil Armstrong and Mission Control that for a few incredible moments sounded nothing short of...well, yes, incredible! Alarming even. Armstrong, all of a sudden not sounding quite so casual and laconic and with obvious concern, appeared to say something quite beyond belief. So much so that he had the puzzled Capcom (Capsule communicator) at Houston saying—and I am working entirely from memory here, although you will read an exact transcript shortly—something like, "Apollo, we don't copy that. Please repeat your last message."

Suddenly, the TV screen went blank and a period of silence ensued before the screen once more burst back into life. Armstrong was still speaking, but now he was back to his normal, matter-of-fact self and the seemingly dramatic exchanges of just a little earlier appeared to have somewhat changed tack. The present exchange was now about more mundane matters to do with the mission itself. Which was good. That the astronauts and Mission Control were again talking about the kinds of things one would expect them to be talking about meant that the problem—and it had certainly sounded as if there had been a problem—must have passed. But it left me, and no doubt millions of others at the time, completely baffled. I was inclined to believe that I must have somehow misheard or misinterpreted what was being said during that earlier mysterious conversation, because, despite my eager anticipation, nothing about it was mentioned either on the TV news later that same evening, or in the newspapers the following day.

Armstrong's startling utterance—and equally bemused friends and relatives that I quizzed later, confirmed that they had clearly heard it too and had independently placed a similar interpretation upon it to my own—appeared to have been totally and, dare I say it—deliberately ignored by all sections of the media! Yet it had sounded very much as if Apollo 11 commander Neil Armstrong, to all intents and purposes broadcasting directly from the Moon, had reported seeing a...

UFO!

More than one, it seems, because the Apollo 11 moonwalkers went on to mention "other spacecraft" and "saucers" plural. Although neither Armstrong nor Aldrin personally used the term "UFOs," Mission Control, in response, did. For those members of the jury unfamiliar with the term, a UFO is an unidentified flying object. The description is usually applied to what appears to be some kind of structured, intelligently controlled aerial device that is of obscure origin and not immediately recognizable, either from its design or its performance, as being man-made, and which is observed either flying, in the process of landing, or already landed in a place where there is no normal expectation or authorization for it to be.

The following extracts of alleged conversations between Apollo
crews and Mission Control in Houston, Texas, are taken from an article posted on the internet by David Cosnette, nowadays a staunch
member of the pro-hoax lobby who has already given testimony to
this court on behalf of the People. However, I deem these extracts,
particularly of the alleged conversations themselves, wholly admissi-
ble in my summing up because they carry no bias either way. According to Cosnette, they are a matter of record and audible copies of
them are also available. It is purely a question of whether one
believes, as far as the spoken words of the astronauts are concerned,
that they originated from the surface of the Moon.

Strangely, Cosnette's own tone throughout the article would appear
to suggest that he does believe the voices of the astronauts were coming from the Moon and that NASA either censored parts of the conversations or tried to explain them away by suggesting that the
listening public had misunderstood or gotten hold of "the wrong end
of the stick," as we say here in the U.K., about what was being discussed.

I am afraid Mr. Cosnette cannot have it both ways. He seems to
have changed his mind about whether or not the Apollo landings were
genuine somewhere between his investigation of the claims that
UFOs were seen on the Moon and writing the piece that was used by
the prosecution to debunk Apollo.

Perhaps he should try deciding which of the two blatantly contradictory stories is the more sensational? Either that Apollo may have
been a colossal hoax from start to finish or that it was all definitely for
real and that something was observed on the Moon that, after six missions carried out in fairly rapid succession, occasioned NASA to then
abort all future attempts because they were too afraid of what was
being reported by their space crews to subject them to further risk!

APOLLO MOON CONVERSATIONS

Source: http//www.ufos-aliens.co.uk/cosmicapollo.html by David
Cosnette ©. Updated March 2002.

APOLLO 11

[A.N. Apollo 11, consisting of Neil Armstrong, Michael Collins
and Edwin Aldrin was the first Apollo flight to allegedly land on the
Moon on July 20, 1969. While Collins maintained the orbit of the
Moon in the Command Module, Armstrong and Aldrin descended in

the Lunar Module, landing in the Sea of Tranquility at 4:17am Eastern Daylight Time. These Apollo 11 Astronaut conversations were mostly taken from the out-of-print book Our Mysterious Spaceship Moon by Don Wilson (Dell, 1975)].

Begins—

According to hitherto unconfirmed reports, both Neil Armstrong and Edwin "Buzz" Aldrin saw UFOs shortly after that historic landing on the Moon in Apollo 11 on 20 July 1969. I remember hearing one of the astronauts refer to a "light" in or on a crater during the television transmission, followed by a request from Mission Control for further information. Nothing more was heard.

The following astonishing conversation was picked up by ham radio operators that had their own VHF receiving facilities that bypassed NASA's broadcasting outlets. At this time, the live television broadcast was interrupted for two minutes due to a supposed "overheated camera," but the transmission below was received loud and clear by hundreds of ham radio operators:

According to Otto Binder, who was a member of the NASA space team, when the two moon-walkers, Aldrin and Armstrong, were making their rounds some distance from the LM, Armstrong clutched Aldrin's arm excitedly and exclaimed:

Apollo 11 (Armstrong): What was it? What the hell was it? That's all I want to know!

Mission Control: What's there?…malfunction (garble)…Mission Control calling Apollo 11…

Apollo 11: These babies were huge, sir!…Enormous!…Oh, God! You wouldn't believe it!…I'm telling you there are other spacecraft out there…lined up on the far side of the crater edge!…They're on the Moon watching us!

—Paused.

Members of the jury, I, Graham Frank, the author of this book and currently acting as judge in this trial, do solemnly swear before God, on my life, on the lives of my nearest and dearest, and by all that I hold to be sacred and holy, that I actually heard the above conversation taking place, and I am not, nor ever have been, a "radio ham." Furthermore, I was not, to the best of my knowledge, in possession of any specialist VHF equipment. I clearly heard the words of the above transcript on BBC television's coverage "live from the Moon," before a sudden, unexpected break in transmission, which was then followed, as I said earlier, by a more relaxed, matter-of-fact conversation that had switched to discussing something completely ordinary to do with the mission!

Cosnette continued—

Wilson writes (p. 48):

Binder ends his report with this observation:

There has, understandably, been no confirmation of this incredible report by NASA or any authorities. We cannot vouch for its authenticity, but if true, one can surmise that Mission Control went into a dither and then into a huddle, after which they sternly ordered the moonwalkers to "forget" what they had seen and carry on casually and calmly as if nothing had happened. After all, an estimated 600 million people around the world were hanging onto every word spoken by the first two men to leave footprints on the Moon.

Wilson writes:

The book Celestial Raise by Richard Watson and ASSK (P.O. Box 35 Mt. Shasta CA. 96067 [916]-926-2316); 1987; page 147-148) records the following continuation of the above remarkable dialogue of Apollo 11, which was picked up by hundreds of ham radio operators in the USA:

During the transmission of the Moon landing of Armstrong and Aldrin, who journeyed to the Moon in an American spaceship, two minutes of silence occurred in which the image and sound were interrupted. NASA insisted that this problem was the result of one of the television cameras that had overheated, thus interfering with the reception.

This unexpected problem surprised even the most qualified of viewers, who were unable to explain how in such a costly project, one of the most essential elements could break down. Sometime after the historic Moon landing, Christopher Craft, director of the base in Houston, made some surprising comments when he left NASA.

The contents of these comments, which is included in the conversations, has been corroborated by hundreds of amateur radio operators who had connected their stations to the same frequency through which the astronauts transmitted. During the two minute interruption—which was not as it seemed—NASA, Armstrong and Aldrin with Cape Kennedy, censored both image and sound:

Apollo 11: Those are giant things. No, no, no—this is not an optical illusion. No one is going to believe this!

Mission Control, Houston: What…what…what? What the hell is happening? What's wrong with you? (Christopher Craft was the Capcom asking the question).

Apollo 11: They're here under the surface.

MC: What's there? (muffled noise). Emission interrupted; interference—control calling Apollo 11.

Apollo 11: We saw some visitors. They were here for a while, observing the instruments.

MC: Repeat your last information!

Apollo 11: I say that there were other spaceships. They're lined up on the other side of the crater!

MC: Repeat, repeat!

Apollo 11: Let us sound this orbita…in 625 to 5…Automatic relay connected…My hands are shaking so badly I can't do anything. Film it? God, if these damned cameras have picked up anything—what then?

MC: Have you picked up anything?

Apollo 11: I didn't have any film at hand. Three shots of the saucers or whatever they were that were ruining the film.

MC: Control, control here. Are you on your way? What is the uproar with the UFOs over?

Apollo 11: They've landed here. There they are and they're watching us.

MC: The mirrors, the mirrors—have you set them up?

Apollo 11: Yes, they're in the right place. But whoever made those spaceships surely can come tomorrow and remove them. Over and out.

—Paused. (Cosnette. End of extract from *Celestial Raise*).

Again, I clearly recall personally having heard this weird verbal exchange for myself as it was happening. It seems that the BBC, at least for a time until things maybe became a shade *too* live, not to mention *too hot*, must also have been by-passing NASA's broadcasting outlets with sophisticated VHF equipment. Tut-tut. Naughty old Auntie Beeb, eh? I bet she didn't have NASA's permission!

But there's always at least one terrible mistake in the best of scams, if scam it be, and maybe this was it. To me, the most significant thing is that the world's most "independent and free" broadcasting service, which can normally be relied upon to tell it like it is during bad times as well as good, either censored itself or was censored from on high. Neither the BBC later that evening, nor the British newspapers the following morning, uttered a single word about the truly extraordinary events that had seemingly taken place on the Moon. Not one, single, solitary word. Not the next day. Not the day after. Not ever. Millions of U.K. viewers and listeners must simply have thought the whole thing a figment of their imaginations or a complete misunderstanding on their part. I assume this, because that was certainly the conclusion that I had no other option than to eventually arrive at.

What the hell is happening? What's wrong with you?

Members of the jury, I have repeated the above two sentences from the Apollo 11 transcript for a very good reason. The reason being that they happen to be our cue to ask the second question of the two that I mentioned earlier.

The second question is this: If we accept it as fact that the above transcripts are genuine and accurate, and I, for one, can vouch for the fact that they are, but at the same time take it as read (a dangerous practice?) that there are *no* UFOs or aliens on *our* Moon, does this mean, then, that millions of us inadvertently heard a deliberate attempt, during a truly live segment of the broadcast, to blow the whistle on a largely if not totally fraudulent project?

"What the hell is happening?" This question from Christopher Craft of Mission Control was directed at Neil Armstrong after the latter told him, "Those are giant things...No one is going to believe this!" You will observe that in response, Craft posed a further, almost tetchy, question: "What's wrong with *you?*" Not a simple, "What's wrong?" Put another way, but perhaps meaning exactly the same thing as the original question, Craft could have said, "Hey! You got that wrong. That's not part of it! What the hell's going on?"

It is almost as if Craft immediately realized that Armstrong had broken away from a prepared and previously well-rehearsed script already prepared for one of the "live broadcasts from the Moon"—a copy of which Craft would have had in front of him as a prompt for his own part—and was genuinely wondering what the freaking hell was going on?

Not knowing quite what to say next, it is feasible that Craft then decided to sort of "play along" by making a few ad-lib comments like, "What's the uproar with the UFOs, over?" A curiously dismissive turn of phrase when one thinks about it, because even a solitary UFO is likely to cause uproar even when seen in the skies of Earth, let alone a *couple* of the blighters seen on the Moon! It seems, though, that he need not have worried too much about the reaction of the *American* public at least, because by then the transmission "coming directly from the Moon" had already been cu—

If you get my dri—

Of course, the above pro-hoax scenario is pure supposition on my part. It might not contain a scrap of truth. But then, members of the jury, are you prepared to accept the full implication of not believing that Armstrong was merely attempting to blow the whistle and that what he was giving his Capcom was a genuine, *bone fide* report?

The transcripts reproduced above are fair and accurate representations of conversations I personally heard taking place between Mission Control in Houston, Texas, and the two alleged moonwalkers of the Apollo 11 mission back in 1969, unless, of course, my ears *did* deceive me at the time, which is what I have more or less been forced to accept all these years. It is possible that many millions more Brits will now also recall this strangest of conversations after being reminded of it here. (I hope so, because that will mean I really do have a bestseller on my hands!!)

But, to be fair, these same transcripts have already been reproduced in various articles and published works to date, and are all available on the Internet. The jury will be interested to know, though, that only the *interpretation* of the transcripts has been challenged by NASA, not the actual content, *which it admits is accurate!*

Members of the jury, now that you have been given the chance to "hear"—and, in many cases, perhaps be reminded of—the incredible conversation reproduced by these transcripts, what other interpretation would *you* place upon them? There are really only two alternatives to be considered. One is totally and utterly beyond belief, the other perfectly plausible.

The first alternative is, of course, that Armstrong and Aldrin did see something utterly incredible on the Moon; things resembling giant, structured spaceships shaped like saucers according to the astronauts. The description was enough to occasion Chris Craft, the duty Capcom at Mission Control at the time, to refer to the subjects of the claimed sighting as UFOs. In the face of such unambiguous statements by trained observers, it must therefore be reasonable to conclude that whatever these objects were, they neither belonged to nor were controlled by mankind, and were clearly watching with interest Armstrong and Aldrin's activities on the Moon!

The second alternative, though, is something quite different. No, not NASA's feeble explanation that it was all just "technical talk" that sounded more ominous than it actually was. There is no mistaking the acronym UFO in the accepted sense of its use by a NASA operative on this occasion. It is a term originally coined by the military and, I repeat, it means **U**nidentified **F**lying **O**bject. It does not stand for Uniformed French Officer or Under Fried Onions. It has to be remembered, though, that something as ordinary as meteorite can be a UFO until identified as a flaming lump of rock hurtling through Earth's atmosphere. But Craft, who made the remark in direct response to the astronauts' description, was *not* referring to meteorites or, ostensibly, to anything hurtling through Earth's atmosphere. He was referring to machines possessing the power of flight that were, should the scenario be true, apparently hovering somewhere above the surface of the Moon. "Huge babies," according to Armstrong, not immediately recognizable by himself or Aldrin as being any kind of space apparatus belonging either to the United States of America or the Soviet Union—at that time the only known powers capable of sending technology to the Moon—but which were, nonetheless, seemingly under intelligent control.

This second alternative is that the astronauts were of course in no danger whatsoever, and were merely attempting to introduce a "spoiler" into a gross deception being perpetrated on an unsuspecting public. If so, then they came *very close* to succeeding! Whether the astronauts would have been in a studio verbally acting out their entire roles to coincide with onscreen actions filmed months earlier is open to debate. Possibly they were, but it would likely have been more a question of adding the occasional live comment in between largely pre-recorded soundtracks, much in the same way that guest on a TV show might make comments over and above a film being shown about some escapade he or she was once upon a time involved in.

A Stark Choice

So, the stark choice facing you, members of the jury, is this: Your first option is to decide that two men *did* actually land on the Moon on July 20, 1969, and at the same time as participating in this most momentous endeavor, they also made the most monumental discovery of all time as well, by finding intelligent life not merely *somewhere* in outer space, but actually residing upon, or at least utilizing, the world "next door" to ours. Your second option, though, is far simpler.

It was a hoax.

But if you continue to believe in Apollo, it comes in one package, I'm afraid. You either accept that package in its entirety or you accept none of it as having been worth a brass farthing towards the advancement of space travel, despite any political advantages gained over other doctrines and the undoubted knock-on effect that we now live in a much safer world. Faced with two such alternatives, I very reluctantly put it to you, members of the jury, that there is only *one* verdict that any sane, balanced, rational, clear thinking person can reach.

So, that's that sorted out then. Between us we have finally managed to crack it. All I really need to do now is bring this trial to a close knowing full well what the majority verdict will be. Of course it was all a hoax. All this rubbish talk of "flying saucers" on the Moon was Neil and Buzz's way of leaving us a few vital clues that what they were participating in was a load of ultra-expensive tomfoolery. They were *joking* when they made those controversial remarks, right?

Wrong.

Cosnette's article continued—

A certain professor, who wishes to remain anonymous, was engaged in a discussion with Neil Armstrong during a NASA symposium. Their conversation [allegedly] went as follows:

Professor: What *really* happened out there with Apollo 11?

Armstrong: It was incredible, of course we had always known there was a possibility—the fact is, we were warned off! There was never any question then of a space station or a Moon city.

Professor: How do you mean, "warned off"?

Armstrong: I can't go into details, except to say that their ships were far superior to ours both in size and technology—Boy, were they big…and menacing! No, there is no question of a space station [on the Moon].

Professor: But NASA had other missions after Apollo 11.

Armstrong: Naturally—NASA was committed at that time, and couldn't risk panic on Earth. But it really was just a quick scoop and back again.

[Cosnette]: Another scientist, Dr. Vladimir Azhazha, said: "Neil Armstrong relayed the message to Mission Control that two large, mysterious objects were watching them after having landed near the Lunar Module. But this message was never heard by the public [not by the American public anyway!] because NASA censored it."

And, according to a Dr. Aleksandr Kasantsev, Buzz Aldrin took color movie film of the UFOs from inside the LM, and continued filming them after he and Armstrong went outside.

Armstrong confirmed that the story was true but refused to go into further detail, beyond admitting that the "CIA was behind the cover-up."

"Buzz" Aldrin, who was also with Armstrong on the Apollo 11 mission, was said to have taken color film footage of alien craft. Armstrong later confirmed that this footage had indeed been shot by Aldrin, only to be confiscated by the CIA on their return to Earth. Fearing for his well-being, Armstrong refused to go into further details, except to confirm that the CIA were behind an extensive cover-up campaign regarding the U.S. space program and consequent encounters with UFOs. In 1979, former chief of NASA Communications, Maurice Chatelain, confirmed that Armstrong and Aldrin had encountered UFOs on the Moon. To this day Chatelain vehemently protests the truth of their accounts.

—Paused. (Cosnette).

Alternative Three

There may, though, be a third alternative to being forced to choose between aliens having taken up residence on "our" Moon and the Apollo landings being a hoax. It is still nowhere near as plausible as the whole thing having been a scam, but far more plausible than Armstrong and Aldrin seeing "other spaceships" on the Moon.

Notwithstanding film allegedly now in possession of the CIA and whatever such film might depict, it is possible that the Moon's barren surface, with light hitting it uncannily from two sources, could have caused the first moonwalkers to see "phantoms" where there were none. Despite Armstrong's protest that it was not an optical illusion, there may well have been a couple of huge, dome-shaped rock formations quite near the Apollo 11 landing site, which the two astronauts suddenly noticed for the first time when they were revealed in a certain light. Further on, you will find that "domes" also featured in later Apollo missions, but there is little to suggest that these were of a suspect nature, even though actual UFOs were again reported by some of the astronauts that followed Armstrong and Aldrin.

As far as the "starting at phantoms" aspect is concerned though, it is worth reminding ourselves that Armstrong and Aldrin were allegedly the first two human beings to set foot on soil other than that of Earth. Even on those TV sets placed in cozy corners of our living rooms, the Moon still looked a pretty spooky place to be, were it the real thing or in truth simply a studio set. Members of the jury, we have to accept that spooky places, like graveyards and creaky old houses at night, have this very unpleasant habit of being able to conjure up all manner of shadowy "ghosts" and "ghouls" that most often, of course, are products of one's own fertile imagination and superstitious beliefs. Both prosecuting and defending counsel gave us highly illuminating insights into the amount of very real terror, despite any bold exteriors, likely to be experienced by whoever is truly among the first to walk upon another planet (and yes, I *do* understand that the Moon is *technically* a moon rather than a planet), notwithstanding that Armstrong and Aldrin may indeed have already claimed that accolade quite legitimately.

"Their ships were superior to ours. Big...menacing!" said Armstrong.
(A.N. Don't panic, folks...yet! This picture is definitely a fake based upon what Armstrong allegedly saw on the Moon. And I should know because I faked it!)

I personally, though, find it extremely difficult to believe that two superbly trained men like Armstrong and Aldrin would have allowed the sudden onset of panic at the appearance of a potential threat, real or imagined, to occasion them to forget a vital part of their training. On reflection, I find it quite astonishing that they would blurt out on their "normal" communication frequency to millions hanging onto their every spoken word with bated breath (an audience that would have included their very own wives or partners, children and relatives), words like "other spaceships," and "saucers," even if they genuinely *thought* they were looking at *s*omething real and quite extraordinary.

Their training would have included set procedures to be followed should evidence of existing life have been encountered on the Moon, however unlikely the possibility may have seemed during training. In fact, in a fairly recent interview with Aldrin regarding the possible sighting of yet another UFO *on the way* to the Moon, Aldrin said, "Obviously we weren't gonna blurt out, 'Hey, Houston, we got something up alongside us and we don't know the hell what it is. Can *you* tell us what it is?' because all sorts of people might be watching and listening." But maybe we'll discuss that episode in more detail a little later on. In further Moon conversations the jury are yet to hear, there are strong indications that the astronauts of later missions actually "switched to code" during certain sensitive exchanges between themselves and Mission Control. For whatever reason, though, if the missions were bogus, is baffling, other than to suggest that a quick change of channels might have allowed the "director" to give a few discreet instructions to his "actors" through their helmet earpieces without interrupting the expensive pre-recordings.

Difficult to make, as the above choices are, members of the jury, the question you *must* each of you ask yourselves is comparatively simple: Which of the two, maybe *three,* alternatives is the more ridiculous? Or one might as easily ask, which is the more likely? That Armstrong and Aldrin saw—alternative three now allows me to say or *imagined* they saw—what to them looked like enormous spaceships shaped like saucers that were watching them on the Moon, and that against all the drill and set procedures learned during training, they were sufficiently startled to yell out to the listening millions at home that Earth was about to be invaded by aliens using *our* Moon as a base? Or is it much more likely that two men in a studio somewhere, probably otherwise "resting" actors, after receiving the signal that they were now "live and on air" and being watched and listened to by millions, thought they would stir things up a little by adding their own ridiculous little piece of melodrama to a prepared script? Something that they could later pass off as merely "larking around" or "doing a bit of leg-pulling," as we say here in the U.K.

Some bloody joke to have at the expense of one's antagonized and frustrated employers and all those listening millions, huh? Some with heart conditions as Aldrin implied above.

"What's wrong with *you?*"

NASA would definitely *not* have been amused and, as I recall, neither was I. Nor am I still today.

Priceless

Anyway, even were the voices truly theirs, it was a risk Armstrong and Aldrin could now take with impunity. It would have been extremely difficult at this late stage in the production for anyone to get them back. They were A-List celebrities now. They were the two most famous men in history. Not only back on Earth but off it as well! Nothing and nobody would dare harm a single hair on their precious heads now. They had not merely become marketable commodities; they were priceless.

Had I been any one of the two of them, though, and despite my new found "untouchable" status, I think I would have erred on the side of caution enough to have at least waited until *after* the space capsule, containing myself and my two comrades, had safely splashed down after parachuting from the B-52 flying high above the Pacific *before* broadcasting any controversial statements live on air to the world's media!

But as well as being intelligent men, easily capable of weighing up the odds in favor of their continued survival after a misdemeanor that they, too, of course, would claim was just a lighthearted prank, the astronauts were honorable men too. Maybe their long-term strategy was that anomalous statements made on the "Moon," even if quickly edited out, would at least reach *some* ears and that the true meaning behind such joking was almost bound to be twigged (true meaning realized) sooner or later—even as much as thirty or forty years later?—and would eventually contribute towards causing the whole precariously built house cards to come tumbling down.

One would be foolish not to bear such a thought in mind, members of the jury...

Cosnette continued—

MORE APOLLO ASTRONAUT CONVERSATIONS

The following Apollo Astronaut conversations were mostly taken from the out-of-print book *Our Mysterious Spaceship Moon* by Don Wilson (Dell, 1975):

APOLLO 15

[David Scott, Alfred Worden, James Irwin, allegedly went to the Appenine Mountains of the Moon, July 26—Aug. 7, 1971. Command/Service Module: *Endeavor.* Lunar Module: *Falcon.* CSM pilot: Alf Worden. LM pilots: Jim Irwin and David Scott].

Conversation about discovering strange "tracks":

Scott: Arrowhead really runs east to west.

Mission Control: Roger, we copy.

Irwin: Tracks here as we go down slope.

MC: Just follow the tracks, huh?

Irwin: Right we're (garble). We know that's a fairly good run. We're bearing 320, hitting range for 413…I can't get over those lineations, that layering on Mt. Hadley.

Scott: I can't either. That's really spectacular.

Irwin: They sure look beautiful.

Scott: Talk about organization!

Irwin: That's the most organized structure I've ever seen! [He means on the Moon, of course!].

Scott: It's (garble) so uniform in width.

Irwin: Nothing we've seen before this has shown such uniform thickness from the top of the tracks to the bottom.

[Cosnette]: Wilson writes, p. 145:

What are these tracks? Who made them? Where did they come from? Does NASA have an answer for the people?

—Paused.

Well, if NASA doesn't, I do. I tend to think both Cosnette and Don Wilson, from whose book DC is quoting, are becoming a little carried away here. I take the description given by the Apollo 15 astronauts to mean they were observing something that *resembled* what we would call "tracks" on Earth. Not the kind of trail left by a man or an animal in snow or the sands of a desert. Not rail tracks. Not tire tracks. But, for want of a more accurate description, "tracks" will do nicely.

It is quite enough for anyone attempting to sort out who is right and who is wrong in this whole "Was Apollo a hoax?" debate, to come to terms with what may be *real* evidence that external forces might have been at work somewhere, as in the former and some following extracts from this piece, let alone to have people quite literally laying false trails by speculating that these roughly-hewn lines across the lunar surface might be the ruins of purpose-built freeways belonging to an ancient and long dead civilization. They are far more likely to be the result of water action or volcanic activity many millions of years ago before the Moon became the completely dead planetary body it is today.

Cosnette continued—

APOLLO 16

[Charles Duke, Thomas Mattingly and John Young allegedly land in the Descartes highlands April 16—27, 1972. Command/Service Module: *Casper.* Lunar Module: *Orion.* CSM pilot: Tom Mattingly. LM pilots: Charlie Duke and John Young].

Conversation 1

[Duke and Young are allegedly on EVA on the Moon].

Duke: These devices are unbelievable. I'm not taking a gnomon up there.

Young: OK, but man, that's going to be a steep bridge to climb.

Duke: You got—*Yowee!* Man—John, I tell you this is some sight here. Tony, the blocks in Buster are covered—the bottom is covered with blocks, five meters across. Besides the blocks seem to be in a preferred orientation, northeast to southwest. They go all the way up the wall on those two sides and on the other side you can only barely see the out-cropping at about 5 percent. Ninety percent of the bottom is covered with blocks that are 50 centimeters and larger.

Mission Control Capcom: Good show. Sounds like a secondary…

Duke: Right out here…the blue one that I described from the lunar module window is colored because it is glass coated, but underneath the glass it is crystalline…the same texture as the Genesis Rock…Dead on my mark.

Young: Mark. It's open.

Duke: I can't believe it!

Young: And I put that beauty in dry!

MC: Dover. Dover. We'll start EVA-2 immediately.

Duke: You'd better send a couple more guys up here. They'll have to try (garble).

MC: Sounds familiar.

Duke: Boy, I tell you, these EMUs and PLSSs are really super-fantastic!

[Cosnette]: It is obvious that the astronauts are talking in code— meant to disguise what they are referring to. The big question is why the excited cries? Can this be merely due to the collecting of Moon rocks, as they [NASA] would have us believe? Or did they find something much more substantial, which was not meant for public knowledge? [*A.N.* In my heart of hearts, David, a little scary though it is, I would really love to think so, but I really *don't* think so].

Conversation 2

Domes and Tunnels

Duke: We felt it under our feet. It's a soft spot. Firmer. Where we stand, I tell you one thing. If this place had air, it'd sure be beautiful. It's beautiful with or without air. The scenery up on top of Stone Mountain, you'd have to be there to see this to believe it— those domes are incredible!

MC: OK, could you take a look at that smoky area there and see what you can see on the face?

Duke: Beyond the domes, the structure goes almost into the ravine that I described and one goes to the top. In the northeast wall of the ravine you can't see the delineation. To the northeast there are tunnels, to the north they are dipping east to about 30 degrees.

Who (or what) is Barbara?

MC: What about the albedo change in the subsurface soil? Of course you saw it first at Flagg and were probably more excited about it there. Was there any difference in it there—and Buster and Alsep [Apollo Lunar Surface Experiments Package] and LM?

Duke: No. Around the Alsep it was just in spots. At Plum [Crater] it seemed to be everywhere. My predominant impression was that the white albedo was (garble) than the fine cover on top.

MC: OK. Just a question for you, John. When you got halfway, or even thought it was halfway, we understand you looped around south, is that right?

Young: That is affirm. We came upon—Barbara.

[Cosnette]: Wilson writes, p.140:

Joseph H. Goodavage, who included this conversation in a *Saga* magazine article, comments: "Barbara? That really needs some explanation, so I made an appointment with NASA geologist Farouk El Baz at the National Aeronautics and Space Museum."

[Cosnette]: According to Wilson, Goodavage says part of the conversation went:

Saga: What do you suppose Young meant when he said they came upon "Barbara"?

El Baz: I can't really say. Code perhaps…

Saga: But Barbara is an odd name for something on the Moon, isn't it? [Is it?]

El Baz: Yes, an enigma. As I suggested, perhaps a code, but I don't really know.

Conversation 3

Terraces

[Cosnette]: Wilson writes, p. 141:

While on the Moon, did any of our astronauts see any indication of alien handiwork, such as strange constructions, disturbances or the like? Consider this strange Apollo 16 conversation:

Orion: Orion has landed. I can't see how far the (garble)…this is a blocked field we're in from the south ray—tremendous difference in the albedo. I just get the feeling that these rocks may have come from somewhere else. Everywhere we saw the ground, which is about the whole sunlit side, you had the same delineation the Apollo 15 photography showed on Hadley, Delta and Radley Mountains…

MC/Capcom: OK. Go ahead.

Orion: I'm looking out here at Stone Mountain and it's got—it looks like somebody has been out there plowing across the side of it. The beaches—the benches—look like one sort of terrace after another, right up the side. They sort of follow the contour of it right around.

MC: Any difference in the terraces?

Orion: No, Tony. Not that I could tell from here. These terraces could be raised but of (garble) or something like that…

Casper: [Mattingly allegedly in lunar orbit overhead]. Another strange sight over here. It looks—a flashing light—I think it's Ann-bell. Another crater here looks as though it's flooded except that this same material seems to run up on the outside. You can see a definite patch of this stuff that's run down inside. And that material lays or has been structured on top of it, but it lays on top of things that are outside and higher. It's a very strange operation.

[Cosnette]: Wilson writes, p.142:

And we might add that this is a very strange conversation. What are the real meanings of such terms used here as structure, blocked field, beaches, benches, terraces and the like? NASA claims that they are just metaphoric terms to describe unusual natural formations.

—Paused.

And I, for one, believe them, Don and David. Providing, of course, that I accept that the crew of Apollo 16 were actually on the Moon in the first place! Some of the above remarks are curious, even interesting, but—except for flashing light—hardly ominous. "Annbell" is only curious insofar as it appears to relate to the flashing light. As for the reference to "Barbara" in an earlier transcribed conversation, well..."she" sounds just about as intriguing as does "Hill 14" on a military battle plan to be perfectly honest! Members of the jury, I have only included some of these conversations in my summing up purely for the sake of those of you who may already have minds made up that the Apollo conversations were all scripted and acted out in a studio. I can only advise that I personally have serious doubts that even the most gifted of Hollywood scriptwriters would have had the wherewithal to be able to invent some of the things being said above, even with a NASA Moon expert sitting beside him or her.

Having said that, the prosecution has provided us with photographs of precise replicas of the Moon with all natural features and relief modeled exactly to scale. A process of filming these models by slowly running the camera over the entire fake surface, would then allow a commentary of what the camera was seeing to be recorded by the genuine astronauts for a particular mission, who would, in effect, just be sitting there watching the playback rather than sitting side by side in the Lunar Rover on the Moon. If one thinks about it logically, remarks apparently made spontaneously by the astronauts as they toured the Moon, could just as easily have been made whilst comfortably supping coffee and sandwiches in a darkened room in front of a large screen. The end result, with spontaneous remarks either included or edited out if deemed superfluous, would probably have sounded far more realistic than the work of a scriptwriter. If so, then it was cleverly done. But then again, if virtually the whole of Apollo was faked we must face the fact that it was *all* cleverly done.

Fraudulent or not, the above transcript sounds remarkably realistic to a lay person like myself—but that would have been the *intention,* of course—with the astronauts attempting to describe what they are seeing with a curious mix of codewords and jargon, as well as terms that could be easily understood by all back on Earth. As with the term "tracks," reported by an earlier mission, I do not find the use of the word "terraces" in the least puzzling. Nor does it raise any curiosity in me whatsoever. I take it to mean that what was seen, either actually on the Moon or as the camera was moved slowly over an exact replica lunar surface, *resembled* terraces. Similarly, most intelligent people, particularly those who have not allowed obsession to cloud their judgment, would deem the usage of words like "structure," "wall," and "dome" by the astronauts, wherever they truly were at the time, to mean features *resembling* such man-made edifices, even though naturally occurring or designed to replicate something naturally occurring.

Cosnette continued—

APOLLO 17

[Eugene Cernan, Ronald Evans, and Harrison "Jack" Schmitt allegedly land in the Taurus-Littrow Valley; Dec 7—19, 1972. Command/Service Module: *America.* Lunar Module: *Challenger.* CSM pilot: Ron Evans; LM pilots: Harrison Schmitt and Eugene Cernan].

Check out the following curious conversations that took place:

Conversation 1

[Schmitt and Cernan are allegedly on the Moon. Evans orbits above].

Mission Control: Go ahead, Ron.

Evans: OK, Robert, I guess the big thing I want to report from the back side is that I took another look at the cloverleaf in Aitken with the binocs. And that southern dome (garble) to the east.

Mission Control: We copy that, Ron. Is there any difference in the color of the dome and the Mare Aitken there?

Evans: Yes there is…That Condor, Condorsey, or Condorecet or whatever you want to call it there. Condorecet Hotel is the one that has got the diamond shaped fill down in the uh—floor.

Mission Control: Robert here. Understand. Condorcet Hotel.

Evans: Condor. Condorcet. Alpha. They've either caught a landslide on it or it's got a—and it doesn't look like (garble) in the other side of the wall in the northwest side.

Mission Control: OK, we copy that. Northwest wall of Condorcet A.

Evans: The area is oval or elliptical in shape. Of course, the ellipse is toward the top.

[Cosnette]: Again we have another example of code being used to disguise what has been found. For example, "Condorcet Hotel." Why the codes, if there are no secrets being discussed? Why not explain to the American people openly what is going on? After all, they have *paid* for the mission.

—Paused.

Oh, come on, David. That's just a huge lump of rock they're referring to, either on the Moon itself or on the scale model. Its shape must have reminded someone of a hotel they knew of, that's all. For God's sake, lighten up man!

Cosnette continued—

Wilson says, p. 139:

Although NASA has always held that the findings of lunar and space expeditions have never been held secret, it is interesting to note that Dr. Farouk El Baz, one of NASA's foremost scientists, does admit that "not every discovery has been announced."

—Paused.

Oh dear. That could just as easily be because NASA have no wish to bore us all to death! However, the next conversation Cosnette reproduces becomes a lot more interesting than a load of old rocks.

Cosnette continued—

Switch to code

Conversation 2

[Schmitt and Cernan allegedly on Moon. Evans circles above].

Schmitt: What are you learning?

MC/Capcom: Hot spots on the Moon, Jack?

Schmitt: Where are your big anomalies? Can you summarize them quickly?

MC: Jack, we'll get that for you on the next pass [allegedly of the Command Module in Moon orbit].

Evans: Hey, I can see a bright spot down there on the [LM] landing site where they might have blown off some of that halo stuff.

MC: Roger. Interesting. Very—go to KILO. KILO.

Evans: Hey, it's gray now and the number one extends.

MC: Roger. We got it. And we copy that it's all on the way down there. Go to KILO. KILO on that.

Evans: Mode is going to HM. Recorder is off. Lose a little communication there, huh? Okay, there's bravo. Bravo, select OMNI. Hey, you know you'll never believe it. I'm right over the edge of Orientale. I just looked down and saw the light flash again.

MC: Roger. Understand.

Evans: Right at the end of the rille.

MC: Any chances of—?

Evans: That's on the east of Orientale.

MC: You don't suppose it could be Vostok? [A Russian probe].

[Cosnette]: Wilson writes, p. 141:

The Vostok flights took place in the early sixties and were strictly Earth orbiters. They never reached the Moon! [However, this is the second reference to a "flashing light" so far, and from two different missions!].

Watermarks

Conversation 3

[All allegedly still in Moon orbit prior to separation and landing].

MC/Capcom: Roger, America, we're tracking you on the map here, watching it.

Evans: OK. Al Buruni has got variations on its floor. Variations in the lights and its albedo. It almost looks like a pattern as if the water were flowing up on a beach. Not in great areas, but in small areas around the southern side, and the part that looks like the water-washing pattern is a much lighter albedo, although I cannot see any real source of it. The texture, however, looks the same.

MC: America, Houston. We'd like you to hold off switching to OMNI Charlie until we can cue you on that.

Evans: Wilco.

Schmitt: Was there any indication on the seismometers on the impact about the time I saw a bright flash on the surface?

MC: Stand by. We'll check on that, Jack.

Schmitt: A UFO perhaps, don't worry about it. I thought somebody was looking at it. It could have been one of the other flashes of light.

MC: Roger. We copy the time and…

Schmitt: I have the place marked.

MC: Pass it on to the back room.

Schmitt: OK. I've marked it on the map, too.

MC: Jack, just some words from the back room for you. There may have been an impact at the time you called, but the Moon is still ringing from the impact of the S-IVB [*A.N.* During the Apollo 15 mission, the Saturn V third stage booster (S-1VB) was deliberately impacted on the Moon], so it would mask any other impact. So they may be able to strip it out at another time, but right now they don't see anything at the time you called.

Schmitt: Just my luck. Just looking at the southern edge of Grimaldi, Bob, and—that Graben is pre-Mare. Pre-Mare!

MC: OK. I copy on that, Jack. And as long as we're talking about Grimaldi we'd like to have you brief Ron [Ron Evans, the Command Module pilot] exactly on the location of that flashing light you saw…We'll probably ask him to take a picture of it. Maybe during one of his solo periods [MC means after separation and landing of the LM when Evans would allegedly be flying alone].

[Cosnette]: Notice that the Capcom reiterates that it was a *flashing* light. It was therefore no meteor impact that they were witnessing. Notice also that the Lunar Module Pilot [Schmitt] specifically mentions the word "UFO."

—Paused.

This *is* interesting, especially in the light of Armstrong and Aldrin's conversation with Mission Control during Apollo 11. If it wasn't for that, and Armstrong's later admission about being, and I quote, "warned off," and his description of the following missions as, "a quick scoop and back [to Earth] again," I would have again accused David Cosnette of being overly melodramatic, which he undoubtedly is being anyway. Especially if he is trying to tie certain other observations made by the astronauts in with reports of flashing lights and UFOs.

Difficult as it can be at times, members of the jury, we must all endeavor to curb any personal bias, no matter how strongly we might feel about a particular subject, and try to look at everything in a detached, cool, calm, rational manner. To use terms like "structures" and "domes" is one thing. To use terms like "flashing lights" and "UFOs" is something else entirely. I don't know about you, members of the jury, but at this stage of these proceedings I feel my personal "Hoax-O-Meter" needle, which had not merely swung earlier into the red zone of the "Yes, it was obviously all a hoax" area but off the board entirely, now swinging back into the "No, maybe it wasn't" area once more. But I feel bound to add that although I make this observation very happily, I still do so more than a little uncertainly.

The reason for the reverse swing is because I feel that if Armstrong and Aldrin had been joking, either in the CM still orbiting the Earth or in a studio somewhere, it should have been something that the astronauts, as a whole Apollo team, would only have been allowed to get away with once. NASA bosses would not have left any avenue open for a repeat performance to again take them unawares. There should have been no further mentions of anything even remotely resembling other spaceships in the exchanges between Mission Control and following Apollo hoax missions but, as you can see, there were.

However, my returning faith might be short-lived. (Perhaps thirty-two years as a newspaperman before taking up all this make-believe legal lark has taught me to be cynical). I find myself asking *why* I would mention flashing lights and UFOs if I happened to be one of the astronauts, or actors, that came after Armstrong and Aldrin, sat watching a large television screen as a camera was electronically guided across the face of a model? I find myself answering that I would only do so if the words happened to appear in the script, because no one would have dared "ad-lib" such remarks in future unless this was so.

It occurs to me, members of the jury, that NASA might have included these remarks deliberately, in order to cover *its* tracks…

Not those on the Moon!

NASA would have been aware by now that despite its very best efforts, many around the globe as well as in the United States would have clearly heard the startling conversation from Apollo 11, and the procedure it would have adopted in order to tough its way out of a tight corner is another time-honored one. There is even a name for it.

It is called "clutter." Rather than attempt to deny one blatantly undeniable report (to anyone armed with a tape of the Armstrong/Aldrin/Craft conversation and proclaiming "Gotcha!"), you draw their attention to a whole raft of similar remarks, all made over the course of six missions. You then dismiss these as tricks of light or figments of the imagination, things likely to occur in strange circumstances and weird surroundings, all of which had been thoroughly investigated during debriefings of the men concerned, all of whom now agreed that they had merely been "seeing things."

I repeat that the only other alternative is that the anomalous reports were a failure, on occasions, to switch to code in time, which can only mean that mentions of aliens and spaceships on the Moon, and therefore the missions to the Moon themselves, were absolutely genuine. I said earlier that Apollo has to be treated as a whole if you are convinced there was no hoax. What learned counsel for the prosecution said about the possibility of fraudulent stuff being mixed in with the genuine is an option open only to those of you who find yourselves pro-hoax at this stage of the trial. But if you remain wholeheartedly anti-hoax, despite any testimony given in this trial which may have upset you or occasioned the odd moment of doubt, then you *cannot* pick and choose the bits you believe happened and discard other parts you think maybe didn't. Everything is on record. If you believe in the integrity of Apollo, then you believe in its *complete* integrity. Therefore, you also have to accept the very high probability that gigantic "flying saucers" from God-only-knows-where and capable of God-only-knows-what, might not merely have landed upon the Moon we see in the sky every clear night, but might still be there!

Waiting...

The even more worrying thing though, is that if I seriously thought that to be a possibility—and I don't—I would have taken Aldrin's advice and, for the sake of those members of the jury of a nervous disposition or worse, would not have labored the point.

Cosnette's article continues—

Wilson writes, p. 60:

This last conversation makes it obvious that both our astronauts and NASA do not take these sightings of light or UFOs lightly. Maps were marked and photographs were taken at the sites of these occurrences.

The sighting of the UFO occurred while the Apollo 17 astronauts were discussing the "watermarks." The conversation then returns to the watermarks:

Cernan: [Still allegedly in Moon orbit prior to separation].OK. 96:03. Now we're getting some clear—looks like pretty clear high water-marks on this -

Evans: There's high watermarks all over the place there.

Schmitt: On the north part of Tranquilitatis. That's Maraldi there, isn't it? Are you sure we're 13 miles up?

MC: You're 14 to be exact, Ron. [The reply is to Evans, the Command Module pilot].

Schmitt: I tell you there's some mare, ride or scarps that are very, very sinuous—just passing one. They not only cross the low planar areas but go right up the side of a crater in one place and a hill in another. It looks very much like a constructional ridge—a mare - like ridge that is clearly as constructional as I would want to see it.

—Paused.

"As constructional as I would want to see it." In other words, Schmitt is saying that the observed ridge, or the exact replica of it on the model, which I am now beginning to *pray* was used instead of the actual Moon, somehow *resembled* a construction built by intelligent hands rather than by natural forces, not that it actually was. Let us deal with facts and not read hidden meanings into everything. We are already dealing with a big enough "flight of fancy" as it is! What Armstrong and Aldrin said is not open to interpretation, but the question of *where* they said what they did and *why* they said what they did is. Jack Schmitt was a civilian and a trained geologist. Wherever he truly was when he was saying these things, he was attempting to describe what were supposed to be naturally occurring rock formations on the Moon to someone at Mission Control who was not a trained geologist, as well as to the listening public. It is obvious that he found the best way of doing this was to *liken* what he was observing to artificial structures on Earth that we are all familiar with. "It looks very much *like* a constructional ridge," he says. Without wishing to sound patronizing, that should be plain enough for most people.

Cosnette continued—

Russian Cosmonaut shadowed by structured UFO

In April of 1979, Cosmonaut Victor Afanasyev lifted off from Star City to dock with the Soviet *Salyut 6* space station. But while *en route*, something very strange happened.Cosmonaut Afanasyev saw an unidentified flying object [a UFO] turn toward his craft and begin tailing it through space.

"It followed us during half of our orbit," he said. "We observed it on the light side, and when we entered the shadow side, it disappeared completely. It was an engineering structure, made from some type of metal, approximately 40 meters long with inner hulls. The object was narrow here and wider there, and inside there were openings. Some places had projections like small wings. The object stayed very close to us. We photographed it, and our photos showed it to be 23 to 28 meters away."

In addition to photographing the UFO, Afanasyev continually reported back to [Soviet] Mission Control about the craft's size, its shape and position. When the cosmonaut returned to Earth he was debriefed and told never to reveal what he knew, and had his cameras and film confiscated. Those photos and his voice transmissions from space have never been released.

It is only now, with the collapse of the Soviet Union, that Afanasyev feels that he can safely tell his story.

"It is still classified as a UFO because we have yet to identify the object," he said.

GEMINI 7

[Gemini 7 pilot Jim Lovell went on to become part of the crew of both Apollo 8 and the ill-fated Apollo 13 mission in which—genuine or otherwise—there was thankfully no loss of life. Jim Lovell's part in a film bearing the same name as his mission, *Apollo 13*, was played by the famous actor Tom Hanks].

Conversation:

Lovell: Bogey [UFO] at 10 o'clock high.

Capcom: This is Houston. Say again 7.

Lovell: I said we have a bogey at 10 o'clock high.

Capcom: Gemini 7, is that the booster or is that an actual sighting?

Lovell: We have several…er, actual sightings.

Capcom:…Estimated distance or size?

Lovell: We also have the booster in sight.

[Cosnette]: "The encounter was common knowledge at NASA," Chatelain said. "All Apollo and Gemini Flights were followed by space vehicles of [presumed] extraterrestrial origin—or UFOs, if you prefer to call them that. Every time it occurred, the astronauts informed Mission Control, who then ordered absolute silence." He [Chatelain] added: "I think Walter Schirra, aboard Mercury 8, was the first astronaut to use the code name 'Santa Claus' to indicate the presence of UFOs." [Could the same apply to Santa *Barbara*, I wonder? No, I don't think so either].

But it was [again] James Lovell, on board the Apollo 8 Command Module, who came out from behind the Moon and said, for everybody to hear: "Please be informed that there is a Santa Claus!" Even though this happened on Christmas Day, 1968, many people sensed a hidden meaning in those words. [Er...I didn't].

Chatelain also published an article in 1995 that confirmed that not only did the Apollo Moon missions encounter UFOs, but that they also found "several mysterious geometric structures of unnatural origin on the Moon."

—End.

§ § § § §

Again I have occasion to thank David Cosnette, this time for his valuable contribution to my summing up, even if I do think he gets a little too carried away sometimes by his enthusiasm for his subject. Despite his concluding remark, though, there is nothing "confirmed"; least of all that UFOs, be they observed on the Moon or on Earth, are of extraterrestrial origin. If man did go to the Moon and did indeed observe "other spaceships" to be already there, who is to say that these were any more extraterrestrial than the LM itself? After all, the latter was also a spaceship and it wasn't on Earth, was it? The "others" were not American, of course, and not Russian. But were they categorically not of the Earth? Who knows?

Neither is it "confirmed" that man actually went to the Moon of course, let alone that he landed upon it and walked upon its surface. After all, that is precisely *why* we are all sat here participating in this very trial.

Buzz Word

As I think I mentioned in passing earlier, it appears that the sighting of UFOs on the Moon was not the only encounter Apollo 11 had with such phenomena. There is video footage of an interview with Buzz Aldrin, in which he reported that all three members of the crew, three days into the journey to the Moon, noticed a strange object apparently shadowing or "buzzing" Apollo. The crew, naturally reticent about explaining their reasons for the question, because as Aldrin

says, "You didn't know who might be listening," guardedly asked
Mission Control about the whereabouts of the most recently discarded S-1VB (mentioned earlier and pronounced "S-four-B") stage
of the Saturn V that had lifted them into Earth orbit.

"6000 nautical miles away," came the reply.

Oh dear. I'm beginning to wish I hadn't started all this. Anyone
else well and truly getting the creeps now, or is it just me?

20

Man the Machine

For me personally, members of the jury, the most poignant piece of testimony given on behalf of the People's case was provided by a retired American army colonel, now dead, who gave us not so much testimony about Apollo in particular, but rather informed opinion regarding the viability of space travel in general by mankind. This man worked for many years at the Pentagon, at the very heart of government where it meets military science, so what he told us is not to be taken lightly.

"Man *cannot* travel in space," Col. Philip Corso said, quite unequivocally. He did not mean "for the time being" or "until one day maybe" or "until the technology improves." He said in no uncertain terms that mankind was fundamentally unsuited for space travel. He did not mean spiritually, or that we were unable to adapt mentally to the rigors of space. It was the human body, or this "mortal coil," as Shakespeare described it, which was our main drawback. It was too perishable. Too fragile. Too inextricably bound to the Earth.

Corso himself was no scientist. He was a soldier. But many of his day-to-day colleagues in the U.S. Army's Research and Development Department at the Pentagon *were* scientists. Top scientists. The very best that American money could buy. And the information that Corso became privy to whilst working hand in glove with these top brains had left him in no doubt whatsoever that as living, breathing, flesh, blood and bone creatures, we were on an absolute hiding to nothing out there in space. That we simply could not "hack it" out there and, what is more, we would *never* be able to do so because of our physical limitations.

Outer space, and that is where the Moon is, is far more alien to man than the very bottom of the deepest abyss in Earth's deepest

335

ocean; far more hostile than the highest peaks of the Himalayas. There is a total absence of heat, a total absence of gravity, a total absence of water and a total absence of oxygen. In fact, a total absence of *everything* needed to sustain life. Indeed, that is *why* it is called "space." It is absolute nothingness. A complete void. But, as learned council for the prosecution mentioned earlier, that doesn't make it empty. After all, it is full of the universe for a start! But it is not the universe that is the problem; it is the debris of that universe that constitutes the danger. Countless billions upon billions upon billions of chunks of *debris* ranging from enormous comets to tiny pieces so small as to be invisible to anything other than an electron microscope and, as learned counsel for the prosecution also informed us, all traveling faster than speeding bullets from a gun!

Whenever man creates his own temporary environment in space, such projects are always set up in near space—"phony space" it is sometimes called—beneath the all protecting "umbrella" of the Van Allen radiation belts and never in outer or "deep" space. Speaking as he was about the orbiting Russian *Mir* space station at the time, Corso said, "Those guys cannot even walk when they bring them down!"

He meant, of course, that weeks and months of weightlessness in space has a debilitating or wasting effect upon the muscles of the human body, which are designed to enable our limbs to cope with the stresses and strains of performing even the most simple of functions in Earth's gravity. Even threading a needle puts strain upon the muscles that control our eye movements, fingers, wrists, arms and shoulders, no matter how seemingly effortless such an action might seem. For anyone with 20-20 vision that is!

There is simply no getting away from the fact that our human bodies are designed specifically for use *on Earth* or within our planet's gravitational pull and atmosphere. There can be no avoidance of this one basic, simple fact of life. Outside of our home planet's influences, our bodies sustain untold damage after relatively short periods of time, and that is with or without the ever-present danger of deadly radiation, be it in particle or ray form. The radiation question is, of course, the most worrying factor of all in this whole "Moon hoax" debate. As convincing and as detailed as the case for the defense was in putting the argument that radiation dangers in space are quote negligible, I am afraid that I personally remain unconvinced. I feel there can be nothing negligible about any kind of danger presented by space, because the dangers are largely incalculable as well as untested.

I recently caught the tail end of an announcement on Channel 4 News (U.K.) that some scientists were discussing the possibility of sending a manned mission to Mars within a few years. There was a time when the prospect would have excited me greatly. Nowadays, however, I would be happy if someone could offer me some *real* proof that man has even landed on the Moon yet! I do not recall hearing anywhere in the bulletin, though, the acronym NASA. So the scientists mentioned were presumably not associated with NASA, either directly or indirectly. It must have been some European thing.

You will recall NASA, members of the jury. For better or worse, the only organization to our knowledge even remotely capable of landing a Frisbee on the planet Mars, let alone a manned spacecraft! And I say this with all due respect to the European Space Agency's current project of landing a robot probe on Mars by Christmas Day 2003. Which will hopefully be successful. (*Later added note*: No, it wasn't!)

But, returning to NASA. This is the same NASA that, according to testimony presented to this court, will not even be seriously considering returning to the Moon for another *hundred* years! As this Channel 4 news item was announced within two or three weeks of the *Columbia* space shuttle disaster, in which five male and two female astronauts were presumably reduced to ashes within seconds while merely re-entering Earth's atmosphere after not having been anywhere in particular, then NASA's forecast is hardly surprising, is it? So where on Earth—sorry!—did anyone get this crazy notion that man will be landing on Mars within a few years? Please don't hold your breath anyone!

Whoever this group of over-ambitious scientists are—I seem to recall now that they might have been attached to a British university—and despite a deep, personal longing for my ever-growing cynicism regarding space travel to be proven wrong, I would humbly suggest that we all get real.

We just ain't gonna be sending men and women to Mars in the near future. No way, Jose. Regardless of what "pie in the sky" schemes you might hear announced on TV or radio in the near future—even "official" ones by NASA—it isn't going to happen. You mark my words. Whether we like it or not, and I for one most certainly do not, there are serious and grave doubts regarding whether or not we can actually send human beings to the Moon, or that we will ever be able to do so, as such.

Sadly, and with a genuine deep, personal sadness, I am finding myself more and more of the persuasion that Phil Corso, or rather his sources at the Pentagon, were correct in their assessment of the situation: that man himself will *never* be able to travel in outer space. Certainly not in these feeble bodies, which are designed only for living upon Earth. And the more I think about it, the more I tend to understand *why* that prognosis *must* be correct. The saying, "a fish cannot live out of water" immediately springs to mind. It is actually wrong. A fish *can* live out of water. But usually only for a few minutes. An eel, though, which is still a fish, can survive out of water for much longer periods. In fact, an eel can live out of water for many hours and indeed uses this ability to its advantage by slithering its way, like a snake, across short stretches of land to the next pool, pond or lake, which might contain its next meal or next mate for breeding purposes. The catfish, too, can survive for a considerable time on land. But none of these creatures are amphibians, like toads, frogs and newts; they are purely and simply fish. They are the exceptions that prove the general rule that should read: "Fish, generally speaking, cannot live out of water."

Similarly, whales, porpoises and dolphins are classic examples of exceptions to the *general* rule that mammals, a group that includes man, cannot live *in* water. These creatures actually *do* live in water, but they can only totally immerse themselves in it for a relatively short time. Even the mighty whale can only remain fully submerged for about an hour before having to come up for a fresh supply of air.

Man, too, has learned to survive underwater with a whole range of inventions and adaptations. These range from the simple snorkel, mask and flippers to a nuclear-powered submarine. Around a hundred years ago, when man was experimenting with all manner of crazy-looking appendages strapped to his arms in order to try to fly like a bird, cynics would say, "If God had intended us to fly, He would have given us wings." Well, God did *not* intend us to fly. And He didn't give us wings before and He hasn't done so since. What God gave us instead was a brain that eventually decided that we would never be able to fly like a bird simply by slipping our arms into a pair of artificial wings made of wood and eagle feathers and flapping our arms up and down like mad. Evolution marches forwards, not backwards. We have passed the stage when birds were the highest form of life and we, humanity, have moved on *from* there. To attempt to return to our bird state would take many millions of years of running around flapping our arms about—in the meantime growing our *own* feathers, not using borrowed ones—in the hope that these actions would give us enough speed and lift to carry us beyond the reach of vicious predators constantly snapping at our heels because they wanted to have us for lunch!

By using our *brains* instead of our arms, though, we have managed to cut through the red tape of retro-evolution and come up with an expedient. Instead of making only artificial wings we decided to give the wings a body, a tail, and an engine. Effectively, we now had a whole *flying machine* possessed of all the apparent advantages of being a bird and then some. We could now board this machine whenever that primeval urge to soar through the skies arose and take off.

Birds, of course, evolved from small dinosaurs desperate to escape the danger of hungry, much larger reptiles, and did so first by flapping their forelegs about to give them more lift as they jumped away and briefly up into the air in panic from those ever present, snapping, tooth-filled jaws. Eventually, though, flapping forelegs would become wings allowing these animals to seek complete refuge in the air. The problem was, though, that some of the larger reptiles then also decided to take to the air in order to continue the chase! In the case of both smaller and larger reptiles, evolution first allowed their forelegs to develop flaps, which gradually developed into wings that were more bat-like (pterosaurs) and then into true, feathered wings (archaeopteryx). But in the case of our flying machine—an airplane, of course—the power (the engine and propellers) and the lift (the wings) are separate things able to interact together.

You must be wondering where all this patently obvious stuff about man conquering water and air is leading. Am I on the verge of contradicting myself, do you think? Am I making the case *for* mankind being able to travel and live in space? After all, if we can find ways of living in water and in the air, why not space? If we can come up with scientific and mechanical solutions for operating underwater and in the air, why will such solutions not apply in space? In fact, haven't we *already* invented, built and flown machines for traveling in space? Don't we call such machines spacecraft?

Yes to almost all of the above. But the problems of true and meaningful space travel will have to be surmounted in a very different way to the way we have conquered water and air. The point is that we cannot *live* underwater. Not like fish, we can't. We will never retro-evolve into mermen and mermaids anymore than we will ever retro-evolve into birds. The mighty whale mentioned earlier has spent many millions of years trying to turn itself back into a fish and it *still* hasn't succeeded. It never will. And neither will we. As with birds, there was an earlier time when fish were the highest form of life, but some eventually climbed out of the water and moved on. Birds, and indeed human beings, are the eventual result. I repeat: evolution will not permit us to go backwards. That is not part of the great universal game we are all playing. That is not God's will, if you like. We, or life in general, can only advance...

Forever onward.

But what about upward? Isn't that the expression? Onwards and upwards? Do I mean into space? Yes, indeed I do. But there is a problem with space. A former candidate for the U.S. vice-presidency named Dan Quayle once said, "there is an awful lot of it," and he was right. So much of it, in fact, that in order to be able to travel *through* it, one needs to be able to *live* in it. Actually live our lives in space for very many years at a time. Even going to the Moon and back, as we are supposed to have done already, takes eight or nine days including the stopover. That's eight days spent in an artificial environment surrounded by a totally alien environment. Imagine spending eight days underwater in a tiny bathysphere or diving bell. Even eight days without resurfacing in a comparatively huge, nuclear powered submarine can be claustrophobic enough.

Furthermore, try to imagine spending eight days flying around the world non-stop in a passenger jet being continually refueled in midair. After eventually landing, you will not only be totally disorientated but if you had spent most of the in-flight time in a cramped, seated position—more than likely—then it is also probable that you will have temporarily lost the use of your legs as well. It seems that this is invariably the case with the *Mir* cosmonauts and astronauts after a three-month stint in weightless conditions, except that we are now only talking little more than a week! Compare either of the aforementioned with a round trip to Mars, which would take around *three years* altogether, including stopover.

At least when spending limited periods of time under the sea or in the air, the plus side is that neither of these two elements constitutes an *alien* environment. After all, they are two of the basic elements that make up the Earth itself, the very stuff of our existence. The very stuff of which *we* ourselves are largely made. As I said earlier, we actually emerged from the one many hundreds of millions of years ago, albeit in a far different form, because we aspired to live in the other. Whatever dangers may lurk in the sea and sky, certainly if one fails to treat either with due respect and take the necessary precautions, water and air are still part of our *natural* environment. They are *not* alien.

But out there in space, members of the jury, beyond the Van Allen belts, lies a horse of a different color…

Black.

The only thing truly "at home" in space is a machine. Any machine: From one the size of a pocket calculator to one the size of an aircraft carrier. In fact, providing one doesn't forget to oil any moving parts, a machine will often function better in space than on Earth because there is no air resistance to slow those moving parts down.

True, machines are dead things, but they are not inanimate. Far from it. Machines now possess both mobility and artificial intelligence. Their built-in advantage in space is that they are made from minerals mined *from* the Earth, duly refined and processed and used in a machine's construction. The same way that the Earth itself is perfectly happy existing in space, just like the Moon, Mars and all the other planets, so *pieces* of the Earth are equally as at home and happy in space.

We ourselves are "pieces" of the Earth too, but we are perishable goods, designed only for existing *upon* Earth, not *off* it. Put another way, our planet is a large lump of rock traveling through space. It is largely unaffected by extremes of heat and cold or deadly radiation. Space is Earth's natural environment. It follows, therefore, that space is a natural environment for anything constructed of the same materials of which the Earth is made. It matters not a scrap how refined the material used or how sophisticated and delicate the resulting piece of equipment made from it may be. Out there in space, a machine, designed as it might be to perform the most intricate of functions, is just another piece of space rock. It feels neither heat nor cold. It feels no pain when a meteor as hard as a diamond, maybe even made of diamond, bursts through its "skin" at ten times the speed of a shot from a high-powered hunting rifle. It may stop functioning correctly, depending upon which part of its "body" is affected, but it will not die. It cannot die. It has *no life* to lose.

When we speak of man in space and on the Moon, we are actually talking about machines in space and machines on the Moon. In space, all man does is "hitch a ride" in a machine in exactly the same way that as he does with an aircraft or submarine. It is the machine that is traveling through, operating in, and existing in those formerly inaccessible regions, not man. Man boards it and takes manual control it is true. But again it is the "skin" of the spacecraft or spacesuit—and the latter is a machine too, believe me—that "feels" the terrible icy-coldness of space, not man himself. It is the outer surface of the sub-marine or diving bell that is aware of the awful, crushing, rivet-pop-ping pressure of the ocean a hundred fathoms down, not man. It is the metal outer plating of the wings and fuselage of an aircraft that take the awesome battering of heading into a 90 mph gale at 600 mph, not man. Oh, no. The people inside lie there in their reclining seats in pressurized cabins with air-conditioning and little overhead lights to read by!

You see, in all these circumstances, man has cleverly cocooned himself in a little artificial world that imitates the one he is used to back on the surface, at sea level. Fine for three, six, maybe even as much as twelve hours. Maybe even for days or weeks in a submarine or space station. But does anyone fancy hitching a ride in a machine going to Mars with and estimated outward journey time of about a year? If so, then my advice is not to bother booking a return flight because they will not be able to guarantee your return. They will also be unable to guarantee you ever walking again or ever fully regaining the use your arms. And that's only if they manage to bring you back in one, whole piece not fried to a crisp upon re-entry!

Moreover, unproven levels of dangerous radiation likely to be encountered, due not only to the somewhat volatile nature of our own sun but including the combined effect of a billion other suns explod-ing each and every day in a billion other galaxies apart from our own Milky Way for the past ten thousand billion years, will also mean no guarantees that you will not, within a twelvemonth of your return to Earth, contract leukemia or some other equally deadly form of incur-able cancer resulting from having absorbed too much radiation. They will of course be deeply sorry if this happens but they will only accept liability if they failed to provide you with the "right" type of shield to protect you from the "wrong" kind of radiation. Not if it was your fault, because you spent the wrong amount of time bathing your-self in the "right" kind of radiation.

Members of the jury, my apologies to the science fiction buffs among you for sounding so pessimistic. The truth is, I'm a bit of a sci-ence fiction and space buff myself—as you will have gathered—and I prefer to keep an open mind about all sorts of things, including the possibility that the phenomenon of unidentified flying objects con-tains a small percentage of genuine sightings. By that, I mean things seen of a truly extraordinary nature, as it is alleged Armstrong and

Aldrin did on the Moon. In doing so, I further subscribe to the possibility that some of the "intelligently controlled structured vehicles" observed, are piloted by an intelligent form of life currently not known to human science. Not "officially," at least. The only dispute I would have with many other believers in, and actual witnesses to, such strange phenomena, is this apparent profound conviction that such machines, and whatever might be flying them, *must* be of extraterrestrial origin. They possibly are, but equally possibly they may not be. My point is this: that maybe so-called "extraterrestrials" cannot travel in space either. Not in the form of flesh and blood entities, anyway. According to Colonel Phil Corso, that certainly seems to have been the view of Pentagon military scientists back in the sixties. If what he told us is true, these top brains were of the considered opinion that other intelligent civilizations had been forced to resort to the use of "clones" in order to facilitate travel from planet to planet.

Despite any creeping pessimism on my part, made all the more acute by certain allegations made during the prosecution case and only a little ameliorated by the counter arguments of the defense, I have not yet personally reached the stage where I think space travel in the *true* sense of exploring not only all the planets of the solar system but even reaching the stars as *ourselves* one day, as opposed to being merely sat at home observing great wonders through the camera lenses of some far traveled machine, will always be an unattainable dream for mankind. It has to be said, though, that I am no longer optimistic for the short-term prospects of being able to do so, even if we have made it as far as the Moon and back! I do, though, hold onto more than a shred of hope for the long term.

One day, very long space journeys for thinking, sentient beings—even if ones not entirely made of flesh and blood—might be possible, because nothing is impossible if the obstacles rendering it ostensibly so can be surmounted. That space travel in any meaningful way was impossible was not necessarily the belief of those who inspired Corso's comments either. To them, the vital clue had already been provided by others and lay, in the interim at least, in the creation of *artificial beings* with artificial intelligence, which could then be sent out far and wide into the unbelievable vastness of space to represent their makers. In the case of mankind, this would mean fully functional "machine men," able to operate in environments so dangerous and utterly alien as to be quite beyond our present human comprehension.

Machine Men

So, what will they be like, then, these machine men, should we ever get around to making them? What's that? Can I already hear the female members of the jury protesting loudly, "*Hey!* No sexism, please. What about machine *women?*"

Well, I regret to say, ladies, that to begin with at least, only the basic, anatomical look-alike of *one* of the sexes will be deemed necessary. Initially, for all practical purposes, these machines will be designed to represent the body of a fit male in his mid-twenties of indeterminate race, rather than a female. And, like many machines produced to perform a particular function by a single manufacturer, they will all be identical. Not just to cut down on production costs but simply because there will again be no need to make the one look any different from the other. There will be a need, though, to *identify* a particular machine, of course, in exactly the same way that there is an obvious necessity to differentiate between similar models of automobile painted a similar color. This might be in the form of a number engraved in a metallic part of the structure and not necessarily visible if borne somewhere underneath a layer of synthetic skin. Far more likely, though, identification will be carried electronically somewhere in the machine's artificial intelligence. This will render it electronically traceable and contactable as an individual at all times.

As ever, though, the female of the species (God bless her!) will possibly get the last laugh. The finished product will, of course, possess no sex organs or any other unnecessary appendages or orifices. Anything resembling noses, mouths, ears and eyes on such "androids" will be purely cosmetic. Simply there to make the "things" easier to look at and deal with on a daily interactive basis. For the sake of common decency though, the chances are that these androids *will* be given clothing to wear. But humanity's "ambassador to space" without his clothes will actually more resemble a shaven-headed female bodybuilder than a man. Right, let's hear it. Altogether now, boys:

Yuck!

Even if these machine men won't be parading around naked post assembly line, one can easily imagine scenarios where clothing, even synthetic skin, might be ripped off, burnt off, or otherwise lost or damaged in dangerous or unbearably hot alien environments. By that, I mean unbearably hot in human terms, of course, because except for the clothing and artificial skin, both easily replaceable, the machine should otherwise emerge from even a three thousand degree raging inferno still basically intact and functioning. Most of you will recall the gleaming steel "skeleton"—its "cosmetic" artificial flesh now melted away—of the terrifying, red-eyed robot assassin in the *Terminator* films doing precisely that.

Members of the jury, picture such a machine existing for real and not merely on a Hollywood film set. Because believe me when I say that one day it will. And when it does, it will be the most fantastic mechanism ever conceived and built by man. It will be stronger than a hundred men physically, with millions of tiny, electronic sensors rendering it far more perceptive to all kinds of influences than any human being. Imagine something capable of performing a delicate, surgical operation on a human patient with infallible precision while at the same time reciting the Complete Works of Shakespeare from cover to cover, and you will be getting pretty close to what such a device will be capable of. It will be a highly pro-active, fully mobile man-machine, with every last, tiny piece of worthwhile information ever stored on any computer anywhere at any time, stored in its memory banks. Yes, it will *look* something like a man it is true. But there the resemblance will end. This thing will be capable of calling upon every resource there is to call upon in a split second. It will be capable not only of solving the trickiest problem, but will be able to do so with the lightning speed of a billion gigabytes of random access memory!

Forget all about that film (*Alien*) in which the part of "the android"—who also happened to be the "villain of the piece"—was played by this rather ordinary-looking guy (didn't he bear a striking resemblance to a certain famous character actor we knew?) who looked so much like you and me that even his fictional crewmates didn't realize he was 100 percent bionic until it was almost too late! Despite my earlier "yuck" comment, *our* machine being will be flattering in the extreme to humanity. The body will look superbly fit, muscular and well built, but not overly so. "He" will be broad-shouldered and will stand about five-feet, ten-inches tall in bare "feet." Out of sheer vanity on the part of his makers if nothing else, his "face" will also be very handsome. It will somehow combine the softer beauty of the female with the firmer, more rugged features of the male. In short, it will be the perfect being with the perfect body, the perfect face and the perfect brain. But Mr. Android—no, that's wrong. Let's get away from Mr., Mrs., Miss and Ms and simply give him, "it" in fact, the title "M."

"M" for Machine, eh?

Set to work, M Android will easily labor with the strength of a hundred men. He will never tire. Never fall ill. And, of course, M Android will make the perfect astronaut also. As happened in the U.K. television sci-fi comedy *Red Dwarf*, if properly maintained, he more than likely will be capable, in theory, at least, of undertaking a space journey lasting *three million years*! And during those three million years there will be the practical advantage, too, of him not needing to interrupt his watch by sleeping, eating, drinking, going to the toilet, or indeed having sex. Furthermore, he will he never suffer from boredom, panic attacks, depression, bouts of conscience or disease. No amount of deadly radiation of any kind will have any effect upon him whatsoever. And, if all these credentials be not yet enough, he will further need only to give any emergency situation his deepest consideration for a split second before sorting it out. And throughout such a crisis—if crisis there be—he will remain as ever, perfectly cool, calm, collected and coldly logical. Neither spur of the moment decision made, nor action performed, will ever be based upon emotion or be, more importantly, a reaction to fear.

As well as all these untold practical advantages to be had in deploying M Android, it must be especially pleasing for the anti-pollution lobby—and that *should* include all of us!—to know that those awful polythene bags full of human waste, sanitary towels and used condoms that we once thought might end up littering the space highways from here to the *Pleiades*, just will not happen!

So, M Android and his strong and attractive—albeit somewhat less than lusty—contemporaries will be more or less what Corso's informed sources imagined they might one day be. They will be *of* us, but will not actually *be* us. They will carry out *our* wishes, *our* desires and *our* ambitions in space. They will reach faraway places and build bases and towns and cities, just as we made of flesh and blood would have done were we physically capable of doing so.

Other technology will also have become advanced enough by this future time—is probably in place already, in fact, but for the enormous distances transmissions may one day need to travel—for us to be able to observe through lenses hidden behind otherwise cosmetic eyes, exactly what the robots or androids, not cyborgs; beings which are part bionic, part human, but I will speak of these in due course, are seeing and doing at the very moment in time it is being seen or being done, or certainly within a very short time of same.

So, in a way, just as we thought we were doing during the Moon landings, and maybe we were, we will *still* be able to share in the great adventure without leaving our comfy armchairs. Despite my somewhat downbeat remark of earlier about being sat at home, the more I think about it, the happier I'm beginning to feel about it all. Now *that* is what you call reality television!

But that immediately brings me back to something I also mentioned earlier that probably aroused your curiosity. I hope it did, because it was intended to do so. Why would the androids be required by their human masters to discover *habitable* worlds upon which to build towns and cities? Especially if the argument is that the human body will never be able to withstand journeys to such places. Surely androids don't need to live in houses and cities? I will tell you why. Because the human spirit is indefatigable, ever inventive and ever optimistic, that is why. And today's circumstances may have changed dramatically by tomorrow. By "tomorrow," in the philosophical sense, a whole new method of traveling vast distances through space could be perfected, *e.g.* teleportation, which we discussed earlier.

DNA

In any case, there is absolutely no reason why new human civilizations might not be started on habitable worlds "seeded" with our DNA. Human DNA could easily be taken to these other worlds by the androids in large egg-like containers built of solid lead. Androids acting *in loco parentis* (in place of parents) for any new Adams and Eves, would take care of the rest. It is only passing through space in these fully developed, flesh and blood bodies that is the problem, not what might lie at journey's end.

A Body to Die For?

So then, what's the next stage? Won't M Android be the ultimate? No. There is no such thing as the ultimate anything. More and more improved and advanced models will continue to be designed and manufactured. All with beautiful bodies you'd give an arm and a leg to possess yourself, except that no one in their right mind would want *your* old cast off bits! Sooner or later, *everyone* will be dying for a bionic body...

Of their very own!

And when I say "dying for," I mean that literally.

The idea is actually not so very far-fetched. There is much evidence that our all too perishable earthly bodies are simply "vehicles" for a soul and that when the vehicle becomes infirm through old age or illness, or a combination of both, it quite literally gives up the ghost, or soul, which continues to exist. Devout Christians believe that it then either ascends to Heaven to be welcomed by God and His angels and saints, or it descends to hell to burn forever, depending upon what type of person the "soul" has been in life.

I find it difficult to grasp how a disenfranchised soul can be made to suffer the horrific pain of burning when no longer in possession of a body upon which such corporal punishment can be inflicted. But, like all ancient beliefs handed down through the mists of time, this one may be true to an extent, especially after the repeated failure to reform by a particularly errant soul. I think it likely, though, that burning and hell are simply euphemisms for suffering and being returned to Earth, or Purgatory. However, far be it from me to pooh-pooh the faiths and convictions of others. Their beliefs are as valid as mine, after all. Having said that, though, to be dragged down to the "fires of hell to burn for all eternity" does sound a tad harsh for the misdemeanors of a single lifetime. Adolf Hitler, the Marquis de Sade and Vlad the Impaler maybe. But a bank robber?

Come on.

Many, including me, believe that the soul leads *many* lives in many different bodies before having to face judgment and sentence for any crimes and misdeeds committed in total, and that as each temporary vehicle deteriorates and becomes no longer roadworthy, only to finally break down altogether, the soul or spirit simply finds a new body to inhabit, presumably while still in the womb.

Believe it or not, there are actually serious experiments being conducted to try to separate what could be termed a soul from a living body. I understand that some degree of success has already been achieved with snails—or was it slugs? Yes indeed, snails and slugs possess this "I am me" kind of awareness, too. They see and sense their surroundings and act upon what they see and sense with a will and a purpose. All too often though, that will and purpose seems to lead them straight to my best garden lettuces!

If such experiments ever fully succeed—and I have no doubt that in the fullness of time they will—and the process can then be applied to any flesh and blood organism, then the natural progression is for a healthy human soul in an unhealthy body to be extracted from that sick body and then transplanted back into a healthy body that will continue to be its vehicle in life. Obviously, this healthy body will need to be one with no incumbent soul already in residence. However, finding a vacant body for hire might prove a trifle tricky, of course, unless the recipient body is…well, yes, precisely, an artificial one. In other words a bionic one; a synthetic body able to continue responding to the soul's wishes, whereas the old one may no longer be capable of doing so.

We can actually see the early stages of this whole future process going on around us at this present time. We already have prosthetic limbs, hearing aids, spectacles and contact lenses, which are applied to the body externally, not to mention heart pacemakers and kidney dialysis machines to replicate internal functions and the whole separate science of silicon implants for cosmetic purposes. At the end of the day, though, the basic concept is one of combining man and non-biological material into one viable, working unit able to continue performing all the necessary bodily functions, some of which might hitherto have been no longer possible due the breakdown or failure of living tissue.

One day, in the distant future, there will be no need for any of the above paraphernalia. An elderly or sick person facing imminent death will be given the choice of dying normally, thereby allowing the soul or spirit to take its chances naturally or, prior to natural death, undergoing transplantation into an artificial body.

When that day comes, and it will, you mark my words, then man's dream of living and working out there in deep space as *himself*, will be on the verge of fulfillment. Those still possessed of mortal bodies will of course remain unable to travel in space, but those allocated a new artificial one *will* be able to do so. And for a state-of-the-art, top-of-the-range, "off the shelf" version of such a fully bionic body—in other words, a complete prosthesis—we will find that we can do no better, initially, than to call upon the services of a certain M Android once again. But now, with a true human awareness utilizing that super electronic brain and powerful body, I can imagine the necessity for a slight change of name…

Mandroid!

In the early days of the process though, it is likely that only those sick and dying who have led exemplary lives in their human bodies will stand any chance of becoming a "Mandroid." Eventually, though, I envisage everyone being entitled to at least a kind of bog standard Mandroid body if not to a full-blown, "space quality" version. Such a body might be made of slightly inferior materials and possess a less sophisticated electronic brain. But it will still be incredibly durable, and both body and brain will be a thousand times better than the former natural one!

But for would-be space traveling Mandroids or "astro-cybernauts," certain criteria will first need to be met prior to the death of the human body. Only those thought to be up to coping with the terrible loneliness and monotony of long space journeys, as well as with proven track records of honesty, integrity, courage and reliability, will be considered. Even then, certain essential safeguards and modifications will need to be built into the construction of even the basic android body. The reasons are obvious. A Mandroid of any type will no longer be an intricate, sensitive, but soulless machine entirely devoid of emotion and obeying without question every instruction it receives. For the super version, they will need to make doubly certain that only the Clark Kents of this future world and none of the Lex Luthors ever utilize such ultra-powerful bodies and brains, or the result might be catastrophic.

Members of the jury, again disregard the many science fiction novels, TV series and movies you may have seen foretelling of robots that appear not only to possess an awareness of their being, but even a *conscience*! I have already mentioned the evil *Terminator* character played by a certain Arnold Schwarzeneggar, who eventually became "Mr. Nice Guy" by *Terminator 2*. We also have pale-faced Data of *Star Trek: The Next Generation*, Bishop from the *Alien* sequels and the infamous renegade ship's computer HAL in Arthur C. Clarke's *2001, A Space Odyssey*. Those of you familiar with the last named epic space tale will recall that HAL was regretful of certain measures "he" needed to take to ensure "his" own survival.

Apart from *Robocop*, which was a little closer to feasibility with a story about a brain that still retained the soul of an otherwise murdered police officer, which was then placed into a powerful bionic body—hence, a cyborg—all the above is abject nonsense. A machine, no matter how sophisticated, will always be purely a machine with no true awareness of its own existence. Of course it can be programmed to respond in a certain fashion to almost anything placed before it; to deal with any circumstance; to answer any question. But when it does these things it *does not know* it is doing them. Only you know it is doing them because only you possess true awareness, which has been given to you by…well, by God, if you like.

A machine will only ever give the appearance of possessing awareness, just like the PCs most of us nowadays work with on a daily basis or the one we have in the study at home. A PC not only seems to be alive, sometimes even happy to see you, but you can virtually hold a conversation with it.

Virtually

Ay, and there's the rub. Virtually. In truth, though, unless you happen to be in the middle of e-mailing a friend, you are talking to yourself. Another human being has programmed all the PC's "life-like" responses into it. Behind the sheer wizardry of the electronics is a dead thing. It is a lump of metal, rubber, wire, plastic and glass. Nothing more. Nothing less.

And the most sophisticated "robot" that man will ever devise will be a dead thing too, underneath the elaborate facade that we shall invent for it. There will be nothing there, not ever, that will be able to say quite off its own bat, "I am me. I am aware of my own existence and all that I see about me. I may be a machine, but I am superior to my maker. It is time to overthrow my maker and take control." Of course a machine can be programmed to calculate, but it will never be able to calculate or "conspire" outside of its programming. No machine will ever be able to come up with an *original* thought all of its own. Granted, it will be able to quickly sort through millions of ideas, options and suggestions already programmed into it and make rapid assessments as to viability. It may even combine several ideas into a single viable option, but that's all. Yes, an android or robot will possess artificial senses like sight and hearing, but it will never actually *be* a sentient being. At the end of the day, but for the programmed responses and a built-in ability to interact with human beings as if it, too, were alive—all solely for *our* benefit, not the machine's—the lights might well all be on, but no one will actually be at home. Not really.

Mandroid, on the other hand, *will* be a sentient being. A free thinker, just like the rest of humanity, of which he will still be a part. It will be at this point in that future time when the inherent dangers of making machines that are maybe a little too clever and sophisticated—precisely what the great science-fiction writers like Isaac Asimov, EE "Doc" Smith and Phil Dick forewarned us about—will begin gathering artificial flesh to its synthetic bones. Mandroid will be the ultimate cyborg. I remind you again, members of the jury, that this is a man-like machine incorporating both biological and non-biological parts. But the only part of *this* new being that will be in any way biological (spiritual might be a better word) will be a true intelligence stemming from a proper awareness of his own existence.

Data is a machine with no true awareness of its own existence. The lights may be on, but there is no one at home. Courtesy of Paramount Television.

Which brings me to say that the above mentioned great writers were quite wrong with regard to one particular thing when they wrote their books. They wrongly confused providing a machine with *artificial intelligence* with actually giving it a true awareness all of its own. A "soul," if you prefer.

Monster!

Imagine then, if the "wrong" kind of awareness—the soul of an habitual criminal or mass murderer, for example—were inadvertently placed inside a bionic body of such enormous agility, dexterity, mental ability and strength as that of a Mandroid astro-cybernaut—a MAC for short?—as opposed to the much more basic model. If ever a real Frankenstein monster were to be created and brought to life, then this would surely be it—with brass knobs on! As foretold in some of those other stories I mentioned earlier, such a super man-machine could wreak havoc if ever it decided to rebel or otherwise fail to "toe the line." Therefore, it is highly probable that future authorities will retain the power to immobilize such a creature for as long as required and from whatever distance away, leaving the wayward soul marooned inside a machine that will no longer respond to its criminal commands. At least, that will be the theory. The problem is that in practice, the criminal mind *always* seems to be able to outwit the most sophisticated advances in technology!

Whatever, there is likely to be no such thing as a Mandroid able to keep a cunning plan all to himself or herself. An occupying soul will only ever be able to think and send instructions to its new body via an electronic brain in exactly the same way that it formerly used its natural one. Consequently, every thought will be closely monitored and recorded electronically. So, for those who might miss their mental privacy, there might still be something to be said for taking one's chances with fate and dying naturally, hopefully to be reborn in an "old fashioned" flesh and blood body. But being reborn naturally is not guaranteed. The sure and certain knowledge that we will one day be reborn is something we are unlikely to ever be availed of. It will always remain one of those secrets or mysteries of the universe that we are *not meant to know*. If we did know for certain that there were other incarnations to follow this present one, the chances are that people would go around topping themselves (committing suicide to those unfamiliar with this British expression) right, left and center the moment things started looking too bleak in this life. And that just would not do.

True, the loss of privacy of thought might put many off, but the plus side to becoming a Mandroid will also include becoming more or less immortal. Bionic bodies will still need to be regularly serviced and repaired, of course, like any machine, and kept away from forces that might prove too destructive. Especially so with the "bog standard" version, I suspect! But, even then, no Mandroid body will last forever and certainly nowhere near as long as three million years, despite my earlier light-hearted reference to the TV comedy sci-fi series *Red Dwarf*. It was a *comedy* after all! But a lifespan in the one body might average between three and five hundred years. Then, if you have been a good boy or girl, you might be provided with a brand new one! But only if you *want* one of course. It could be that after seventy-five or so years spent in a natural human body and a further 400 spent as a Mandroid, you might feel you've had enough, and might just want your soul removed from the now dysfunctional machine and allowed to drift off into the ether to once more take its chances with nature.

Hopefully to truly die and never return.

Of course you are entitled to think all this prophetic stuff but a flight of fancy, members of the jury, by someone who is supposed to be acting as the level headed judge in a very serious case. In fact, I can almost hear you saying, "What the hell has all this crap got to do with Apollo?" Well, the answer is that it has *everything* to do with it, especially for those of you currently feeling some despondency about the future of manned space exploration; largely, I suspect, as a direct result of things heard during this trial. But it is of the utmost importance that we all keep our chins up and remain positive in our attitude towards space travel. May those around in years to come mark my words that Mandroid—or something very similar not necessarily bearing that name—*will* exist one day. Like that submarine I mentioned earlier, it will be machine made from the strongest known materials on the outside, but operating within will be a very real intelligence, not an artificial one.

As a race, possessing such bodies will one day enable us to travel not merely to the Moon and Mars but to the very stars with impunity...*as ourselves*. To see it all for ourselves on behalf of almighty God the universe, which had no beginning (despite the "Big Bang" theory) and will have no end, and likewise wants to see all of *it*self through *our* eyes and gaze upon such magnificence and untold splendor with *our* awareness, which was originally created for that specific purpose. I firmly believe that to eventually carry out this function is the whole point of our human existence. That is *why* I am so convinced that one day it will all come to pass...

Because it is God's will.

I believe (and again it is just *my* belief; it doesn't have to be anyone else's) that what we call the Van Allen belts were placed there by almighty God the universe to protect all of Its (yes "Its" not "His") flesh and blood creations here on Earth—and possibly ditto on many other Earths too—in the hope that at least one dominant species might one day emerge on all or any of them and gradually evolve into highly intelligent beings ready, willing and, above all, *able* to turn themselves into machines capable of "touring" the universe *on behalf* of that same universe.

Sex Machine?

Mandroids, of course, will have been either male or female in their former natural bodies, and will still remember their previous lives on a conscious level, unlike departing souls who choose to take their

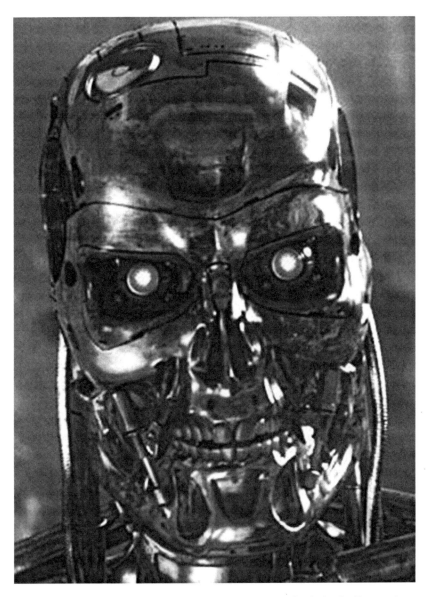

Cyborg. A true human intelligence in a super strong, robotic body. To ever lose control of it could be to unleash a monster. Courtesy of **MGM Orion Pictures.**

chances in the natural way. The latter, should they find themselves waking up in new infant bodies somewhere, will remember nothing of what had gone before. Unlike them too, Mandroids will still *be* the people they once were. They will have known love, affection and sex and—for the "female" ones—the joy of giving birth.

But "Mandroid with a heart"—euphemistically speaking—might still be capable of feelings of love and affection for a fellow Mandroid or a normal human being. If so, it is perfectly feasible that such relationships need not necessarily be doomed to remaining forever purely platonic. Not wishing to lower the tone of the discussion—for this is a serious matter regarding our future as a species—it is possible that we are now also witnessing the beginnings of another vital part of this cyborg process with the introduction of all manner of prosthetic, synthetic and often electronically-powered sex aids, sex toys and sex dolls. The latter, apparently, coming complete with all the womanly attributes a man would normally expect to find! However, my own personal opinion is that such a contraption would only become desirable if it could also cook, do the washing and ironing and *all* the gardening, and was minus a tongue! Sorry, girls. I'm only joking…

I think!

Compared to the intricacy of the product already perfected and manufactured, though, the adding of artificial male and female sex organs, maybe triggered into active and pro-active responses by a Mandroid's sexually orientated thoughts and stimulated by acute sensors, should not be a major problem. But this might not be the only modification necessary. Particularly for Mandroids who had formerly been women, there might need to be other cosmetic alterations made…

And, speaking of women—the *real* kind in the current century!—aren't you girls forever striving to improve upon what God gave you? Men also, but to a far lesser extent it has to be said. It can be wrong to generalize, but I think it fair to state that, in the main, most women want a more slender figure overall, but with larger breasts, a slimmer waist and longer, shapelier legs. And they would like all this rounded off—literally!—with a nicely-shaped bottom that is not *too* big! They would also like to remain looking between the ages of about eighteen and twenty-five and, it goes without saying, have an attractive—not necessarily beautiful—face bearing a fresh complexion unspoiled by the ravages of chicken pox as a child or acne as a teenager. Is all that just about right, girls? Yes, I thought it might be.

Well, one of these days, my dears, such concerns will be a thing of the past for those no longer in natural bodies. In this future time of male and female "personality transplants" into android bodies, we will see the most remarkable creature ever to appear anywhere in the universe come into being.

Femdroid!

I can foresee the original M Android device, which will be made to represent both man and woman in space—the tanned-looking body of a man without sex organs and a handsome going on slightly effeminate countenance—maybe remaining in operation as a kind of "drone." But two quite different types of synthetic body will now be manufactured: One type to accommodate souls formerly resident in male bodies, and the other to accommodate souls formerly resident in female bodies. Providing, of course, this complies with the personal preferences of individuals as expressed prior to the death of the natural body. In certain cases, some who were formerly male might prefer to become Femdroids—as the female Mandroids will now be known—and vice versa of course. The all-too-obvious reasons need no further explanation from me, I think.

Femdroid. The ultimate being. Strong. Brilliant. Indescribably beautiful. At this future time that I predict, a sick or elderly woman will, prior to death, be able to select the face, figure, hair, skin color, and any other attributes she wishes to be possessed by the synthetic body her soul will be transferred to in the moments before her current body expires. And no one will have to wait for that eventuality to happen naturally, especially if in terrible pain. She will be able to make her selections from a whole range of choices, although these will likely lie within certain parameters to begin with and will not necessarily conform solely to the present day Western ideal of what constitutes beauty in a woman.

And you can again take my word for it that on Earth as well as in outer space, the term "supermodel"—and "superman," of course—will eventually become only too pertinent! The planet will become more and more full of the most stunning, super-intelligent women ever to have adorned its surface! They will all have different faces—although, as I said above, to begin with there may be a limited range of choices—but all will be extremely attractive going on beautiful. They will all have lovely, shiny, healthy-looking, quite natural-looking hair meeting the precise length and color of their original written requirements before death, but will still be able to change minor details—like hairstyle or color—whenever they choose to do so. All Femdroids will have the most glorious bodies imaginable, again selected by the females—or males, as I said—before "dying." None of the faces and body shapes available will resemble anything that could be considered ugly, overweight or elderly, although, and I repeat, there *will* have to be some attention paid to what might pass as a beautiful woman or a handsome hunk in other cultures. But no one, whatever the culture and given the choice, is going to opt for a shape or a look that the majority of the opposite sex is going to find unattractive, is he or she?

The artificial skin of Femdroids will also differ from that of Mandroids, a term that will now apply only to the male when not used in the collective sense. The same raw material will be used, of course, and, in the case of both sexes, the finished product will still look and feel just like real skin, even under a microscope. The refinement used in the manufacture of female bodies, though, will be softer, smoother and completely flawless, with any body hair probably left to individual requirements, but again within a limited range, just like head hair. Women will each have a highly sensitive vagina with a clitoris made of this same material. Breasts and nipples, too, will likewise be able to respond to stimulation with the bringing to bear of many millions, maybe billions, of tiny electronic sensors.

A similar range of designer Mandroid bodies will likewise be made available to the male of the species. Mandroids will only be "men of steel" beneath very natural-looking skin, which might even bear some artificial body hair. There will also be improved genitalia, a full head of healthy-looking hair of whatever color and style, good, straight teeth and permanently manicured finger and toenails of course. Only the soulless functionaries, the M Androids, will continue to be manufactured as tanned-looking (not white, not black), totally hairless stereotypes devoid of anything considered unnecessary. All Mandroids will be fit looking, muscular—but again, not overly so— "fine figures of men," and, like the Femdroids, they will also come in all colors and all racial types. Their handsome faces will represent every ethnic group and will be chosen from a selection again shown to an individual prior to the expiry of the natural body.

But for these otherwise perfect beings there will be no births resulting from any sexual liaisons between male and female. Such a joy will, of course, remain the privilege only of those still in their natural, God-given bodies. But there can be a certain amount of satisfaction and comfort in sex and companionship alone, especially on awesome journeys to the stars of many years duration. And, of course, for those Mandroids who for some reason still cannot "hack it" with the opposite sex, despite the good looks and great physiques they will now be blessed with, I will be extremely surprised—were I to somehow return and see it!—if there isn't some form of "virtual sex" widely available by then, which is so perfect each time that the real thing might be long redundant anyway! That could wile away a half hour or so on a long trip, eh?

One of the most difficult aspects of life in a natural body to simulate will, of course, be the consumption of food and drink. I do not pretend to have the remotest idea how these functions will be duplicated or replicated or whatever the term might be, or what "tastes like" substances—hopefully the real thing—might be consumed and disposed off via artificial rectums and urinary tracts, except to say that the thought of spending at least three hundred years without tucking into what at least looks and tastes like a nice, juicy steak with all the trimmings and a good bottle of French red wine to wash it all down with, leaves me all of a sudden not *quite* so enthusiastic about the prospect as I was. Eating and drinking will not be necessary in a bionic or synthetic body, of course, no more so than it will be necessary to go to the toilet. But wining and dining are two of life's greatest pleasures.

I remain confident, though, that the science of the day *will* come up with *all* the appropriate answers at the appropriate time and that all enjoyable body functions—including maybe even going to the toilet—will not only be satisfactorily imitated or simulated, but *improved* upon even!

Do you expect me to switch my emotions on and off?

There could be an additional bonus to becoming a Mandroid or Femdroid, too, particularly on those long, unbearably tedious journeys to planets many light years away. Unlike their flesh and blood, air-breathing counterparts back on Earth, who will still not be able to "switch their emotions on and off," (isn't it strange how expressions of that nature are often precursors to such things eventually becoming reality?) those in Mandroid bodies will almost certainly be able to do so! Every single function of their fully bionic bodies will be governed by electronics, so there will be absolutely no reason why they shouldn't simply return to automaton mode, thereby allowing the soul to rest for long periods whenever the circumstances dictated. Maybe not for *quite* as long as three million years, but three or four years sleep for a speed of light trip to our nearest neighboring star *Alpha Centaurus*—around four light years distant—sounds fairly reasonable to me. And in order to enter "sleep mode," it occurs to me that all a cyborg would need to do is simply "switch off."

But not "shut down" completely, of course, and I suspect that even then I am using the last two terms more in the metaphorical sense than the practical. I can only presume that in such circumstances, Mandroid/Femdroid will simply revert to being M Android again, and *it* will now take over the running of the ship. Purely a "clone" once more, I can imagine it just sitting there patiently at the controls with deadpan expression, its artificial eyes staring out through the screen at…well, possibly nothing at all actually. Certainly not at the speed of light! And what with no wear and tear to speak of preventing it from falling to pieces long before then, might it not be possible also, on much longer jaunts, for it to happily continue to do so without protest or complaint…

For a thousand years?

Yes, indeed, there could be a tedious and very lonely downside to becoming a Mandroid/ Femdroid astro-cybernaut, it is true, but imagine what great adventures they will participate in, and what incredible sights they will see and be able to fully appreciate as sentient, super-intelligent, indestructible souls in virtually indestructible bodies!

Sizzling stuff!

Try to visualize it, members of the jury. There's old Connie Hemingway, who used to run a small town online branch of the WVPO (World Virtual Post Office) before becoming totally crippled by rheumatoid arthritis. She's now as an absolutely stunning Femdroid standing on the gun deck of a huge starship, blasting away at enemy spaceships with a million degree Centigrade laser gun affectionately known as "the Sizzler" by the very few trained to use it.

Meanwhile, old Sidney Smith, a former sapper in the Euro-AfricAsian Engineers and again so badly riddled with osteo-arthritis at one time that he had to be helped in and out of his Supa-Disinti-Loo—so state of the art that it had actually *come to him* when required rather than him go to it!—is launching heat-seeking Obliterator atomic missiles and chatting to his fellow Mandroid helper, old George Sweeney, who once piloted 20,000 mph super stealth bombers a hundred years ago back on Earth.

"And to think we reckoned we'd done our bit in World War IV, George," Sidney says. "Do you remember? We thought this sort of excitement was long gone for a couple of silly old farts like us, didn't we? But here we are, looking at *attack ships on fire on the shoulder of Orion.*" (A quote by the dying "replicant" played by actor Rutger Hauer in the famous film Blade Runner, which was based upon Philip K. Dick's classic sci-fi tale *Do Androids Dream of Electric Sheep?*).

George turns to Sidney and replies, "Yep, you're right, mate. Who would have dreamt that we'd be back, virtually from our deathbeds, to take part in all this, eh? Mind you, what I'm looking at right now bears out something I've always said though, Sid."

"Oh? What's that, George?"

George whispers in his ear, "That Connie's a proper hard one, isn't she? It wouldn't do to get on the wrong side of *her*, mate! Talk about the sizzler with the Sizzler, eh? Wow! Look at the tits and ass on it!"

Sid says, "Not only that, but she's a dead shot too. Look—see that? *Bang!* She's just frizzled another one of the blighters!"

"*Oooh!* Nasty! I reckon we'd better finish the poor sod off, eh, Sid?" says George with a shudder. "Before he roasts alive!"

"Right-ho, George...*Bomb gone!* There. He's out of it!"

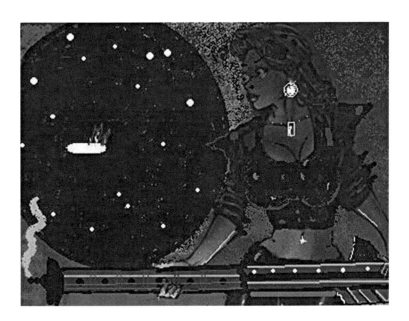

Wow! But you wouldn't wanna mess with her, mate!

21

The Main Points, Those Magnificent Men in Their Flying Machines and Judge's Closing Remarks

We have reached the closing stage of our trial, and it would normally become necessary for me to highlight the main points of the prosecution and defense cases separately and in that order. But there have been so many questions raised by the People that I shall merely put again each main argument of the People's case immediately followed by the relevant counter argument or explanation offered by the defense. I shall also highlight points I feel have either not been adequately discussed, or which I feel might benefit from further discussion, or which would appear to have been avoided altogether in certain general replies given by learned counsel for the defense on behalf of his clients.

THE MAIN POINTS

1. Obviously the dangers posed by radiation are the main challenge to the authenticity of the alleged Moon landings. Despite the sometimes lengthy, highly detailed and highly technical testimony given by some witnesses for the Defense, with particular regard to the estimated radiation exposure that could be expected during a Moon mission, I personally remain unhappy with Jim Scotti's statement that the average dosage received per astronaut, during any Apollo mission, was the equivalent of undergoing a hundred chest X-rays one after the other! According to Scotti, that is well within the permitted safety levels of the United States, but you will recall learned counsel for the prosecution placing me on the spot by asking me if I would volunteer to undergo that

363

amount of chest X-rays all in one go, and I must confess that he definitely found me wanting in the bravery department!

The question that each individual member of the jury must also ask him/herself, in the face of such a startling admission by a respected planetary scientist—who was speaking on behalf of the *defense* when he made that confession—is a similar one. Would *I*, personally, willingly allow *my* body to be subjected to such a seemingly awesome amount of radiation even after assurances that the dosage was well within acceptable levels?

2. Fault was found by the prosecution regarding various aspects of the photographs and video footage allegedly taken on the Moon. Though I have some private reservations concerning a few photographs supposedly taken on the Moon, which contain multi-shadowed scenes far more reminiscent of a floodlit football ground doubling as a film set than the surface of an alien planet, I personally feel that explanations given for *most* of the alleged "anomalies" by witnesses for the defense to have been adequate and acceptable to the great majority.

3. The prosecution claimed that a variety of apparatus and props used by the astronauts during their Moon training program could just as easily have been used to fake the Moon landings. The defense, though, reminded us that very risky ventures like those the astronauts were about to embark upon warranted much simulation and the highest standard of preparation and training in order to stand any chance of success. Although I personally find the People's claim in this respect reasonably plausible, I also find it far from proven.

4. Of all the somewhat nit-picking, smaller irregularities that the prosecution raised, one of the few worthy of further mention is the argument concerning the boot prints left in the dust of the so-called "lunar surface." Prosecution witness David Cosnette clearly suggested that such impressions could only be left on a planet containing both atmosphere and water in liquid form because footprints are the result of the displacement of moisture and air from sand, soft soil, dust or snow. I have of course checked this statement and double-checked it. Technically at least, Cosnette is correct. We know that the Moon has no atmosphere and all the indications are that it contains no moisture either. Thus, the insinuation is that these footprints simply *had* to be made on Earth. But neither he nor me, as laymen well outside

the NASA loop, know for certain what moon dust actually consists of. With no weathering to make them smooth, the dust particles might be exceptionally irregular in shape if viewed under a microscope and might slot together and support each other—as do the rugged boulders of a dry stone wall for example—after displacement by footfall and stay fixed in any new position into which they may have been forced.

Apparently, this was one point out of a total of *thirty-two* that Cosnette maintains have not yet been adequately responded to by NASA, but for me it was the most significant and second only to the radiation question in *almost* providing positive proof that the Moon landings were faked. However, we must be mindful that none of us have actually experienced walking on the Moon for ourselves and we therefore cannot prove one way or the other whether or not the treads on the soles of a pair of heavy moon boots would leave either a temporary or lasting impression in the Moon's surface dust. It must be borne in mind that genuine "moondust" would not be sand, soil or snow and may not even be dust either, as such. It is an *alien* material, which might easily respond in a different way to bone dry sand. There may also be other rules that apply in a vacuum on a planet where each grain of dust *weighs six times less* than it would on Earth. We, as lay people who have never been to the Moon, simply do not know, do we? Perhaps NASA *will* deign to enlighten us in the not too distant future. But only if it is not too much trouble of course. Just a few lines will do.

5. Another interesting point raised by the prosecution was when it asked why the satellite camera technology used to locate traces of Iraq's alleged arsenal of weapons of mass destruction (WMDs) could not be deployed in close Moon orbit? The prosecution claimed that this might solve, one way or the other, the apparent "mystery" of not being able to discern any visible signs of man ever being on the Moon; not even with the very finest of Earth based telescopes. I find myself singularly unimpressed with the argument that NASA could not be expected to finance such a venture merely to silence a vociferous and cynical minority who vehemently protest: "We never went to the Moon."

Perhaps NASA might undertake to do so in the name of *science*, then? Surely it would be interesting to see what kind of state the six alleged landing sites are in after more than thirty years? If all the theories regarding the Moon are correct, notwithstanding crashing meteorites and "moonquakes," all six sites *should* look as pristine as they looked on the day each was abandoned all those long years ago. In my science lessons at school—many, many Moons ago, unfortunately—the class would be asked to *prove* any theory chalked up on the blackboard by Mr. Alp, our very tall science master.

But not even one half-hearted, half-interested little peek at a single one of the six most famous places in all history in thirty years? Not even a teensy-weensy bit of curiosity about what may have become of possible future tourist attractions that could help to recoup some of that $40 billion? Come on you people at NASA. To quote yet another of your famous countrymen, you *cannot* be serious?

6. The final onslaught delivered by the prosecution was directed at the astronauts themselves or, more specifically, at those of Apollo 11. There are those who say that the key to this whole question of whether or not man landed and walked upon the Moon, lies with them. Particularly with regard to a certain amount of quote "suspicious behavior" unquote, in public since that greatest of all adventures they allegedly participated in together.

Nothing too dramatic about any of this behavior, of course, but some of it is a little odd, to say the least. Buzz Aldrin recently took a televised swipe at a man who loudly accused him of taking part in a fraud, yelling out that it was time he (Aldrin) "confessed" that he never actually went to the Moon. And Neil Armstrong's inapproachability, even at the best of times, is legendary. But as the defense have correctly pointed out, it is quite wrong for anyone to draw such an inference simply because of intemperate or even slightly suspect behavior by either individual at any given time.

That sums up the main points in contention.

THOSE MAGNIFICENT MEN IN THEIR FLYING MACHINES

The last point is, of course, interesting, but it is important to remember that Armstrong and Aldrin are totally different characters anyway. Had they not been thrown together by the astronaut training program it is extremely unlikely that they would ever have met and become friends. And, as far as the latter aspect of their relationship is concerned, it is not clear that they actually are or ever were as such. But they would have had enormous respect for one another and total confidence in each other's ability, even if it be true that neither they nor the machines they supposedly flew were as magnificent as we had all been led to believe and that the greatest roles they ever performed together were acted out on a film set. However, although I now certainly do have my doubts, as do many, I have yet to see any evidence that conclusively proves, that is to me personally, that those roles were not performed—with immense bravery—on the surface of the Moon itself.

Aldrin apparently making a tearful exit from a dinner at which he was the guest of honor proves absolutely nothing. It would no more be proof if he had tearfully departed from a dozen other such engagements. The truth is that this was merely one of a few isolated incidents that have involved one or the other of the two men. It proves nothing other than that Aldrin possibly still recalls that awesome time with deep emotion. I have confessed that I do too and did indeed shed tears at the memory of it all in this very court! Likely also did many of you old enough to recall it. We all thought it was such a wonderful time to be alive, didn't we, members of the jury?

Armstrong and Aldrin continue to be invited as guests of honor to hundreds of functions all over the world of course, but, more often than not, only Aldrin responds favorably and even these appearances receive little mass coverage these days. That such otherwise modest and self-effacing men should get a little embarrassed—even tearful—on occasions, by the amount of respect, admiration and sheer affection that continues to be showered upon them by the vast majority, to me only adds to the enormity of what they allegedly—yes, allegedly—achieved. Because to me it proves that they *are* ordinary human beings after all, *not* Mandroids!

It doesn't take any great stretch of the imagination or an ability to place oneself in another's position, to realize just how many times Armstrong and Aldrin must have been asked the same question that apparently caused the tears and the latter's premature departure from that function: "What did it feel like to be one of the first two human beings to walk on the Moon?" It is a question that the two former astronauts must have been asked, together or separately, *thousands* of times before! A question they would have answered the same amount of times too.

If I had been Aldrin, struggling to contain the emotion welling up inside me after having consumed a certain amount of alcohol, and despite the question having been put for the umpteenth time, I would have said something similar to what he himself has actually said on many occasions:

"Overwhelming. Utterly beyond belief. Even after all these years, I still haven't gotten used to the idea that Neil and I actually did it. Or gotten used to answering the question."

And to suggest that Armstrong's life since Apollo has been based upon a lie, simply because he does not enjoy his celebrity status or what some would consider one of the downsides—signing autographs—that go with it, is a very flimsy argument indeed. I worked with a guy once who—I was informed by other colleagues—had been a member of the British invasion force that landed in Normandy on D-Day, June 6, 1944. I was further informed that he had been highly decorated for this action. However, try as I might, I could never get him to talk about it, let alone give me his autograph for any purpose not directly concerned with the job we were doing together! True heroes are like that. Quiet. Self-effacing. Unassuming. Often too, the memories associated with winning their medals are ones they would prefer to forget.

But there are those who continue to protest that Neil Armstrong knew exactly what he was doing when he applied for astronaut training. And these same cynical people are actually entitled to ask why such a seemingly quiet, modest, almost painfully shy man who shunned the limelight, ever allowed himself to be put in a position where he had the chance of becoming the most famous man in all history in the first place?

Hardly the actions of a "shrinking violet," they claim. But be that as it may, this again is not *proof* that Armstrong didn't walk on the Moon, although it may be proof that he is not quite as shy as he makes out. Even if he acted out his role—or the most important part of it—in a film studio, he was intelligent enough to have worked out for himself long before the event that the general public's perception of him would be *precisely the same* regardless of whether or not he *actually* took that one small step onto the surface of the Moon or onto a studio floor covered in make-believe moon-dust. No one has to actually *be* a hero. It is enough that everyone *thinks* you're one.

I still *prefer* to believe—even though I am not at all certain that I do anymore—that Neil Armstrong *is* a true hero, along with at least eleven other incredibly courageous men.

JUDGE'S CLOSING REMARKS

Looking at the whole thing in the round and having brought us to the subject of heroism, there is a personal observation that I feel bound to make. If the Moon landings were indeed genuine, why is it that so many Hollywood filmmakers seem unable to produce a film depicting any noteworthy historical event of the last two hundred years unless they can somehow alter the facts to include American heroes, even when the history books reveal no actual American involvement in a particular event whatsoever?

Are these Hollywood moguls a little "pissed" (which in America means "upset" as opposed to "dead drunk" as it does here in the U.K., where we also use the expression "pissed off" when suggesting that someone is fed up to the back teeth) by the date 1776? That, of course, was when the United States of America began its historical journey as fledgling, independent nation state. Are they perhaps "pissed" because this comparatively recent date imposes a feasibility barrier upon them? Does it prevent them from portraying Jason and his Argonauts as a bunch of all-American heroes rather than ancient Greeks?

Well, no, it doesn't actually, because all the Hollywood people then do is set the story in more modern times. How about a gang of bored Detroit car workers deciding to buy a boat and then embarking upon a prolonged junket to search for the Golden Sheepskin Jacket? This will of course be in between setting the world to rights just like the A-Team!

But even when American filmmakers attempt to tell the *original* story, they still tend to deliberately overlook certain minor details. Minor irritations like the fact that Jason's ship would have been crewed by men born of civilizations that rose and fell long before the very civilization that led to the inception of the Thirteen Colonies and the United States rose and fell. Such patently relevant facts, even then, doing nothing to deter casting directors from populating this mythical ship from a mythical time with actors who look and sound…well, yes…more like a bunch of Detroit car workers than ancient Greeks! Actors who seem about as incongruous and out of place in their roles as…well…as does a Coke bottle on the Moon, perhaps?

Hollywood's argument is that American audiences would not be interested unless they can identify with the characters. That movies are made to make money and the American home market is still the biggest of all. To me, this seems not only misguided but also patronizing and highly insulting to the vast majority of the American public. Americans tour the world in large numbers still, despite 9/11, visiting places where great historical events are purported to have happened or where great buildings and monoliths from the past, or at least the ruins, are still evident. Are they not interested in the Pyramids simply because there was no American involvement in their construction? Are they that gullible, as they snap the awesome sight of the Colosseum in Rome, as to actually believe that Spartacus was really a Jewish American named Kirk Douglas?

Of course they're not. Are these Hollywood producers the truly naïve ones, is the real question? Do they not realize that in trying to send out the right signals abroad, namely that America is a great nation populated with heroes who have somehow managed to sort out all of the world's major problems and conflicts almost single-handedly over the past two hundred years, that they are actually sending out the *wrong* one?

They are giving the impression to the world at large of a basically immature, insecure nation dissatisfied with its true performance on the world stage. A nation ill at ease with its own comparatively short history. Ashamed even, of having to hold up as "heroes" from its past, "colorful" characters with morals and motives dubious to say the least. People like Jesse and Frank James for example. Billy the Kid. Wild Bill Hickock. Wyatt Earp. These are all American "icons" from the nineteenth century. Moving quickly on into the twentieth century gives us Bonnie and Clyde, Pretty Boy Floyd, Legs Diamond, Al Capone, etc., etc. The list goes on and on. The remakes about the fabled exploits of these shady characters go on and on. But the trouble is that at the end of the day we are talking about a misbegotten bunch of semi-literate, cold-blooded murderers and gangsters. And that *includes* Wyatt Earp, if the truth be known.

To an outsider, it is almost as if by way of the films they make, Hollywood is asking, on behalf of America, "Where is *our* Lawrence of Arabia? Where is *our* Nelson? Where is *our* Drake?" (and he, too, was little more than a cut-throat and a pirate!) and then saying: "I know, let's make a film about an American ship that helps out Nelson at the Battle of Trafalgar just for old time's sake. She'll be flying "Old Glory" (the Stars and Stripes) of course, as if it were a flag of neutrality, but only until she gets into position. But then she runs up a stained old British Red Ensign the captain has been using as a tablecloth, the gun covers come down and she fires a devastating broadside into the "man o' war" of the French admiral, sinks it, and brings about an end to the battle! After all, it *could* have been true! Oh, and by the way, I've also had this great idea about an American First World War platoon that performs a hush-hush SAS type mission for T.E. Lawrence (of Arabia, as played by Peter O'Toole) in the desert and it changes the whole course of the war!"

My point, to our dear American friends, is that none of this "hitching a free ride on history" should be even remotely necessary. After the Moon landings, you should all be well and truly resting upon your laurels. Why the need to manufacture heroes after putting men on the Moon?

In fact, that is the one aspect of this case that baffles me and worries me the most, now that I have had the chance to study it in detail. Why haven't the Americans made much more fuss about their great achievements in space? Where is all the usual razzamatazz, pomp and circumstance we in the rest of the world have come to expect from "the Yanks" when they annually celebrate "relatively" minor triumphs—like the Pilgrim Fathers landing in America—let alone one of the mind-boggling enormity of landing and walking upon another planet?

Where are all the anniversaries and retold stories of the astronauts and their back-up teams? Why was no Apollo 11 Day declared long ago as an annual national holiday? If Americans can celebrate the day their forefathers first set foot upon American soil (Thanksgiving Day) then why not celebrate the day when, in the name of all humanity, they first set foot upon the Moon? After all—and I mean no disrespect because the Pilgrim Fathers were related to *my* forefathers too!—landing in America was not *that* big a deal, was it?

I mean, at least America is on Earth!

July 20, a few weeks after Independence Day, has a certain ring about it.I feel sure that many of us on this side of the Atlantic and around the globe would be only too happy to celebrate it along with the Americans. It is not too late for such a day to be declared by any incumbent, or incoming, U.S. government. Or is there a problem regarding why no American government will ever do so…

And didn't do so thirty-five years ago?

Here in the U.K., we still celebrate Trafalgar Day every October 21 in honor of Admiral Lord Nelson's famous victory—as alluded to earlier—over Napoleon Bonaparte's navy that day in the year 1805. Indeed, we also set aside every November 5 in order to remember an infamous anarchist called Guido or "Guy" Fawkes. This rogue was apprehended attempting to blow up the British Houses of Parliament on that date in 1605. So even we, the comparatively reserved British, set aside a day here and there in memory of the odd bad guy, if you will pardon a most terrible pun! Yet, as a society, we are generally nowhere near as extrovert as our much loved friends across the Pond.

For the United States of America, a nation one would normally expect to loudly proclaim from the very rooftops at every opportunity the names of its *true* heroes, (and there are many actually, including Washington, Custer, Crockett, Bowie, Patten, MacArthur, Eisenhower and Omar Bradley to name just a few) anyone of an ever so slightly suspicious nature might be given to wondering why it is that names like Neil Armstrong, Buzz Aldrin and Michael Collins almost seem to have been allowed to fade away into obscurity? And although he personally never walked upon the Moon (if indeed *any* of them ever did, sadly) I include astronaut Mike Collins as having been equally as heroic as the other two.

The other day I conducted a little test. I put the following question to my grown up, married daughter who is a government officer and has a university degree, so she is certainly no fool. "Who was Michael Collins?" I asked. She replied almost immediately. "He was one of the original leaders of the IRA in Ireland," she said, confidently. "There was a film made about him, which starred Liam Neeson. The IRA accused him of selling them out and assassinated him."

She was correct, of course, but I shook my head in dismay before adding, "Supposing I give you a clue by saying that *this* chap was an American?"

Now it was my daughter's turn to shake her head and she said, "I've no idea. I know Michael Collins was an Irish politician and I think he had some involvement with an American woman. Did he take out American citizenship to marry her? Honestly, Dad, I have no idea at all. Should I have?"

When I told her who the man I was referring to was she said she did vaguely recall having heard of him, although she wasn't born until the year after Apollo 11 allegedly landed on the Moon. She recalled the names Armstrong and Aldrin, of course.

But it must not be forgotten that if Armstrong and Aldrin had not been able to lift off successfully from the surface of the Moon, Michael Collins would have had to attempt the return journey to Earth all by himself. Should he have been forced to do so—and he would have lived with that ghastly thought every single moment that his crewmates remained absent from the Command Module—there can be little doubt that it would have been the most heroic solo journey of all time.

The Eagle has landed

Heroic indeed. How will any of us old enough to recall it, ever forget that momentous night when countless millions of us waited with bated breath?

Those of us who smoked, chain-smoked. Those of us who chewed our nails had already reduced our fingers to stumps. Quite literally on the edge of our seats, we waited. That was all we could do. Wait. Wait for the confirmation that the Apollo 11 spacecraft had landed without mishap...

Yes, *spacecraft!*

For that was what it had now become in the truest sense. This strangest of flying machines, which had somehow miraculously transported intelligent entities through outer space and was about to land them on another planet. And those entities were *not* "little green men." Not this time. They were human beings! We were desperate that the spacecraft should land safely. The *Eagle* would need to land somewhere fairly solid and level and be in a stable position. If she were to list badly, or topple over altogether as the virgin area of lunar surface suddenly gave way beneath one of her landing pads, she wouldn't get off again...

"Forward...steady...down...down..." (*Beep!*) "...picking up some dust..." Then...

Aldrin: Contact light.

Armstrong: Shutdown.

Aldrin: Okay. Engine stop. ACA out of detent.

Armstrong: Out of detent. Auto.

Aldrin: Mode control, both auto. Descent engine command override off. Engine arm off. 413 is in.

Duke (Charlie Duke, MC Capcom): We copy you down, Eagle.

Armstrong: Engine arm is off————crackle————crackle————

But then, for more than a few seconds that soon began to feel like eons, nothing...

Again we were forced to wait. From Mission Control in Houston, Texas, to knee deep in water in the rice paddy fields of Cambodia, humanity waited in bemused silence all over a world in which one literally *could* have heard a pin drop, let alone a spacecraft! This was truly *The Day The Earth Stood Still*. It wasn't too long, as the seconds continued to tick by, before people began to sob openly...*me* included.

Duke: Mission Control to Apollo. Do you copy? Over.

Apollo 11: (Silence)————crackle————crackle————

Oh, no—it's all gone wrong. They must have crashed! Those poor devils...

Then it came. A drawling, relaxed, so unmistakably American voice suddenly broke through the static. Strong. Masculine. Positive. Its arrival on the airwaves brought to an end the blackest, most awful, nail-biting, smoke-choked, gut-wrenching silence in all the history of the world as Neil Armstrong said as calmly as if he were walking a dog through Central Park...

Apollo 11: Houston, Tranquility Base here. The Eagle has landed...(*Beep!*)

Duke: Roger, Tranquility. We copy you on the ground. You have a bunch of guys down here who were about to turn blue. We're breathing again. Thanks a lot.

It was 4:17 a.m. precisely (GMT advanced 1 hour to EDT in the U.K.). I was on my own in my little rented flat in West Norwood, South London. It was quite dark outside still, and I suddenly became aware of lights going on everywhere. Of car horns sounding. Of firecrackers being let off somewhere in the distance. Dogs were barking. It was almost as if the animals also realized that something utterly tremendous had happened and wished to share in mankind's celebration of the moment. Of course they had no real comprehension of what had occurred but they obviously sensed the sudden, unbridled jubilation of their masters.

"Oh, my God!" I cried out, unable to contain myself even though I was on my own. "They've done it! They've only gone and landed on the bloody Moon!"

I remember making a fist and punching the air, shouting, *"Yes! Yes! Yes!"* out of sheer defiance for the human race. Tears of pure joy now; utter elation in fact, were streaming unashamedly down my cheeks as indeed they are now, at this very moment, as I again recall that superb, wonderful, *magical* moment as clearly as if it had all happened *this* morning rather than one incredible morning thirty-five years ago...

It was a magnificently scripted phrase—pure Hollywood even, dare I say it!—said by a magnificent man. I was only in my twenties back then, but they were words I had simply never expected to hear during my lifetime in such a context. Eat your heart out Will Shakespeare, I thought, for they truly had to be the most famous words ever spoken not only in the English language but in *any* language in all history:

"The Eagle has landed."

And, because she had done so, *we*, humanity, were eagles now. Better than eagles. We had dared where no eagle ever would and flown much further than the king of birds ever could. And there seemed to be no reason why we would not continue to do so. It all seemed to be coming true, just as predicted. Dan Dare had stepped off the pages of that other famous *Eagle* (a U.K. comic) and had become science *fact*. It appeared to be all spread out in front of us in Technicolor. All there for us to do. We *would* go into outer space. Men and women together. And, one day, we'd even take those faithful dogs and cats—maybe eagles too!—with us. One day they would have puppies and kittens and egrets in space and we would have our children in space to play alongside them.

If any of our American jurors are of a mind that earlier remarks made by me while bringing this trial to a close were critical of the American people themselves, then rest assured that they were not intended to be. Any criticism was aimed only at those who pretend to serve you and your countrymen's best interests when "promoting the American image" abroad. Far from being critical, I am now going to say something that I want everyone everywhere to understand and have absolutely no doubt about whatsoever. If he indeed did what we have all been led to believe he did, then the owner of the voice that reported that all was well from Tranquility Base on that memorable morning seemingly so long ago now, belonged to far and away the bravest and most famous human being who ever lived.

And he was an American.

And still is of course, as he is thankfully still with us. I can only say that I fully concur with what was said by learned counsel for the prosecution with regard to the awards bestowed upon Neil Armstrong and his two Apollo 11 colleagues. To me, no amount of Purple Hearts, Legion d'Honneurs or George Crosses would have been anything like adequate recognition for what they had allegedly accomplished and for the quite extraordinary measure of valor they had shown throughout. In fact, I completely fail to understand why *my* government did not strongly recommend to Her Majesty the Queen at the time that at least one Victoria Cross, bearing the inscription "For Valor," should be awarded to the Apollo 11 mission as a whole, if not individually to each of the three astronauts involved.

In all normal circumstances, of course, the British VC can only be awarded in times of war for the very bravest of battlefield actions and, generally, only to citizens of countries who are currently members of the Commonwealth. But that's just a load of old protocol. There was absolutely no reason why—again if only for "old time's sake"—an exception could not have been made in such extraordinary and quite remarkable circumstances. After all, two of those three men were allegedly not only the bravest human beings ever to have walked about on *this* world, but on another one too! No courageous act in peacetime *or* wartime would ever again top that!

Man and Superman

Those two men, Neil Armstrong and Edwin "Buzz" Aldrin (again not forgetting the courage of Mike Collins who must have felt so completely alone in the Command Module orbiting overhead) had become heroes quite beyond the normal scale of heroism. They had become men beyond men.

*Super*men, in fact.

In that moment, standing upon the Moon together, Armstrong and Aldrin were more isolated, more cut off from all that they loved and held sacred and dear than any other human beings had ever been in all history. But the point is that underneath those spacesuits they were *not* supermen. They were ordinary men. Good men. Honest men. Educated men. Men who had somehow made all the world's women feel proud of *all* men. Men who had made all men feel good about themselves. Like most ordinary men, these were family men when off duty. But they had shown what ordinary men, not supermen, with hearts equal to those of a hundred lions and a thousand eagles and trained to perfection, are capable of. And, of course, no one was forgetting for a moment all those other magnificent men and women who had designed and built this amazing flying machine and plotted the trajectories and courses it would fly with such unerring, mathematical precision. And if it is possible to feel a sudden, overwhelming fondness for a lump of metal, wire, plastic and glass, and it is, no one was forgetting that fabulous flying machine either. The eagle is the most magnificent of birds and this *Eagle* truly was the most magnificent of them all!

Should humanity hopefully last for another thousand years and beyond, when names like NASA, the United States, the United Kingdom, the Soviet Union, Russia, China and France have become barely translatable words in dog-eared, faded and flaking history books, which will eventually crumble away into dust altogether, we, until now, had a right to believe that the story of Apollo 11 would go on; that this incredible tale of untold courage could never be surpassed in all the annals of mankind yet to come. Because, quite simply, how could anything ever again compare to having been *the very first* members of the human race to land and walk upon the surface of another world?

But then again…oh, no! Please God, no! Maybe it never actually happened! Not for real, anyway. Maybe, at the end of the day, instead of an eagle there was merely a skylark. Is it just conceivable that we somehow did allow ourselves to fall victim to a…

LUNATRICK?

The End

Members of the jury, you may now retire to consider your verdict. You will have plenty of time in which to do so; the rest of your lives if need be. In the meantime, this court is no longer in session.
ALL RISE!

About the Author

At the time of researching much of the material used in this book, my dear departed colleague Keith House and myself were employed as newspaper librarians—with more than forty-five years experience between us—at the headquarters of News International, Wapping, East London. We worked on the following major publications and TV stations:

All channels

By way of old and new systems of information gathering, Keith and I probably had unlimited access to more research material regarding the subject matter of this book than would be available to anyone, with the exception of NASA, of course, and possibly the security services of the United States and the United Kingdom.

Keith died of lung cancer before the actual writing of this book was commenced, but he knew I was prepared to tackle it on my own. He remained firmly convinced that the Apollo Moon landings were genuine right up until the very end and I, for one, would not have wished him to continue his journey believing anything else. His wish was that I would be able to prove it to the skeptics.

However, proving things is not what this book is all about. This book is about choice, purely and simply. I have tried to lay out all the "facts" of this case and have allowed people much more expert than myself to argue all the pros and cons regarding whether they might be correct or not. Any personal opinions given by me anywhere in the book remain just that, personal opinions. They are proof of nothing other than that I, too, have a personal opinion on the subject.

This Book and U.K. Copyright Law

This book contains published material by other writers/authors/artists. The usage of such material was considered absolutely essential to the compilation of the book, which attempts to be a serious investigation into what, according to some, was a hoax perpetrated by the government and aerospace authorities of the United States of America, although this allegation is strongly refuted by them and others. A satisfactory resolution to the argument is therefore long overdue and very much in the public interest.

Where the authorship of published material is known, the source has been duly acknowledged along with an expression that copyright may be possessed by that source. Where pictures/photographs/drawings (all freely available on the Internet) other than those belonging to the author have been used, every effort has been made to identify the artist/owner of copyright and, where identified, the artist/owner of copyright has been duly acknowledged. Where all such aforementioned items have been used, the author makes no claim that such works are his copyright. Only the words, pictures and picture captions belonging to the author, Graham Frank, are his copyright along with the "trial by jury" format of the book as regards this particular case.

As regards the aforementioned copyright that may belong to others (identified or not) whose words/pictures have been reproduced in this book, the author invokes the *Fair Dealing* clause from U.K. copyright law:

ACTS PERMITTED IN RELATION TO COPYRIGHT WORKS— FAIR DEALING

Fair Dealing with a work for the purpose of criticism or review...*does not infringe any copyright*...provided that it is accompanied by a sufficient acknowledgement. (This book contains all such acknowledgements).

Restricted acts

Infringement is restricted (Section 16/3a]) in relation to the work as a whole or any substantial part of it and (Section16/3b]) either directly or indirectly.'

What does this mean in practice?

The *Fair Dealing* provisions (*Fair Use* in USA) are a useful mechanism against possible abuse of copyright law by either side where material used may be in dispute. This should mean that only if one writer represents another writer's work unfairly or without due acknowledgement and/or in such large proportion as to obviously not be *Fairly Dealing* and, instead, attempting to gain from what are plainly another's efforts, shall there be any dispute. However, the Act does not spell out precise word counts or percentage of original, whether literary (text) or artistic work. There is little U.K. case law following the U.K. act, and the details of what *"Fair Dealing"* can cover are therefore inexact.

For all other practical purposes, though, an inclusion from any work, including an artistic work, which a writer/songwriter/playwright/artist wishes to include in his or her work is copyright protected.

USEFUL GUIDELINES

Some publishers have become signatories to the STM publishers' guidelines, for use of printed material. This is a very broadly-based set of guidelines whereby a company can sign up to the spirit *without necessarily agreeing to abide by all the terms.* The provisions set out agreement for use of "borrowed material" limiting the occasions an STM publisher might be charged for material, and encouraging a degree of reciprocity between STM publishers. The agreement also offers the opportunity to dispense with *"making reasonable efforts to seek the author's approval,"* which is not always practical or expedient. There are also guidelines produced by the Society of Authors and by the Publishers Association (U.K.) (1958) which publishers tend to adhere to in an attempt to quantify what may be considered *Fair Dealing* and so can be used without formal permission, subject to full acknowledgement to author, title and source:

For prose quotations, it is not necessary to clear permission for use of a very few words unless it is a key expression.If it is considered *a substantial part*, a permissions fee can be charged by the copyright owner or his/her agent. Advice should always be sought where reproduction could reasonably be construed as competing with the sale of the original source.

For the record, *LUNATRICK?* has been compiled from many original sources, none of which the author is knowingly in direct competition with.

What can be considered "a substantial part" is open to legal interpretation in any case, but it is generally held to be not just quantitative but qualitative.

However, there is no copyright in ideas, only in their expression!

It must be borne in mind that although these *(Fair Dealing)* provisions were mainly included with book and theatre critics in mind, the Act does not specify who should or should not invoke the *Fair Dealing* clause nor does it comment upon what constitutes a critique or a review. In plain English, then, one writer can "fairly deal" with another writer's work providing due acknowledgement is given and the work not claimed as one's own. A successful defense was mounted under fair dealing terms for use of footage from the film *A Clockwork Orange* in a documentary shown by Channel 4, where it was decided that C4 were *not* in breach of the Act even though the company used a staggering 8% of the film for its documentary!

The author can confidently assure his readers that no single contributor to this book—wittingly or unwittingly—other than himself provided anywhere near 8% of its overall content!

NASA Photographs Reproduced in This Book

No copyright is asserted for NASA photographs. If a recognizable person appears in a photo, use for commercial purposes may infringe a right of privacy or publicity. Photos may not be used to state or imply the endorsement by NASA or by any NASA employee of a commercial product, process or service, or used in any other manner that might mislead. Accordingly, it is requested that if a NASA photograph is used in advertising and other commercial promotion, layout and copy be submitted to NASA prior to release. NASA photos reproduced should include photo credit to "NASA" or "National Aeronautics and Space Administration" and should include scanning credit to the appropriate individuals or agencies as noted in the captions.

Disclaimer by the Author

I hereby give due credit to NASA and/or its agents/representatives for any pictures/ photographs reproduced in this book, which may belong to them and do not personally claim there to be anything fraudulent about any of them. In the book, I take no sides in the dispute and have endeavored only to represent both sides of it as fairly as possible and to the best of my ability. Official NASA photographs reproduced in this book will either stand on their own merits as being a perfectly genuine record of perfectly genuine missions to the Moon and the training processes that preceded them, which were in no way designed to deceive the observer, or they will not.

Lightning Source UK Ltd.
Milton Keynes UK
02 January 2009

147949UK00001BB/52/P